Handbook of Statistical Distributions with Applications

Second Edition

Handbook of
Statistical Distributions
with Applications
Second Edition

K. Krishnamoorthy

University of Louisiana

Lafayette, Louisiana, USA

CRC Press
Taylor & Francis Group
Boca Raton London New York

CRC Press is an imprint of the
Taylor & Francis Group, an **informa** business

A CHAPMAN & HALL BOOK

CRC Press
Taylor & Francis Group
6000 Broken Sound Parkway NW, Suite 300
Boca Raton, FL 33487-2742

First issued in paperback 2020

© 2016 by Taylor & Francis Group, LLC
CRC Press is an imprint of Taylor & Francis Group, an Informa business

No claim to original U.S. Government works

ISBN-13: 978-1-4987-4149-1 (hbk)
ISBN-13: 978-0-367-65870-0 (pbk)

Visit the Taylor & Francis Web site at
http://www.taylorandfrancis.com

and the CRC Press Web site at
http://www.crcpress.com

In Memory of My Parents

Contents

List of Figures

List of Tables

Symbols

Symbol Description

α Level of significance

μ Population mean

σ^2 Population variance

Σ Population covariance matrix

Φ Standard normal distribution function

ϕ Standard normal pdf

pmf   Probability mass function

pdf   Probability density function

cdf   Cumulative distribution function

df   Degrees of freedom

$\Gamma(x)$ Gamma function

ρ Population correlation coefficient

r Sample correlation coefficient

$\bar{x}$ Sample mean

UPL   Upper prediction limit

CI   Confidence interval

TI   Tolerance interval

PI   Prediction interval

s Sample standard deviation

Q_α 100α percentile of Q

μ_k' kth moment about the origin

μ_k kth moment about the mean

m_k' Sample kth moment about the origin

m_k Sample kth moment about the mean

χ_m^2 Chi-square random variable with m degrees of freedom

$\widehat{p}$ Sample proportion

$F_{m,n}$ F distribution with numerator df $= m$ and denominator df $= n$

$t_m(\delta)$ Noncentral t random variable with df $= m$ and noncentrality parameter δ

ψ Digamma function; $\psi(x) = \frac{\Gamma'(x)}{\Gamma(x)}$

Preface to the First Edition

Statistical distributions and models are commonly used in many applied areas, such as economics, engineering, social, health, and biological sciences. In this era of inexpensive and faster personal computers, practitioners of statistics and scientists in various disciplines have no difficulty in fitting a probability model to describe the distribution of a real-life data set. Indeed, statistical distributions are used to model a wide range of practical problems, from modeling the size grade distribution of onions to modeling global positioning data. Successful applications of these probability models require a thorough understanding of the theory and familiarity with the practical situations where some distributions can be postulated. Although there are many statistical software packages available to fit a probability distribution model for a given data set, none of the packages is comprehensive enough to provide table values and other formulas for numerous probability distributions. The main purpose of this book and the software is to provide users with quick and easy access to table values, important formulas, and results of the many commonly used, as well as some specialized, statistical distributions. The book and the software are intended to serve as reference materials. With practitioners and researchers in disciplines other than statistics in mind, I have adopted a format intended to make it simple to use the book for reference purposes. Examples are provided mainly for this purpose.

I refer to the software that computes the table values, moments, and other statistics as *StatCalc*. For rapid access and convenience, many results, formulas and properties are provided for each distribution. Examples are provided to illustrate the applications of *StatCalc*. The *StatCalc* is a dialog-based application, and it can be executed along with other applications.

The programs of *StatCalc* are coded in C++ and compiled using Microsoft Visual C++ 6.0. All intermediate values are computed using double precision so that the end results will be more accurate. I compared the table values of *StatCalc* with classical hard copy tables, such as *Biometrika Tables for Statisticians*, *Handbook of Mathematical Functions* by Abramowitz and Stegun (1965), *Tables of the Bivariate Normal Distribution Function and Related Functions* by the National Bureau of Standards (1959), *Pocket Book of Statistical Tables* by Odeh et al. (1977), and the tables published in various journals listed in the references. Table values of the distributions of Wilcoxon rank-sum statistic and Wilcoxon signed-rank statistic are compared with those given in *Selected Tables in Mathematical Statistics*. The results are in agreement wherever I checked. I have also verified many formulas and results given in the book either numerically or analytically. All algorithms for random number generation and evaluating cumulative distribution functions are coded in Fortran, and verified for their accuracy. Typically, I used 1,000,000 iterations to evaluate the performance of random number generators in terms of the speed and accuracy. All the algorithms produced satisfactory results. In order to avoid typographical errors, algorithms are created by copying and editing the Fortran codes used for verification.

A reference book of this nature cannot be written without help from numerous people. I am indebted to many researchers who have developed the results and algorithms given in the book. I thank my colleagues for their valuable help and suggestions. Special thanks are due to Tom Rizzuto for providing me numerous books, articles, and journals. I am grateful to computer science graduate student Prasad Braduleker for his technical help at the initial stage of the *StatCalc* project. It is a pleasure to thank P. Vellaisamy at IIT–Bombay who thoroughly read and commented on the first fourteen chapters of the book. I am thankful to my graduate student Yanping Xia for checking the formulas and accuracy of the software *StatCalc*.

K. Krishnamoorthy
University of Louisiana at Lafayette

Preface to the Second Edition

This revision maintains the organization of chapters and presentation of materials as in the first edition. Although I cover very much the same topics, the book has been largely revised by adding some popular results, including tolerance intervals, prediction intervals, confidence intervals, and many new examples. In the following, I summarize the major changes by chapter.

Chapter 3 on binomial distribution and Chapter 5 on Poisson distribution have been revised by adding tolerance intervals, prediction intervals, confidence intervals for the risk ratio, odds ratio, and for a linear combination of parameters. Fiducial methods and some recent new methods for finding interval estimates are added in these two chapters. I have included one- and two-sample problems involving coefficients of variation, and one-sample inference with censored data in the chapter on normal distribution. Inferential methods for estimating/testing are added to chapters on exponential, gamma, lognormal, logistic, Laplace, Pareto, Weibull, and extreme value distributions. Pivotal-based exact methods, likelihood methods, and other accurate approximate methods are now described in some of these chapters. The chapter on gamma distribution now includes likelihood methods for testing means and parameters, parametric bootstrap confidence intervals for parameters, mean, and for comparing means. In addition, tolerance intervals, prediction intervals, comparison of several gamma distributions with respect to parameters, and means are also covered in the chapter on gamma distribution. The chapter on Weibull distribution has been enhanced by adding confidence intervals, prediction intervals, and tolerance intervals. Some new materials, such as the Fisher z-transformation, comparison of two correlation coefficients, and comparison of two correlated variances, are provided in the chapter on bivariate normal distribution.

The PC calculator *StatCalc* has also been enhanced by incorporating the new materials from the book. Calculation of confidence intervals, tolerance intervals, prediction intervals, sample size calculation, and power calculation can be carried out using *StatCalc* in a straightforward manner. Furthermore, *StatCalc* now comes with help topics, and the topic for a dialog box can be accessed by clicking the help button in the dialog box. The purpose of *StatCalc* is to aid users to carry out calculation accurately with ease, although it may lack graphical appearance. R functions are given for the cases where *StatCalc* is not applicable. To avoid entering codes manually, an electronic version of the R codes is provided along with *StatCalc* software. This R program file

"HBSDA.r" is located in the *StatCalc* installation directory. *StatCalc* software is available at www.ucs.louisiana.edu/~kxk4695.

I have been very much encouraged by positive comments from the reviewers of the first edition of this book, reviewers of the sample chapters from the current edition, and many readers. I am grateful to those readers who have passed on errors found in the first edition.

K. Krishnamoorthy
University of Louisiana at Lafayette

1

StatCalc

1.1 Introduction

The software accompanying this book is referred to as *StatCalc*[1], which is a PC calculator that computes various statistical table values. More specifically, it computes table values of all the distributions presented in the book, necessary statistics to carry out some hypothesis tests and to construct confidence intervals, prediction intervals, tolerance intervals, sample size to carry out a test within specified accuracy, and much more. Readers who are familiar with some statistical concepts and terminologies, and PC calculators may find *StatCalc* simple and easy to use. In the following, we explain how to use this program and illustrate some features.

The dialog boxes that compute various table values are grouped into three categories, namely, continuous, discrete, and nonparametric, as shown in the main page of *StatCalc* in Figure 0.1(a). Suppose we want to compute binomial probabilities, percentiles, or moments; if so, then we should first select "Discrete dialog box" (by clicking on the radio button [Discrete]) as the binomial distribution is a discrete distribution (see Figure 0.1(b)). Click on [Binomial], and then click on [Probabilities, Critical Values and Moments] to get the binomial probability dialog box. This sequence of selections is indicated in the book by the trajectory [StatCalc→Discrete→Binomial→Probabilities, Critical Values and Moments]. Similarly, if we need to compute factors for constructing tolerance intervals for a normal distribution, we first select [Continuous] (because the normal distribution is a continuous one), and then select [Normal] and [Tolerance and Prediction Intervals]. This sequence of selections is indicated by the trajectory [StatCalc→Continuous→Normal→Tolerance and Prediction Intervals]. After selecting the desired dialog box, input the parameters and other values to compute the desired table values.

StatCalc is a stand-alone application, and many copies (as much as the screen can hold) of *StatCalc* can be opened simultaneously. To open two copies, click on *StatCalc* icon on your desktop or select from the start menu. Once the main page of *StatCalc* opens, click on the *StatCalc* icon again on your desktop. The second copy of *StatCalc* pops up exactly over the first copy, and so using the mouse, drag the second copy to a different location on your desktop. Now, we have two copies of *StatCalc*. Suppose we want to compare binomial probabilities with those of the

---

[1]To download, visit www.ucs.louisiana.edu/~kxk4695 or the publisher's site.

hypergeometric with lot size 5000, and the number of defective items in the lot is 1000, then select binomial from one of the copies and hypergeometric from the other.

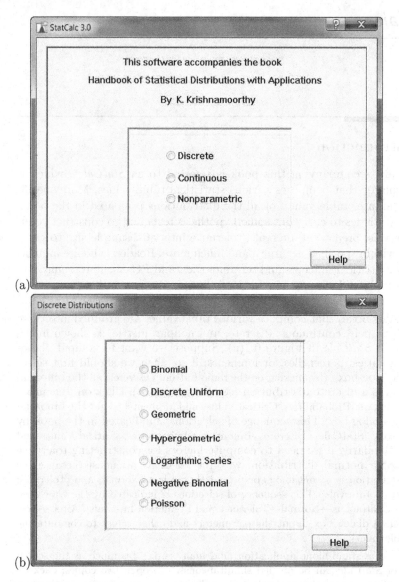

FIGURE 1.1: Selecting the dialog box for computing binomial probabilities

Input the values as shown in Figure 0.2. We observe from these two dialog boxes

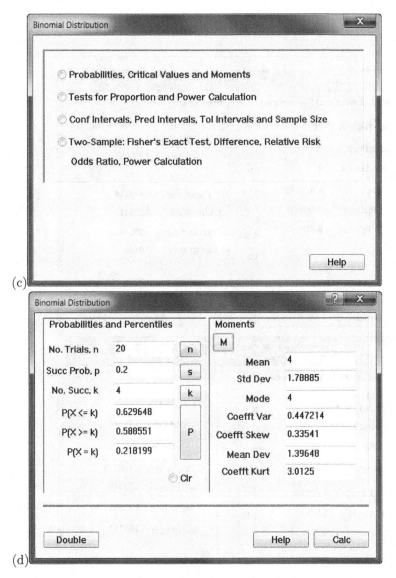

FIGURE 1.1 continued.

that the binomial probabilities with $n = 20$ and $p = 0.2$ are very close to those of the hypergeometric with lot size (population size) 5000 and the number of items with the attribute of interest is 1000. Furthermore, good agreement of the moments of these two distributions clearly indicates that, when the lot size is 5000 or more, the hypergeometric probabilities can be safely approximated by the binomial probabilities.

StatCalc can be opened along with other applications, and the values from the

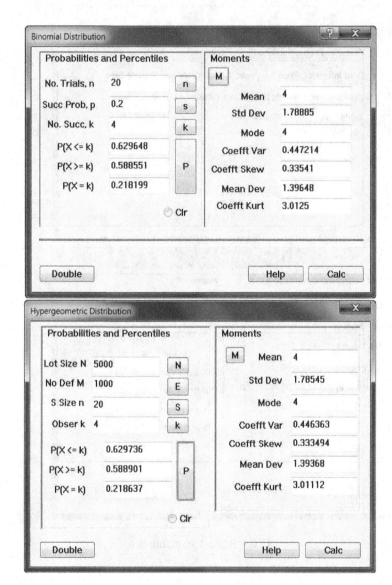

FIGURE 1.2: Dialog boxes for computing binomial and hypergeometric probabilities

edit boxes (the white boxes) can be copied [Ctrl+c] and pasted [Ctrl+v] in a document. *StatCalc* also comes with a help file, and the appropriate help topic for a dialog box can be accessed by clicking the [Help] button in the dialog box.

The following contents list various computations that can be performed using *StatCalc*.

1.2 Contents of *StatCalc*

Discrete Distributions

1. Binomial Distribution

   1.1 Calculation of cumulative probabilities, percentiles, and moments

   1.2 Tests for proportions and power calculation

   1.3 Confidence intervals, prediction intervals, tolerance intervals and sample size calculation

   1.4 Tests for comparing proportions and power calculation: Fisher's test, difference, relative risk, and odds ratio

2. Discrete Uniform Distribution

3. Geometric Distribution

4. Hypergeometric Distribution

   4.1 Calculation of cumulative probabilities, percentiles, and moments

   4.2 Tests for proportions and sample size power

   4.3 Confidence intervals for proportion and sample size for precision

   4.4 Tests for the difference between two proportions and power calculation

5. Logarithmic Series Distribution

6. Negative Binomial Distribution

   6.1 Calculation of cumulative probabilities, percentiles, and moments

   6.2 Test and confidence interval for the proportion

7. Poisson Distribution

   7.1 Calculation of cumulative probabilities, percentiles, and moments

   7.2 Tests and confidence intervals for the mean and power calculation

   7.3 Tolerance intervals and prediction intervals

   7.4 Tests and confidence intervals for comparing two means and power calculation

Continuous Distributions

8. Beta Distributions

9. Bivariate Normal

2

Preliminaries

This reference book is written for those who have some knowledge of statistical distributions. In this chapter, we will review some basic terms and concepts, and introduce the notations used in the book. Readers should be familiar with these concepts in order to understand the results, formulas, and properties of the distributions presented in the rest of the book. This chapter also covers two standard methods of fitting a distribution for an observed data set, two classical methods of estimation, some recent methods of finding approximate confidence intervals, and some aspects of hypothesis testing and interval estimation. Furthermore, some methods for generating random numbers from a probability distribution are outlined.

2.1 Random Variables and Expectations

Random Experiment: An experiment whose outcomes are determined only by chance factors is called a random experiment.

Sample Space: The set of all possible outcomes of a random experiment is called a sample space.

Event: The collection of none, one, or more than one outcomes from a sample space is called an event.

Random Variable: A variable whose numerical values are determined by chance factors is called a random variable. Formally, it is a function from the sample space to a set of real numbers.

Discrete Random Variable: If the set of all possible values of a random variable X is countable, then X is called a discrete random variable.

Mutually Exclusive and Exhaustive Outcomes: A set of outcomes is said to be mutually exclusive if occurrence of one prevents occurrence of all other outcomes. The set is said to be exhaustive if at least one of the outcomes must occur.

Probability of an Event: If all the outcomes of a random experiment are equally likely, mutually exclusive, and exhaustive, then the probability of an event A is given by

$$P(A) = \frac{\text{Number of outcomes in the event A}}{\text{Total number of outcomes in the sample space}}.$$

Probability Mass Function (pmf): Let S be the set of all possible values of a discrete random variable X, and $f(k) = P(X = k)$ for each k in S. Then $f(k)$ is called the probability mass function of X. The expression $P(X = k)$ means the probability that X assumes the value k.

Example 2.1. A fair coin is to be flipped three times. Let X denote the number of heads that can be observed out of these three flips. Then X is a discrete random variable with the set of possible values $\{0, 1, 2, 3\}$; this set is also called the *support* of X. The sample space for this example consists of all possible outcomes ($2^3 = 8$ outcomes) that could result out of three flips of a coin, and is given by

$$\{HHH, HHT, HTH, THH, HTT, THT, TTH, TTT\}.$$

Note that all the above outcomes are mutually exclusive and exhaustive; also they are equally likely to occur with a chance of $1/8$. Let A denote the event of observing two heads. The event A occurs if one of the outcomes HHT, HTH, and THH occurs. Therefore, $P(A) = 3/8$. The probability distribution of X can be obtained similarly and is given below:

k	0	1	2	3
P($X = k$)	1/8	3/8	3/8	1/8

This probability distribution can also be obtained using the probability mass function. For this example, the pmf is given by

$$P(X = k) = \binom{3}{k} \left(\frac{1}{2}\right)^k \left(1 - \frac{1}{2}\right)^{3-k}, \quad k = 0, 1, 2, 3,$$

and is known as the binomial$(3, \frac{1}{2})$ probability mass function (see Chapter 4).

Continuous Random Variable: If the set of all possible values of X is an interval or union of two or more nonoverlapping intervals, then X is called a continuous random variable.

Probability Density Function (pdf): Any real valued function $f(x)$ that satisfies the following requirements is called a probability density function:

$$f(x) \geq 0 \text{ for all } x, \text{ and } \int_{-\infty}^{\infty} f(x)dx = 1.$$

Cumulative Distribution Function (cdf): The cdf of a random variable X is defined by

$$F(x) = P(X \leq x) \quad \text{for all } x.$$

For a continuous random variable X with the probability density function $f(x)$,

$$P(X \le x) = \int_{-\infty}^{x} f(t)dt \ \text{ for all } x.$$

For a discrete random variable X, the cdf is defined by

$$F(k) = P(X \le k) = \sum_{i=-\infty}^{k} P(X = i).$$

Many commonly used distributions involve constants known as parameters. If the distribution of a random variable X depends on a parameter θ (θ could be a vector), then the pdf or pmf of X is usually expressed as $f(x|\theta)$, and the cdf is written as $F(x|\theta)$ or $F_X(x|\theta)$.

Inverse Distribution Function: Let X be a random variable with the cdf $F(x)$. For a given $0 < p < 1$, the inverse of the distribution function is defined by

$$F^{-1}(p) = \inf\{x : P(X \le x) = p\}.$$

Expectation: If X is a continuous random variable with the pdf $f(x)$, then the expectation of $g(X)$, where g is a real valued function, is defined by

$$E(g(X)) = \int_{-\infty}^{\infty} g(x)f(x)dx.$$

If X is a discrete random variable, then

$$E(g(X)) = \sum_{k} g(k)P(X = k),$$

where the sum is over all possible values of X. Thus, $E(g(X))$ is the weighted average of the possible values of $g(X)$, each weighted by its probability.

2.2 Moments and Other Functions

The moments are a set of constants that represent some important properties of the distributions. The most commonly used such constants are measures of central tendency (mean, median, and mode), and measures of dispersion (variance and mean deviation). Two other important measures are the coefficient of skewness and the coefficient of kurtosis. The coefficient of skewness measures the degree of asymmetry of the distribution, whereas the coefficient of kurtosis measures the degree of flatness of the distribution.

2.2.1   Measures of Central Tendency

Mean: Expectation of a random variable X is called the mean of X or the mean of the distribution of X. It is a measure of location of all possible values of X. The mean of a random variable X is usually denoted by μ, and for a discrete random variable X, it is defined by

$$\mu = E(X) = \sum_k kP(X = k),$$

where the sum is over all possible values of X. For a continuous random variable X with probability density function $f(x)$, the mean is defined by

$$\mu = E(X) = \int_{-\infty}^{\infty} xf(x)dx.$$

Median: The median of a continuous random variable X is the value such that 50% of the possible values of X are less than or equal to that value. For a discrete distribution, median is not well defined, and it need not be unique. *Mode*: The most probable value of the random variable is called the mode.

2.2.2   Moments

Moments about the Origin (Raw Moments): The moments about the origin are obtained by finding the expected value of the random variable that has been raised to k, $k = 1, 2, \ldots$. That is,

$$\mu'_k = E(X^k) = \int_{-\infty}^{\infty} x^k f(x)dx$$

is called the kth moment about the origin or the kth raw moment of X.

Moments about the Mean (Central Moments): When the random variable is observed in terms of deviations from its mean, its expectation yields moments about the mean or central moments. The first central moment is zero, and the second central moment is the variance. The third central moment measures the degree of skewness of the distribution, and the fourth central moment measures the degree of flatness. The kth moment about the mean or the kth central moment of a random variable X is defined by

$$\mu_k = E(X - \mu)^k, \quad k = 1, 2, \ldots,$$

where $\mu = E(X)$ is the mean of X. Note that the first central moment μ_1 is always zero.

Sample Moments: The sample central moments and raw moments are defined analogous to the moments defined above. Let $X_1, \ldots, X_n$ be a sample from a population. The sample kth moment about the origin is defined by

$$m'_k = \frac{1}{n} \sum_{i=1}^{n} X_i^k, \quad k = 1, 2, \ldots,$$

and the sample kth moment about the mean is defined by

$$m_k = \frac{1}{n} \sum_{i=1}^{n} (X_i - \bar{X})^k, \quad k = 1, 2, \ldots,$$

where $\bar{X} = m_1'$. In general, for a real valued function g, the sample version of $E(g(X))$ is given by $\sum_{i=1}^{n} g(X_i)/n$.

2.2.3 Measures of Variability

Variance: The second moment about the mean (or the second central moment) of a random variable X is called the variance and is usually denoted by σ^2. It is a measure of the variability of all possible values of X. The positive square root of the variance is called the *standard deviation*.

Coefficient of Variation: Coefficient of variation is the ratio of the standard deviation to the mean, that is, (σ/μ). This is a measure of variability independent of the scale. That is, coefficient of variation is not affected by the units of measurement.

Note that the variance defined above is affected by the units of measurement. As an example, consider the height measurements (in inches) of 10 people:

$$68, \quad 62.5, \; 63, \; 65.5, \; 70, \; 65, \; 69, \; 66.5, \; 67, \; 64.5$$

The mean height is 66.1 inches with the variance 6.1 inches. The coefficient of variation is $\sqrt{\text{var}}/\text{mean} = .0374$. Suppose that the above data are transformed to centimeters. As one inch is equal to 2.54 cm, the mean will be 167.894 and the variance will be $(2.54)^2 \times 6.1 = 39.35$. However, the coefficient of variation is $\sqrt{39.3548}/167.894 = .0374$, the same as the coefficient of variation for the measurements in inches.

Mean Deviation: Mean deviation is a measure of variability of the possible values of the random variable X. It is defined as the expectation of absolute difference between X and its mean. That is,

$$\text{Mean Deviation} = E(|X - \mu|).$$

2.2.4 Measures of Relative Standing

Percentile (quantile): For a given $0 < p < 1$, the $100p$th percentile of a distribution function $F(x)$ is the value of x for which $F(x) = p$. That is, $100p\%$ of the population data are less than or equal to x. If a set of values of x satisfy $F(x) = p$, then the minimum of the set is the $100p$th percentile. The $100p$th percentile is also called the pth quantile.

Quartiles: The 25th and 75th percentiles are called the first and the third quartile, respectively. The difference (third quartile − first quartile) is called the *inter quartile range*.

2.2.5 Other Measures

Coefficient of Skewness: The coefficient of skewness is a measure of skewness of the distribution of X. If the coefficient of skewness is positive, then the distribution is skewed to the right; that is, the distribution has a long right tail. If it is negative, then the distribution is skewed to the left. The coefficient of skewness is defined as

$$\frac{\text{Third Moment about the Mean}}{(\text{Variance})^{\frac{3}{2}}} = \frac{\mu_3}{\mu_2^{3/2}}.$$

Coefficient of Kurtosis:

$$\gamma_2 = \frac{\text{4th Moment about the Mean}}{(\text{Variance})^2} = \frac{\mu_4}{\mu_2^2}$$

is called the coefficient of kurtosis or coefficient of excess. This is a scale and location invariant measure of degree of *peakedness* of the probability density curve. If $\gamma_2 < 3$, then the probability density curve is called *platykurtic*; if $\gamma_2 > 3$, it is called *lepto kurtic*; if $\gamma_2 = 3$, it is called *mesokurtic*.

Coefficient of skewness and coefficient of kurtosis are useful to approximate the distribution of a random variable X. For instance, if the distribution of a random variable Y is known, and its coefficient of skewness and coefficient of kurtosis are approximately equal to those of X, then the distribution functions of X and Y are approximately equal. In other words, X and Y are approximately identically distributed.

2.2.6 Moment Generating Function

Moment Generating Function: The moment generating function of a random variable X is defined by

$$M_X(t) = E\left(e^{tX}\right),$$

provided that the expectation exists for t in some neighborhood of zero. If the expectation does not exist for t in a neighborhood of zero, then the moment generating function does not exist. The moment generating function is useful to derive the moments of X. Specifically,

$$E(X^k) = \left.\frac{\partial^k E(e^{tx})}{\partial t^k}\right|_{t=0}, \quad k = 1, 2, \ldots$$

Characteristic Function: The characteristic function of a random variable X is defined by

$$\phi_X(t) = E\left(e^{itX}\right),$$

where i is the complex number and t is a real number. Every random variable has a unique characteristic function. Therefore, the characteristic function of X uniquely determines its distribution.

Probability Generating Function: The probability generating function of a nonnegative, integer-valued random variable X is defined by

$$P(t) = \sum_{i=0}^{\infty} t^i P(X = i),$$

so that

$$P(X = k) = \frac{1}{k!} \left(\frac{d^k P(t)}{dt^k} \right) \Bigg|_{t=0}, \quad k = 1, 2, \ldots,$$

Furthermore, $P(0) = P(X = 0)$ and $\left. \frac{dP(t)}{dt} \right|_{t=1} = E(X)$.

2.3   Some Functions Relevant to Reliability

Survival Function: The survival function of a random variable X with the distribution function $F(x)$ is defined by

$$1 - F(x) = P(X > x).$$

If X represents the life of a component, then the value of the survival function at x is called the survival probability (or reliability) of the component at x. *Inverse Survival Function*: For a given probability p, the inverse survival function returns the value of x that satisfies $P(X > x) = p$. *Hazard Rate*: The hazard rate of a random variable at time x is defined by

$$r(x) = \frac{f(x)}{1 - F(x)}.$$

Hazard rate is also referred to as failure rate, intensity rate, and force of mortality. The survival probability at x in terms of the hazard rate is given by

$$P(X > x) = \exp \left(- \int_0^x r(y) dy \right).$$

Hazard Function: The cumulative hazard rate

$$R(x) = \int_0^x \frac{f(y)}{1 - F(y)} dy$$

is called the hazard function.

Increasing Failure Rate (IFR): A distribution function $F(x)$ is said to have increasing failure rate if

$$P(X > x|X > t) = \frac{P(X > t + x)}{P(X > t)} \text{ is decreasing in time } t \text{ for each } x > 0.$$

Decreasing Failure Rate (DFR): A distribution function $F(x)$ is said to have decreasing failure rate if

$$P(X > x|X > t) = \frac{P(X > t + x)}{P(X > t)} \text{ is increasing in time } t \text{ for each } x > 0.$$

2.4   Model Fitting

Let $X_1, \ldots, X_n$ be a sample from a continuous population. To verify whether the sample can be modeled by a continuous distribution function $F(x|\theta)$, where θ is an unknown parameter, the plot called Q–Q plot can be used. If the sample size is 20 or more, the Q–Q plot can be safely used to check whether the data fit the distribution.

2.4.1   Q–Q Plot

Construction of a Q–Q plot involves the following steps:

1. Order the sample data in ascending order and denote the jth smallest observation by $x_{(j)}$, $j = 1, \ldots, n$. The $x_{(j)}$s are called *order statistics* or *sample quantiles*.

2. The proportion of data less than or equal to $x_{(j)}$ is usually approximated by $\left(j - \frac{1}{2}\right)/n$ for theoretical convenience.

3. Find an estimator $\widehat{\theta}$ of θ (θ could be a vector).

4. Estimate the population quantile $q_{(j)}$ as the solution of the equation

$$F(q_{(j)}|\widehat{\theta}) = \left(j - \frac{1}{2}\right)/n, \quad j = 1, \ldots, n.$$

5. Plot the pairs $(x_{(1)}, q_{(1)}), \ldots, (x_{(n)}, q_{(n)})$.

If the sample is from a population with the distribution function $F(x|\theta)$, then the Q–Q plot forms a line pattern close to the $y = x$ line, because the sample quantiles and the corresponding population quantiles are expected to be equal. If this happens, then the distribution model $F(x|\theta)$ is appropriate for the data (for examples, see Example 11.1 and Section 17.5).

The following chi-square goodness-of-fit test may be used if the sample is large or the data are from a discrete population.

2.4.2   The Chi-Square Goodness-of-Fit Test

Let X be a discrete random variable with the support $\{x_1, ..., x_m\}$. Assume that $x_1 \leq ... \leq x_m$. Let $X_1, ..., X_n$ be a sample of n observations on X. Suppose we hypothesize that the sample is from a particular discrete distribution with the probability mass function $f(k|\theta)$, where θ is an unknown parameter (it could be a vector). The hypothesis can be tested as follows.

1. Find the number O_j of data points that are equal to x_j, $j = 1, 2, \ldots, m$. The O_js are called observed frequencies.

2. Compute an estimator $\widehat{\theta}$ of θ based on the sample.

3. Compute the probabilities $p_j = f(x_j|\widehat{\theta})$ for $j = 1, 2, \ldots, m-1$ and
$$p_m = 1 - \sum_{j=1}^{m-1} p_j.$$

4. Compute the expected frequencies $E_j = p_j \times n, \quad j = 1, ..., m.$

5. Evaluate the chi-square statistic $\chi^2 = \sum_{j=1}^{m} \frac{(O_j - E_j)^2}{E_j}$.

Let d denote the number of components of θ. If the observed value of the chi-square statistic in Step 5 is larger than the $(1 - \alpha)$th quantile of a chi-square distribution with degrees of freedom $m - d - 1$, then we reject the hypothesis that the sample is from the discrete distribution with pmf $f(k|\theta)$ at the level of significance α.

If we have a large sample from a continuous distribution, then the chi-square goodness-of-fit test can be used to test the hypothesis that the sample is from a particular continuous distribution $F(x|\theta)$. The interval (the smallest observation, the largest observation) is divided into l subintervals, and the number O_j of data values fall in the jth interval is counted for $j = 1, \ldots, l$. The theoretical probability p_j that the underlying random variable assumes a value in the jth interval can be estimated using the distribution function $F(x|\widehat{\theta})$. The expected frequency for the jth interval can be computed as $E_j = p_j \times n$, for $j = 1, \ldots, l$. The chi-square statistic can be computed as in Step 5, and compared with the $(1 - \alpha)$th quantile of the chi-square distribution with degrees of freedom $l - d - 1$, where d is the number of components of θ. If the computed value of the chi-square statistic is greater than the percentile, then the hypothesis will be rejected at the level of significance α.

2.5   Methods of Estimation

We shall describe here two classical methods of estimation, namely, the moment estimation and the method of maximum likelihood estimation. Let

$X_1, \ldots, X_n$ be a sample of observations from a population with the distribution function $F(x|\theta_1, \ldots, \theta_k)$, where $\theta_1, \ldots, \theta_k$ are unknown parameters to be estimated based on the sample.

Moment Estimation

Let $f(x|\theta_1, \ldots, \theta_k)$ denote the pdf or pmf of a random variable X with the cdf $F(x|\theta_1, \ldots, \theta_k)$. The moments about the origin are usually functions of $\theta_1, \ldots, \theta_k$. Notice that $E(X_i^k) = E(X_1^k)$, $i = 2, \ldots, n$, because the X_is are identically distributed. The moment estimators can be obtained by solving the following system of equations for $\theta_1, \ldots, \theta_k$:

$$\frac{1}{n} \sum_{i=1}^{n} X_i = E(X_1)$$
$$\frac{1}{n} \sum_{i=1}^{n} X_i^2 = E(X_1^2)$$
$$\vdots$$
$$\frac{1}{n} \sum_{i=1}^{n} X_i^k = E(X_1^k),$$

where

$$E(X_1^j) = \int_{-\infty}^{\infty} x^j f(x|\theta_1, \ldots, \theta_k) dx, \quad j = 1, 2, \ldots, k.$$

Maximum Likelihood Estimation

For a given sample $x = (x_1, \ldots, x_n)$, the function defined by

$$L(\theta_1, \ldots, \theta_k | x_1, \ldots, x_n) = \prod_{i=1}^{n} f(x_i|\theta_1, \ldots, \theta_k)$$

is called the *likelihood function*. The maximum likelihood estimators are the values of $\theta_1, \ldots, \theta_k$ that maximize the likelihood function.

2.6   Inference

Let $\mathbf{X} = (X_1, \ldots, X_n)$ be a random sample from a population, and let $\mathbf{x} = (x_1, \ldots, x_n)$, where x_i is an observed value of X_i, $i = 1, \ldots, n$. For simplicity, let us assume that the distribution function $F(x|\theta)$ of the population depends only on a single parameter θ. In the sequel, $P(X \leq x|\theta)$ means the probability that X is less than or equal to x when θ is the parameter of the distribution of X.

2.6.1 Hypothesis Testing

The main purpose of the hypothesis testing is to identify the range of the values of the population parameter based on a sample data. Let Θ denote the parameter space. The usual format of the hypotheses is

$$H_0 : \theta \in \Theta_0 \quad \text{vs.} \quad H_a : \theta \in \Theta_0^c, \qquad (2.1)$$

where H_0 is called the null hypothesis, H_a is called the alternative or research hypothesis, Θ_0^c denotes the complement set of Θ_0, and $\Theta_0 \cup \Theta_0^c = \Theta$. For example, we want to test the mean difference θ between durations of two treatments for a specific disease. If it is desired to compare these two treatment procedures, then one can set hypotheses as $H_0 : \theta = 0$ vs. $H_a : \theta \neq 0$.

In a hypothesis testing, a decision based on a sample of data is made as to "reject H_0 and decide H_a is true" or "do not reject H_0." The subset of the sample space for which H_0 is rejected is called the *rejection region* or *critical region*. The complement of the rejection region is called the *acceptance region*.

Test Statistic: A statistic that is used to develop a test for the parameter of interest is called the test statistic. For example, usually the sample mean $\bar{X}$ is used to test about the mean of a population, and the sample proportion is used to test about the proportion in a population.

Errors and Powers

Type I Error: Wrongly rejecting H_0 when it is actually true is called the type I error. Probability of making a type I error while testing hypotheses is given by

$$P(\boldsymbol{X} \in R | \theta \in \Theta_0),$$

where R is the rejection region. The type I error is also referred to as the *false positive* in clinical trials where a new drug is being tested for its effectiveness.

Type II Error: Wrongly accepting H_0 when it is false is called the type II error. Probability of making a type II error is given by

$$P(\boldsymbol{X} \in R^c | \theta \in \Theta_0^c),$$

where R^c denotes the acceptance region of the test. The type II error is also referred to as the *false negative*.

Level of Significance: The maximum probability (over Θ_0) of making type I error is called the level or level of significance; this is usually specified (common choices are 0.1, 0.05, or 0.01) before carrying out a test.

Power Function: The power function $\beta(\theta)$ is defined as the probability of rejecting null hypothesis. That is,

$$\beta(\theta) = P(\boldsymbol{X} \in R | \theta \in \Theta).$$

Power: Probability of not making a type II error is called the power. That is, the probability of rejecting false H_0, and it can be expressed as $\beta(\theta) = P(\boldsymbol{X} \in R | \theta \in \Theta_0^c)$.

Size of a Test: The probability of rejecting H_0 at a given $\theta_1 \in \Theta_0$ is called the size at θ_1. That is, $P(\boldsymbol{X} \in R | \theta_1 \in \Theta_0)$ is called the size.

Level α Test: For a test, if $\sup\limits_{\theta \in \Theta_0} P(X \in R | \theta) \leq \alpha$, then the test is called a level α test. That is, if the maximum probability of rejecting a true null hypothesis is less than or equal to α, then the test is called a level α test.

If the size exceeds α for some $\theta \in \Theta_0$, then the test is referred to as a *liberal* or *anti conservative* test. If the sizes of the test are smaller than α, then it is referred to as a *conservative* test.

Size α Test: For a test, if $\sup\limits_{\theta \in \Theta_0} P(X \in R | \theta) = \alpha$, then the test is called a size α test.

Exact Test: If the distribution of a test statistic under $H_0 : \theta = \theta_0$ (null distribution) does not depend on any parameters, then the test for θ basted on the statistic is an exact test. This implies that, for any specified value θ_0 of θ, the type I error rate is exactly equal to the nominal level α.

The above definition for an exact test is applicable only for continuous distributions. For discrete distributions, type I error rates is seldom equal to the nominal level, and a test is said to be "exact" as long as the type I error rate is no more than the nominal level for all parameter values under the null hypothesis. Such exact tests are usually more conservative, in the sense that the type I error rates are much less than the nominal level, as a result, they have poor power properties.

Unbiased Test: A test is said to be unbiased if $\beta(\theta_1) \leq \beta(\theta_2)$ for every θ_1 in Θ_0 and θ_2 in Θ_0^c.

The Likelihood Ratio Test (LRT):   Let $\boldsymbol{X} = (X_1, ..., X_n)$ be a random sample from a population with the pdf $f(x|\theta)$. Let $\mathbf{x} = (x_1, ..., x_n)$ be an observed sample. Then the likelihood function is given by

$$L(\theta|\mathbf{x}) = \prod_{i=1}^{n} f(x_i|\theta).$$

The LRT statistic for testing (2.1) is given by

$$\lambda(\mathbf{x}) = \frac{\sup_{\Theta_0} L(\theta|\mathbf{x})}{\sup_{\Theta} L(\theta|\mathbf{x})}.$$

Notice that $0 < \lambda(\mathbf{x}) < 1$, and the LRT rejects H_0 in (2.1) for smaller values of $\lambda(\mathbf{x})$.

Pivotal Quantity: A pivotal quantity is a function of sample statistics and the parameter of interest whose distribution does not depend on any unknown parameters. The distribution of $T(\boldsymbol{X})$ can be used to make inferences on θ. The distribution of $T(\boldsymbol{X})$ when $\theta \in \Theta_0$ is called the *null distribution*, and when $\theta \in \Theta^c$ it is called the *non-null distribution*. The value $T(\mathbf{x})$ is called the observed value of $T(\boldsymbol{X})$. That is, $T(\mathbf{x})$ is the numerical value of $T(\boldsymbol{X})$ based on the observed sample $\mathbf{x}$.

P-Value: The p-value of a test is a measure of sample evidence in support of H_a. The smaller the p-value, the stronger the evidence for rejecting H_0. The p-value based on a given sample $\mathbf{x}$ is a constant in $(0,1)$ whereas the p-value based on a random sample $\boldsymbol{X}$ is a uniform$(0, 1)$ random variable. A level α test rejects H_0 whenever the p-value is less than or equal to α.

We shall now describe a test about θ based on a pivotal quantity $T(\boldsymbol{X})$. Consider testing the hypotheses

$$H_0 : \theta \leq \theta_0 \text{ vs. } H_a : \theta > \theta_0, \tag{2.2}$$

where θ_0 is a specified value. Suppose the statistic $T(\boldsymbol{X})$ is a stochastically increasing function of θ. That is, $T(\boldsymbol{X})$ is more likely to be large for large values of θ. The p-value for testing the hypotheses in (2.2) is given by

$$\sup_{\theta \leq \theta_0} P\left(T(\boldsymbol{X}; \theta) > T(\mathbf{x})|\theta\right) = P\left(T(\boldsymbol{X}) > T(\mathbf{x})|\theta_0\right).$$

For two-sided alternative hypothesis, that is,

$$H_0 : \theta = \theta_0 \text{ vs. } H_a : \theta \neq \theta_0,$$

the p-value is given by

$$2 \min\left\{P\left(T(\boldsymbol{X}) > T(\mathbf{x})|\theta_0\right), P\left(T(\boldsymbol{X}) < T(\mathbf{x})|\theta_0\right)\right\}.$$

For testing (2.2), let the critical point c be determined so that

$$\sup_{\theta \in \Theta_0} P(T(\boldsymbol{X}) \geq c|\theta) = \alpha.$$

Notice that H_0 will be rejected whenever $T(\mathbf{x}) > c$, and the region

$$\{\mathbf{x} : T(\mathbf{x}) > c\}$$

is the rejection region.

The power function of the test for (2.2) is given by

$$\beta(\theta) = P(T(\boldsymbol{X}) > c|\theta).$$

The value $\beta(\theta_1)$ is the power at θ_1 if $\theta_1 \in \Theta_0^c$, and the value of $\beta(\theta_1)$ when $\theta_1 \in \Theta_0$ is the size at θ_1.

For an efficient test, the power function should be an increasing function of $|\theta - \theta_0|$ and the sample size. Between two level α tests, the one that has more power than the other should be used for practical applications.

2.6.2   Interval Estimation

Confidence Intervals

Let $L(\boldsymbol{X})$ and $U(\boldsymbol{X})$ be functions satisfying $L(\boldsymbol{X}) < U(\boldsymbol{X})$ for all samples. Consider the interval $(L(\boldsymbol{X}), U(\boldsymbol{X}))$. The probability

$$P((L(\boldsymbol{X}), U(\boldsymbol{X})) \text{ contains } \theta|\theta)$$

is called the *coverage probability* of the interval. The minimum coverage probability, that is,

$$\inf_{\theta \in \Theta} P((L(\boldsymbol{X}), U(\boldsymbol{X})) \text{ contains } \theta|\theta)$$

is called the *confidence coefficient*. If the confidence coefficient is specified as, say, $1-\alpha$, then the interval $(L(\boldsymbol{X}), U(\boldsymbol{X}))$ is called a $1-\alpha$ confidence interval. That is, an interval is said to be a $1 - \alpha$ confidence interval if its minimum coverage probability is $1 - \alpha$.

One-Sided Limits: If the confidence coefficient of the interval $(L(\boldsymbol{X}), \infty)$ is $1 - \alpha$, then $L(\boldsymbol{X})$ is called a $1 - \alpha$ lower confidence limit for θ, and if the confidence coefficient of the interval $(-\infty, U(\boldsymbol{X}))$ is $1-\alpha$, then $U(\boldsymbol{X})$ is called a $1 - \alpha$ upper confidence limit for θ.

Prediction Intervals

Prediction interval, based on a sample from a population with distribution $F(x|\theta)$, is constructed to assess the characteristic of an individual in the population. Let $\boldsymbol{X} = (X_1, ..., X_n)$ be a sample from $F(x|\theta)$. A $1 - \alpha$ prediction interval for $X \sim F(x|\theta)$, where X is independent of $\boldsymbol{X}$, is a random interval $(L(\boldsymbol{X}), U(\boldsymbol{X}))$ that satisfies

$$\inf_{\theta \in \Theta} P\left[(L(\boldsymbol{X}), U(\boldsymbol{X})) \text{ contains } X|\theta\right] = 1 - \alpha.$$

The prediction interval for a random variable X is wider than the confidence interval for θ because it involves the uncertainty in estimates of θ and the uncertainty in X.

Tolerance Intervals

A p content $- (1-\alpha)$ coverage tolerance interval (or simply $(p, 1-\alpha)$ tolerance interval) is an interval based on a random sample that would contain at least proportion p of the sampled population with confidence $1 - \alpha$. Let $\boldsymbol{X} = (X_1, ..., X_n)$ be a sample from $F(x|\theta)$, and $X \sim F(x|\theta)$ independently of $\boldsymbol{X}$. Then, a p content $- (1 - \alpha)$ coverage one-sided tolerance interval of the form $(-\infty, U(\boldsymbol{X}))$ is required to satisfy the condition

$$P_{\boldsymbol{X}} \{P_X [X \le U(\boldsymbol{X})] \ge p|\boldsymbol{X}\} = P_{\boldsymbol{X}} \{F(U(\boldsymbol{X})) \ge p\} = 1 - \alpha, \qquad (2.3)$$

where X also follows $F(x|\theta)$ independently of $\boldsymbol{X}$. That is, $U(\boldsymbol{X})$ is to be determined such that at least a proportion p of the population is less than or equal to $U(\boldsymbol{X})$ with confidence $1 - \alpha$. The interval $(-\infty, U(\boldsymbol{X})]$ is called a one-sided tolerance interval, and $U(\boldsymbol{X})$ is called a one-sided upper tolerance limit. Note that based on the definition of the p quantile q_p (see Section 2.2.4), we can write (2.3) as

$$P_{\boldsymbol{X}}\{q_p \leq U(\boldsymbol{X})\} = 1 - \alpha. \tag{2.4}$$

Thus, $U(\boldsymbol{X})$ is indeed a $1 - \alpha$ upper confidence limit for the p quantile q_p. The $(p, 1 - \alpha)$ one-sided lower tolerance limit is a $100(1 - \alpha)\%$ lower confidence limit for q_{1-p}, the $100(1 - p)$ percentile of the population of interest.

A $(p, 1 - \alpha)$ two-sided tolerance interval $(L(\boldsymbol{X}), U(\boldsymbol{X})]$ is constructed so that

$$
\begin{aligned}
P_{\boldsymbol{X}}\{P_X\left[L(\boldsymbol{X}) \leq X \leq U(\boldsymbol{X})\right] \geq p|\boldsymbol{X}\} &= P_{\boldsymbol{X}}\{F_X(U(\boldsymbol{X})) - F_X(L(\boldsymbol{X})) \geq p\} \\
&= 1 - \alpha.
\end{aligned}
$$

It is important to note that the computation of the above two-sided tolerance interval does not reduce to the computation of confidence limits for certain percentiles.

Equal-Tailed Tolerance Interval

Another type of interval, referred to as equal-tailed tolerance interval, is defined as follows. Assume that $p > 0.5$. A $(p, 1 - \alpha)$ equal-tailed tolerance interval $(L_e(\boldsymbol{X}), U_e(\boldsymbol{X}))$ is such that, with confidence $1 - \alpha$, no more than a proportion $\frac{1-p}{2}$ of the population is less than $L_e(\boldsymbol{X})$, and no more than a proportion $\frac{1-p}{2}$ of the population is greater than $U_e(\boldsymbol{X})$. That is, for $(L_e(\boldsymbol{X}), U_e(\boldsymbol{X}))$ to be a $(p, 1 - \alpha)$ equal-tailed tolerance interval, the condition to be satisfied is

$$P_{\boldsymbol{X}}\left(L_e(\boldsymbol{X}) \leq q_{\frac{1-p}{2}} \text{ and } q_{\frac{1+p}{2}} \leq U_e(\boldsymbol{X})\right) = 1 - \alpha. \tag{2.5}$$

In fact, the interval $(L_e(\boldsymbol{X}), U_e(\boldsymbol{X}))$ includes the interval $\left(q_{\frac{1-p}{2}}, q_{\frac{1+p}{2}}\right)$ with $100(1 - \alpha)\%$ confidence.

2.7 Pivotal-Based Methods for Location-Scale Families

A continuous univariate distribution is said to belong to the location-scale family if its pdf can be expressed in the form

$$f(x; \mu, \sigma) = \frac{1}{\sigma}g\left(\frac{x - \mu}{\sigma}\right), \quad -\infty < x < \infty, \ -\infty < \mu < \infty, \ \sigma > 0, \tag{2.6}$$

where g is a completely specified function. Here, μ and σ are referred to as the location and scale parameters, respectively. As an example, the family of normal distributions is a location-scale family because the pdf can be expressed as

$$f(x; \mu, \sigma) = \frac{1}{\sigma} \phi\left(\frac{x - \mu}{\sigma}\right) \quad \text{with} \quad \phi(x) = \frac{1}{\sqrt{2\pi}} e^{-\frac{x^2}{2}}.$$

2.7.1 Pivotal Quantities

Let $X_1, ..., X_n$ be a sample from a distribution with the location parameter μ and scale parameter σ. Estimators $\widehat{\mu}(X_1, ..., X_n)$ of μ and $\widehat{\sigma}(X_1, ..., X_n)$ of σ are said to be equivariant if for any constants a and b with $a > 0$,

$$\begin{aligned} \widehat{\mu}(aX_1 + b, ..., aX_n + b) &= a\widehat{\mu}(X_1, ..., X_n) + b \\ \widehat{\sigma}(aX_1 + b, ..., aX_n + b) &= a\widehat{\sigma}(X_1, ..., X_n). \end{aligned} \quad (2.7)$$

For example, the sample mean $\bar{X}$ and the sample variance S^2 are equivariant estimators for a normal mean and variance, respectively.

Result 2.7.1. Let $X_1, ..., X_n$ be a sample from a continuous distribution with the pdf of the form in (2.6). Let $\widehat{\mu}(X_1, ..., X_n)$ and $\widehat{\sigma}(X_1, ..., X_n)$ be equivariant estimators of μ and σ, respectively. Then

$$\frac{\widehat{\mu} - \mu}{\sigma}, \quad \frac{\widehat{\sigma}}{\sigma} \quad \text{and} \quad \frac{\widehat{\mu} - \mu}{\widehat{\sigma}}$$

are all pivotal quantities. That is, their distributions do not depend on any parameters (see Lawless, 2003, Theorem E2).

The above result implies that

$$\frac{\widehat{\mu} - \mu}{\widehat{\sigma}} = \frac{(\widehat{\mu} - \mu)/\sigma}{\widehat{\sigma}/\sigma}$$

is also a pivotal quantity.

Pivotal Quantity for an Equivariant Function

Let $h(\mu, \sigma)$ be an equivariant function. That is, $h(b\mu + a, b\sigma) = bh(\mu, \sigma) + a$ for all a and $b > 0$. Then

$$\frac{\widehat{\mu} - h(\mu, \sigma)}{\widehat{\sigma}} = \frac{\widehat{\mu} - [\sigma h(0, 1) + \mu]}{\widehat{\sigma}} = \frac{\widehat{\mu} - \mu}{\widehat{\sigma}} - h(0, 1)\frac{\sigma}{\widehat{\sigma}}$$

is a pivotal quantity. Furthermore, the definition of pivotal quantity implies that

$$\frac{\widehat{\mu} - h(\mu, \sigma)}{\widehat{\sigma}} \sim \frac{\widehat{\mu}^* - h(0, 1)}{\widehat{\sigma}^*}, \quad (2.8)$$

where the notation "$\sim$" means "distributed as," and $(\widehat{\mu}^*, \widehat{\sigma}^*)$ are the equivariant estimators based on a sample from the sampled distribution with $\mu = 0$ and $\sigma = 1$. Therefore, the percentiles of the above pivotal quantity can be obtained either by using a numerical method or by Monte Carlo simulation. For example, if k_1 and k_2 satisfy

$$P\left(k_1 \leq \frac{\widehat{\mu}^* - h(0, 1)}{\widehat{\sigma}^*} \leq k_2\right) = 1 - \alpha,$$

then $(\widehat{\mu} - k_2\widehat{\sigma}, \; \widehat{\mu} - k_1\widehat{\sigma})$ is a $1 - \alpha$ confidence interval for $h(\mu, \sigma)$.

2.7.2   Generalized Pivotal Quantities (GPQ)

Let X be a sample from a population with the parameter of interest θ, and let $\mathbf{x}$ be an observed value of X. A generalized confidence interval for θ is computed using the percentiles of a so-called generalized pivotal quantity (GPQ), say $G(X; \mathbf{x}, \theta)$, a function of X, $\mathbf{x}$, and θ (and possibly the nuisance parameter δ) satisfying the following conditions:

(i) For a given $\mathbf{x}$, the distribution of $G(X; \mathbf{x}, \theta)$ is free of all unknown parameters.

(ii) The value of $G(X; \mathbf{x}, \theta)$, namely its value at $X = \mathbf{x}$, is θ, the parameter of interest.   (C.1)

When the conditions (i) and (ii) in (C.1) hold, appropriate percentiles of $G(X; \mathbf{x}, \theta)$ form a $1 - \alpha$ confidence interval for θ. For example, if G_p is the pth quantile of $G(X; \mathbf{x}, \theta)$, then $\left(G_{\frac{\alpha}{2}}, G_{1-\frac{\alpha}{2}}\right)$ is a $1 - \alpha$ confidence interval for θ.

Numerous applications of generalized confidence intervals have appeared in the literature. Several such applications are given in the books by Weerahandi (1995). It should, however, be noted that generalized confidence intervals may not satisfy the usual repeated sampling properties. That is, the actual coverage probability of a 95% generalized confidence interval could be different from 0.95, and the coverage could in fact depend on the nuisance parameters. The asymptotic accuracy of a class of generalized confidence interval procedures has recently been established by Hannig, Iyer, and Patterson (2005). However, the small sample accuracy of any procedure based on generalized confidence intervals should be investigated at least numerically. As this generalized variable approach is used to obtain inference for several distributions in later chapters, we shall illustrate the method for finding GPQs for a location-scale family of distributions.

GPQs for Location-Scale Families

Consider a location-scale family of distributions with the pdf of the form $f(x|\mu, \sigma)$, where μ is the location parameter and σ is the scale parameter. Let $\widehat{\mu}$ and $\widehat{\sigma}$ be equivariant estimators of μ and σ, respectively. Let $(\widehat{\mu}_0, \widehat{\sigma}_0)$ be an observed value of $(\widehat{\mu}, \widehat{\sigma})$. A GPQ for μ, denoted by G_μ can be constructed as follows.

$$
\begin{aligned}
G_\mu &= \widehat{\mu}_0 - \frac{\widehat{\mu} - \mu}{\sigma} \frac{\sigma}{\widehat{\sigma}} \widehat{\sigma}_0 \\
&= \widehat{\mu}_0 - \frac{\widehat{\mu}^*}{\widehat{\sigma}^*} \widehat{\sigma}_0,
\end{aligned}
\tag{2.9}
$$

where $\widehat{\mu}^*$ and $\widehat{\sigma}^*$ are the equivariant estimators based on a random sample from $f(x|0, 1)$. It is easy to see that when $(\widehat{\mu}, \widehat{\sigma}) = (\widehat{\mu}_0, \widehat{\sigma}_0)$, the expression G_μ simplifies to μ, the parameter of interest. Furthermore, for a given $(\widehat{\mu}_0, \widehat{\sigma}_0)$, the distribution of G_μ does not depend on any parameters. Thus, G_μ is a bona fide GPQ for μ.

A GPQ for the scale parameter σ is given by

$$G_\sigma = \frac{\sigma}{\hat{\sigma}}\hat{\sigma}_0 = \frac{\hat{\sigma}_0}{\hat{\sigma}^*}, \qquad (2.10)$$

where $\hat{\sigma}^*$ are the equivariant estimator based on a random sample from $f(x|0,1)$. It can be easily verified that G_σ is a valid GPQ satisfying the conditions in (C.1).

In general, a GPQ for a function $h(\mu,\sigma)$ can be obtained by replacing the parameters by their GPQs, which gives $h(G_\mu, G_\sigma)$.

Example 2.2. Let us find the GPQs for the mean μ and the standard deviation (SD) σ for a normal distribution. Let $(\bar{X}, S)$ denote the (mean, SD) based on a random sample of size n from the normal distribution. Let $(\bar{x}, s)$ be an observed value of $(\bar{X}, S)$. A GPQ for μ follows from (2.9) as

$$G_\mu = \bar{x} - \frac{\bar{X}^*}{S^*}s, \qquad (2.11)$$

where $(\bar{X}^*, S^*)$ is the (mean, SD) based on a random sample of size n from a standard normal distribution. Note that $\bar{X}^* \sim N(0, 1/\sqrt{n})$ independently of $(n-1)S^* \sim \chi^2_{n-1}$, chi-square distribution with $n-1$ degrees of freedom. As a result,

$$\frac{\bar{X}^*}{S^*} \sim \frac{1}{\sqrt{n}}t_{n-1},$$

where t_m denotes the Student's t distribution with df $= m$. The lower and upper α quantiles of the G_μ in (2.11) form a $1 - 2\alpha$ confidence interval for μ, and is given by

$$\bar{x} \pm t_{n-1;1-\alpha}\frac{s}{\sqrt{n}},$$

the usual t interval for the mean μ.

A GPQ for σ is given by

$$G_\sigma = \frac{s}{S^*},$$

where S^* is the standard deviation based on a random sample from a standard normal distribution. Noting that $S^* \sim \frac{\chi^2_{n-1}}{n-1}$, a $1 - 2\alpha$ confidence interval for σ is given by

$$(G_{\sigma,\alpha},\ G_{\sigma,1-\alpha}) = \left(\frac{s\sqrt{n-1}}{\sqrt{\chi^2_{n-1;1-\alpha}}},\ \frac{s\sqrt{n-1}}{\sqrt{\chi^2_{n-1;\alpha}}} \right),$$

where $\chi^2_{m;q}$ denotes the qth quantile of a χ^2_m distribution. Note that the above confidence interval is exact.

A GPQ for the $100p$th percentile of a $N(\mu,\sigma^2)$ distribution can be obtained by substituting the GPQs for μ and σ. Note that this percentile is expressed as $h(\mu, \sigma) = \mu + z_p \sigma$, where z_p is the p quantile of the standard normal distribution. A GPQ for $h(\mu, \sigma)$ is given by

$$
\begin{aligned}
h(G_\mu, G_\sigma) &= G_\mu + z_p \sqrt{G_{\sigma^2}} \\
&= \bar{x} - \frac{\bar{X}^*}{S^*} s + z_p \frac{s}{S^*}.
\end{aligned}
$$

Noting that $\bar{X}^* \sim Z/\sqrt{n}$, where $Z \sim N(0,1)$, independently of $S^* \sim \frac{\chi^2_{n-1}}{n-1}$, we can write

$$
G_\mu + z_p G_\sigma = \bar{x} + \frac{1}{\sqrt{n}} \left(\frac{Z + z_p \sqrt{n}}{\sqrt{\chi^2_{n-1}/(n-1)}} \right) s.
$$

The term within the parentheses is distributed as $t_{n-1}(z_p \sqrt{n})$, the noncentral t distribution with degrees of freedom $n-1$ and the noncentrality parameter $z_p \sqrt{n}$; see Chapter 20. So the GPQ for $\mu + z_p \sigma$ is given by

$$
\bar{x} + \frac{1}{\sqrt{n}} t_{n-1}(z_p \sqrt{n}) s.
$$

The $100(1 - \alpha)$ percentile of the above quantity is $\bar{x} + \frac{1}{\sqrt{n}} t_{n-1;1-\alpha}(z_p \sqrt{n}) s$, which is a $1 - \alpha$ upper confidence limit for $\mu + z_p \sigma$, and is referred to as the $(p, 1-\alpha)$ upper tolerance limit for the sampled normal population. See Section 11.6.2.

2.8   Method of Variance Estimate Recovery

The Method of Variance Estimate Recovery (MOVER) is useful to find an approximate confidence interval for a linear combination of parameters based on confidence intervals of the individual parameters. Consider a linear combination $\sum_{i=1}^{k} c_i \theta_i$ of parameters $\theta_1, \ldots, \theta_g$, where c_i's are known constants. Let $\hat{\theta}_i$ be an unbiased estimate of θ_i, $i = 1, \ldots, k$. Assume that $\hat{\theta}_1, \ldots, \hat{\theta}_g$ are independent. Further, let (l_i, u_i) denote the $1 - \alpha$ confidence interval for θ_i, $i = 1, \ldots, k$. The $1 - \alpha$ MOVER confidence interval (L, U) for $\sum_{i=1}^{k} c_i \theta_i$ can be expressed as

$$
L = \sum_{i=1}^{g} c_i \hat{\theta}_i - \sqrt{\sum_{i=1}^{g} c_i^2 \left(\hat{\theta}_i - l_i^* \right)^2}, \quad \text{with } l_i^* = \begin{cases} l_i & \text{if } c_i > 0, \\ u_i & \text{if } c_i < 0, \end{cases} \tag{2.12}
$$

and

$$
U = \sum_{i=1}^{g} c_i \hat{\theta}_i + \sqrt{\sum_{i=1}^{g} c_i^2 \left(\hat{\theta}_i - u_i^* \right)^2}, \quad \text{with } u_i^* = \begin{cases} u_i & \text{if } c_i > 0, \\ l_i & \text{if } c_i < 0. \end{cases} \tag{2.13}
$$

Graybill and Wang (1980) first obtained the above confidence interval for a linear combinations variance components, and refer to their approach as the modified large sample method. Zou and coauthors gave a Wald type argument so as to the above confidence interval is valid for any parameters; see Zou and Donner (2008), Zou et al. (2009a, 2009b). These authors refer to the confidence intervals of the above form as the *method of variance estimate recovery* confidence intervals.

2.9 Modified Normal-Based Approximation

There are situations where the problem of finding confidence intervals for a function of parameters simplifies to finding percentiles of a linear combination of independent continuous random variables or percentiles of the ratio of two independent random variables from different families. Approximate percentiles for a linear combination of independent continuous random variables can be obtained along the lines of the MOVER.

2.9.1 Linear Combination of Independent Random Variables

Let $X_1, ..., X_k$ be independent continuous random variables, not necessarily from the same family of distributions. Let $Q = \sum_{i=1}^{k} w_i X_i$, where w_is are known constants. For $0 < \alpha < .5$, let $X_{i\alpha}$ denote the 100α percentile of the distribution of the random variable X_i, $i = 1, ..., k$. Then the 100α percentile of Q is approximated by

$$Q_\alpha \simeq \sum_{i=1}^{k} w_i E(X_i) - \left[\sum_{i=1}^{k} w_i^2 \left[E(X_i) - X_i^l \right]^2 \right]^{\frac{1}{2}}, \qquad (2.14)$$

where $X_i^l = X_{i;\alpha}$ if $w_i > 0$, and is $X_{i;1-\alpha}$ if $w_i < 0$. The upper percentile

$$Q_{1-\alpha} \simeq \sum_{i=1}^{k} w_i E(X_i) + \left[\sum_{i=1}^{k} w_i^2 [E(X_i) - X_i^u]^2 \right]^{\frac{1}{2}}, \qquad (2.15)$$

where $X_i^u = X_{i;1-\alpha}$ if $w_i > 0$, and is $X_{i;\alpha}$ if $w_i < 0$. Furthermore,

$$P(Q_{\alpha/2} \leq Q \leq Q_{1-\alpha/2}) \simeq 1 - \alpha, \quad \text{for } 0 < \alpha < .5.$$

It can be readily verified that the above approximate percentiles in (2.14) and (2.15) are exact for normally distributed independent random variables. These modified normal-based approximations are very satisfactory for many commonly used distributions, such as the beta and Student's t. For more details, see Krishnamoorthy (2014).

2.9.2   Ratio of Two Independent Random Variables

Let X and Y be independent random variables with mean μ_x and μ_y, respectively. Assume that Y is a positive random variable. For $0 < \alpha < 1$, let c denote the α quantile of $R = X/Y$ so that $P(X - cY \le 0) = \alpha$. This means that c is the value for which the α quantile of $X - cY$ is zero. The approximate α quantile of $X - cY$ based on (2.14) is

$$\mu_x - c\mu_y - \sqrt{(\mu_x - X_\alpha)^2 + c^2(\mu_y - Y_{1-\alpha})^2}.$$

Equating the above expression to zero, and solving the resulting equation for c, we get an approximate α quantile for X/Y as

$$R_\alpha \simeq \begin{cases} \dfrac{r - \left\{ r^2 - \left[1 - \left(1 - \frac{Y_{1-\alpha}}{\mu_y}\right)^2\right]\left[r^2 - \left(r - \frac{X_\alpha}{\mu_y}\right)^2\right] \right\}^{\frac{1}{2}}}{\left[1 - \left(1 - \frac{Y_{1-\alpha}}{\mu_y}\right)^2\right]}, & 0 < \alpha \le .5, \\[3em] \dfrac{r + \left\{ r^2 - \left[1 - \left(1 - \frac{Y_{1-\alpha}}{\mu_y}\right)^2\right]\left[r^2 - \left(r - \frac{X_\alpha}{\mu_y}\right)^2\right] \right\}^{\frac{1}{2}}}{\left[1 - \left(1 - \frac{Y_{1-\alpha}}{\mu_y}\right)^2\right]}, & .5 < \alpha < 1, \end{cases}$$

$$(2.16)$$

where $r = \mu_x/\mu_y$.

2.10   Random Number Generation

Inverse Method

The basic method of generating random numbers from a distribution is known as the inverse method. The inverse method for generating random numbers from a continuous distribution $F(x|\theta)$ is based on *the probability integral transformation*: If a random variable X follows $F(x|\theta)$, then the random variable $U = F(X|\theta)$ follows a uniform(0, 1) distribution. Therefore, if $U_1, \ldots, U_n$ are random numbers generated from uniform(0, 1) distribution, then

$$X_1 = F^{-1}(U_1, \theta), \ldots, X_n = F^{-1}(U_n, \theta)$$

are random numbers from the distribution $F(x|\theta)$. Thus, the inverse method is quite convenient if the inverse distribution function is easy to compute. For example, the inverse method is simple to use for generating random numbers from the Cauchy, Laplace, Logistic, and Weibull distributions.

If X is a discrete random variable with support $x_1 < x_2 < \ldots < x_n$ and cdf $F(x)$, then random variates can be generated as follows:

Generate a $U \sim$ uniform(0,1)

If $F(x_i) < U \leq F(x_{i+1})$, set $X = x_{i+1}$.

X is a random number from the cdf $F(x)$. The above method should be used with the convention that $F(x_0) = 0$.

The Accept/Reject Method

Suppose that X is a random variable with pdf $f(x)$, and Y is a random variable with pdf $g(y)$. Assume that X and Y have common support, and random numbers from $g(y)$ can be easily generated. Define

$$M = \sup_x \frac{f(x)}{g(x)}.$$

The random numbers from $f(x)$ can be generated as follows.

1   Generate $U \sim$ uniform(0,1), and Y from $g(y)$
        If $U < \frac{f(Y)}{Mg(Y)}$, deliver $X = Y$
        else go to 1.

The expected number of trials required to generate one X is M.

2.11   Some Special Functions

In this section, some special functions that are used in the following chapters are given.

Gamma Function: The gamma function is defined by

$$\Gamma(x) = \int_0^\infty e^{-t} t^{x-1} dt \quad \text{for} \ \ x > 0.$$

The gamma function satisfies the relation that $\Gamma(x+1) = x\Gamma(x)$.

Digamma Function: The digamma function is defined by

$$\psi(z) = \frac{d\left[\ln\Gamma(z)\right]}{dz} = \frac{\Gamma'(z)}{\Gamma(z)},$$

where $\Gamma(z) = \int_0^\infty e^{-t} t^{z-1} dt$. The value of $\gamma = -\psi(1)$ is called Euler's constant and is given by

$$\gamma = 0.5772\ 1566\ 4901\ 5328\ 6060\cdots.$$

For an integer $n \geq 2$, $\psi(n) = -\gamma + \sum_{k=1}^{n-1} \frac{1}{k}$. Furthermore, $\psi(0.5) = -\gamma - 2\ln(2)$ and

$$\psi(n+1/2) = \psi(0.5) + 2\left(1 + \frac{1}{3} + \cdots + \frac{1}{2n-1}\right), \quad n \geq 1.$$

The digamma function is also called the *Psi* function.

Beta Function: For $a > 0$ and $b > 0$, the beta function is defined by

$$B(a, b) = \frac{\Gamma(a)\Gamma(b)}{\Gamma(a + b)}.$$

The following logarithmic gamma function can be used to evaluate the beta function.

Logarithmic Gamma Function: The function $\ln\Gamma(x)$ is called the logarithmic gamma function, and it has wide applications in statistical computation. In particular, as shown in the later chapters, $\ln\Gamma(x)$ is needed in computing many distribution functions and inverse distribution functions. The following continued fraction for $\ln\Gamma(x)$ is quite accurate for $x \geq 8$ (see Hart et al., 1968). Let

$b_0 = 8.33333333333333E - 2, \quad b_1 = 3.33333333333333E - 2,$
$b_2 = 2.52380952380952E - 1, \quad b_3 = 5.25606469002695E - 1,$
$b_4 = 1.01152306812684, \quad b_5 = 1.51747364915329,$
$b_6 = 2.26948897420496 \quad \text{and} \quad b_7 = 3.00991738325940.$

Then, for $x \geq 8$,

$$
\begin{aligned}
\ln\Gamma(x) \quad = \quad & (x - 0.5)\ln(x) - x + 9.1893853320467E - 1 \\
+ \quad & b_0/(x + b_1/(x + b_2/(x + b_3/(x + b_4/(x + b_5/(x + b_6/(x + b_7))))))).
\end{aligned}
$$

Using the above expression and the relation that $\Gamma(x + 1) = x\Gamma(x)$, $\ln\Gamma(x)$ can be evaluated for $x < 8$ as

$$
\begin{aligned}
\ln\Gamma(x) \quad &= \quad \ln\Gamma(x + 8) - \ln\prod_{i=0}^{7}(x + i) \\
&= \quad \ln\Gamma(x + 8) - \sum_{i=0}^{7}\ln(x + i).
\end{aligned}
$$

The R function 2.1 based on the above method evaluates $\ln\Gamma(x)$ for a given $x > 0$.

R function 2.1. (Calculation of logarithmic gamma function)

```
alng <- function(x){
b <- c(8.33333333333333e-2, 3.33333333333333e-2,
       2.52380952380952e-1, 5.25606469002695e-1,
       1.01152306812684,  1.51747364915329,
       2.26948897420496,  3.00991738325940)
if(x < 8.0){
xx <- x + 8.0
indx <- 1}
else
{indx <- 0
xx <- x}
fterm <- (xx-0.5)*log(xx) - xx + 9.1893853320467e-1
sum <- b[1]/(xx+b[2]/(xx+b[3]/(xx+b[4]/(xx+b[5]/(xx+b[6]
  /(xx+b[7]/(xx+b[8]))))))))
als <- sum + fterm
if(indx == 1){al <- (als-log(x+7.0)-log(x+6.0)-log(x+5.0)
                -log(x+4.0)-log(x+3.0)-log(x+2.0)-log(x+1.0)
                -log(x))
return(c(al))}
else{
return(als)}
}
```

Discrete Distributions

3

Discrete Uniform Distribution

3.1 Description

The probability mass function of a discrete uniform random variable X is given by

$$P(X = k) = \frac{1}{N}, \quad k = 1, \ldots, N.$$

The cumulative distribution function is given by

$$P(X \leq k) = \frac{k}{N}, \quad k = 1, \ldots, N.$$

This distribution is used to model experimental outcomes that are "equally likely." The mean and variance can be obtained using the formulas that

$$\sum_{i=1}^{k} i = \frac{k(k+1)}{2} \quad \text{and} \quad \sum_{i=1}^{k} i^2 = \frac{k(k+1)(2k+1)}{6}.$$

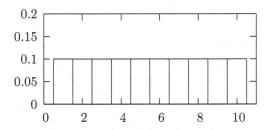

FIGURE 3.1: The probability mass function when $N = 10$

3.2   Moments

Mean:	$\frac{N+1}{2}$
Variance:	$\frac{(N-1)(N+1)}{12}$
Coefficient of Variation:	$\left(\frac{N-1}{3(N+1)}\right)^{\frac{1}{2}}$
Coefficient of Skewness:	0
Coefficient of Kurtosis:	$3 - \frac{6(N^2+1)}{5(N-1)(N+1)}$
Moment Generating Function:	$M_X(t) = \frac{e^t(1-e^{Nt})}{N(1-e^t)}$
Mean Deviation:	$\begin{cases} \frac{N^2-1}{4N} & \text{if } N \text{ is odd,} \\ \frac{N}{4} & \text{if } N \text{ is even.} \end{cases}$

4

Binomial Distribution

4.1 Description

A binomial experiment involves n independent and identical Bernoulli trials, such that each trial can result in to one of the two possible outcomes, namely, success or failure. If p is the probability of observing a success in each trial, then the number of successes X that can be observed out of these n trials is referred to as the binomial random variable with n trials and success probability p. The probability of observing k successes out of these n trials is given by the probability mass function

$$P(X = k | n, p) = \binom{n}{k} p^k (1-p)^{n-k}, \quad k = 0, 1, ..., n.$$

The cumulative distribution function of X is given by

$$P(X \leq k | n, p) = \sum_{i=0}^{k} \binom{n}{i} p^i (1-p)^{n-i}, \quad k = 0, 1, ..., n.$$

Binomial distribution is often used to estimate or determine the proportion of individuals with a particular attribute in a large population. Suppose that a random sample of n units is drawn by sampling with replacement from a finite population or by sampling without replacement from a large population. The number of units that contain the attribute of interest in the sample follows a binomial distribution. The binomial distribution is not appropriate if the sample was drawn without replacement from a small finite population; in this situation, the hypergeometric distribution in Chapter 5 should be used. For practical purposes, binomial distribution may be used for a population of size around 5,000 or more.

We denote the binomial distribution with n trials and success probability p by binomial(n, p). This distribution is right-skewed when $p < 0.5$, left-skewed when $p > 0.5$, and symmetric when $p = 0.5$. See the plots of probability mass functions in Figure 4.1. For large n, binomial distribution is approximately symmetric about its mean np.

4.2   Moments

Mean:	np
Variance:	$np(1-p)$
Mode:	The largest integer $\le (n+1)p$
Mean Deviation:	$2n\binom{n-1}{m}p^{m+1}(1-p)^{n-m}$, where m denotes the largest integer $\le np$. [Kamat, 1965]

Coefficient of Variation: $\sqrt{\frac{1-p}{np}}$

Coefficient of Skewness: $\frac{1-2p}{\sqrt{np(1-p)}}$

Coefficient of Kurtosis: $3 - \frac{6}{n} + \frac{1}{np(1-p)}$

Factorial Moments: $E\left(\prod_{i=1}^{k}(X-i+1)\right) = p^k \prod_{i=1}^{k}(n-i+1)$

Moments about the Mean: $np(1-p) \sum_{i=0}^{k-2} \binom{k-1}{i}\mu_i$

$\qquad -p \sum_{i=0}^{k-2} \binom{k-1}{i}\mu_{i+1}$,

where $\mu_0 = 1$ and $\mu_1 = 0$. [Kendall and Stuart 1958, p. 122]

Moments Generating Function:	$(pe^t + (1-p))^n$
Probability Generating Function:	$(pt + (1-p))^n$

4.3   Probabilities, Percentiles and Moments

The dialog [StatCalc→Discrete→Binomial→Probabilities, Critical Values and Moments] can be used to compute the following.

To compute probabilities: Enter the values of the number of trials n, success probability p, and the observed number of successes k; click [P]. When $n = 20, p = 0.2$, and $k = 4$,

$$P(X \le 4) = 0.629648, \; P(X \ge 4) = 0.588551, \text{ and } P(X = 4) = 0.218199.$$

To compute the value of p: Input values for the number of trials n, the number

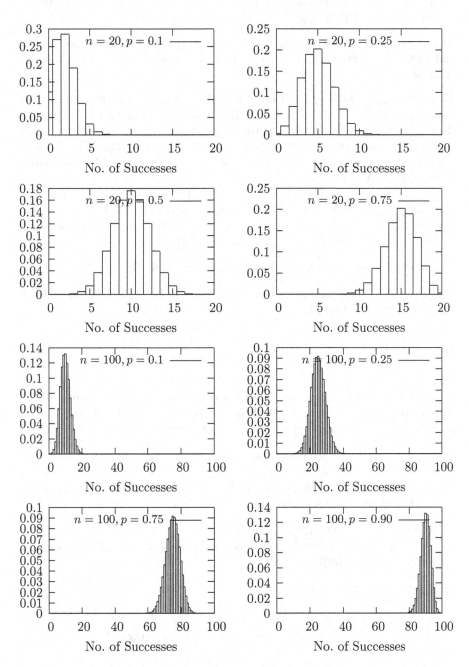

FIGURE 4.1: Binomial probability mass functions

of successes k, and for the cumulative probability P(X <= k); click [s]. When $n = 20$, $k = 4$ and $P(X \le k) = 0.7$, the value of p is 0.183621.

To compute the value of n: Enter the values of p, the number of successes k, and P(X <= k); click [n]. When $p = 0.20$, $k = 6$, and $P(X \le k) = 0.4$, the value of n is 36.

To compute the value of k: Enter the values of n, p, and the cumulative probability P(X <= k); click [k]. If the cumulative probability c is greater than 0.5, then *StatCalc* computes the smallest value of k, such that $P(X \ge k) \le 1 - c$. That is, the value of k is computed so that the right-tail probability is less than or equal to $1 - c$; if $c < 0.5$, then the largest value of k is computed so that the $P(X \le k) \le c$.

   As an example, $p = 0.4234$, $n = 43$, and $P(X <= k) = 0.90$, the value of k is 23. Notice that $P(X \ge 23) = 0.0931953$, which is less than $1 - 0.90 = 0.10$. If $P(X \le k) = 0.10$, then k is 13, and $P(X \le 13) = 0.071458$. Note that $P(X \le 14) = 0.125668$, which is greater than 0.10.

To compute moments: Enter values for n and p; click [M].

Example 4.1. Suppose that a balanced coin is to be flipped 20 times. Find the probability of observing

a. 10 heads;
b. at least 10 heads;
c. between 8 and 12 heads.

Solution: Let X denote the number of heads that can be observed out of these 20 flips. Here, the random variable X is binomial with $n = 20$, and the success probability $= 0.5$, which is the probability of observing a head at each flip.

a. To find the probability, select the dialog box [StatCalc→Discrete→Binomial→ Probabilities, Critical Values and Moments], enter 20 for n, 0.5 for success probability, 10 for k, and click on [P] to get $P(X = 10) = 0.176197$. That is, the chances of observing exactly 10 heads are about 18%.

b. To get this probability, enter 20 for n, 0.5 for p, 10 for k, and click [P] to get $P(X \ge 10) = 0.588099$. That is, the chances of observing 10 or more heads are about 59%.

c. The desired probability is

$$
\begin{aligned}
P(8 \le X \le 12) &= P(X \le 12) - P(X \le 7) \\
&= 0.868412 - 0.131588 \\
&= 0.736824.
\end{aligned}
$$

Example 4.2. What are the chances of observing exactly 3 girls in a family of 6 children?

Solution: Let us assume that the probability of giving birth to a boy = probability of giving birth to a girl = 0.5. Let X be the number of girls in the family. Here, X is a binomial random variable with $n = 6$ and $p = 0.5$. To find the probability, select the dialog box [StatCalc→Discrete→Binomial→Probabilities, Critical Values and Moments], enter 6 for n, 0.5 for p, and 3 for k; click [P] to get $P(X = 3) = 0.3125$.

Example 4.3. Inspection of a random sample of 30 items from a large shipment showed 5 defective items. Let X denote the number of defective items in a sample of 30 items, and let p denote the true proportion of defective items in the shipment.

a. Identify the distribution of X.
b. Find the value of p for which $P(X \geq 5|n = 30, p) = .05$.
c. What can we conclude from the value of p obtained in part b?

Solution:

a. The distribution of X is binomial with $n = 30$ and "success probability" p, which is unknown.

b. Note that the value of p that satisfies $P(X \geq 5|n = 30, p) = .05$ is the same as the one that satisfies $P(X \leq 4|n = 30, p) = .95$. To find the p using *StatCalc*, select the dialog box [StatCalc→Discrete→Binomial→Probabilities, Critical Values and Moments], enter 30 for n, 4 for k, and .95 for $P(X \leq k)$. Click on [s] to get 0.0680556. For this value of p, $P(X \geq 5|n = 30, p = .0680556) = .05$.

c. The result in part b implies that if p were .06806, then only 5% of all possible samples, each of size 30, include 5 or more number of defective items. As a result, we can conclude that the true proportion of defective items is likely to be to be more than .06806.

4.4 Proportion

Suppose that investigation of a sample of n units from a population revealed that X successes (the number of units with an attribute of interest). Let p denote the proportion of individuals in the population with the attribute. The sample proportion $\hat{p} = X/n$ is the maximum likelihood estimate as well as an unbiased estimate of p.

In the following sections, we shall outline an asymptotic test, score test, and an exact test for proportion. These tests are based on X and n.

4.4.1   Tests

The Wald Test

The Wald test is based on the result that

$$Z = \frac{\widehat{p} - p}{\sqrt{\frac{\widehat{p}(1-\widehat{p})}{n}}} \sim N(0,1), \text{ asymptotically,}$$

where $N(0,1)$ denotes the standard normal distribution. Consider testing hypotheses

$$H_0 : p \le p_0 \quad \text{vs.} \quad H_a : p > p_0, \tag{4.1}$$

where p_0 is a specified value, at the level of significance α. Let z_0 be an observed value of

$$\frac{\widehat{p} - p_0}{\sqrt{\widehat{p}(1-\widehat{p})/n}}.$$

The null hypothesis H_0 will be rejected if $z_0 \ge z_{1-\alpha}$, where $z_{1-\alpha}$ is the $100(1-\alpha)$ percentile of the standard normal distribution, or the p-value $1 - \Phi(z_0) \le \alpha$, where Φ is the standard normal cumulative distribution function. For testing

$$H_0 : p \ge p_0 \quad \text{vs.} \quad H_a : p < p_0, \tag{4.2}$$

the null hypothesis will be rejected if $z_0 < z_\alpha$ or if the p-value $\Phi(z_0) \le \alpha$, and for testing

$$H_0 : p = p_0 \quad \text{vs.} \quad H_a : p \ne p_0, \tag{4.3}$$

the null hypothesis will be rejected if $|z_0| > z_{1-\frac{\alpha}{2}}$ or if the p-value

$$2\min\{\Phi(z_0), 1 - \Phi(z_0)\} \le \alpha.$$

The Score Test

The score test is based on the result that

$$Z^* = \frac{\widehat{p} - p}{\sqrt{\frac{p(1-p)}{n}}} \sim N(0,1), \text{ asymptotically.}$$

Let z_0^* be an observed value of

$$\frac{\widehat{p} - p_0}{\sqrt{\frac{p_0(1-p_0)}{n}}},$$

where p_0 is the specified value under H_0 in (4.3). For testing (4.1), the null hypothesis will be rejected if $z_0^* > z_{1-\alpha}$, for testing (4.2), the null hypothesis will be rejected if $z_0^* < z_\alpha$, and for testing (4.3) the null hypothesis will be rejected if $|z_0^*| > z_{1-\frac{\alpha}{2}}$.

An Exact Test

The exact test is based on the exact p-value that can be computed using the binomial(n, p_0) probabilities. Let k be an observed value of X. For testing (4.1), the null hypothesis will be rejected if the p-value

$$P(X \geq k | n, p_0) = \sum_{x=k}^{n} \binom{n}{x} p_0^x (1 - p_0)^{n-x} \leq \alpha,$$

for testing (4.2), the null hypothesis will be rejected if the p-value

$$P(X \leq k | n, p_0) \leq \alpha,$$

and for testing (4.3) the null hypothesis will be rejected if the p-value

$$2 \min \{ P(X \leq k | n, p_0), P(X \geq k | n, p_0) \} \leq \alpha. \tag{4.4}$$

4.4.2 Power and Sample Size Calculation

For a given p, p_0 and the level α, the exact power of a test can be computed using the expression

$$\sum_{k=0}^{n} \binom{n}{k} p^k (1 - p)^{n-k} I \left(\text{p} - \text{value} \leq \alpha \right), \tag{4.5}$$

where $I(.)$ is the indicator function. Note that the p-value is a function of (k, n, p_0). The above expression at $p = p_0$ is the type I error rate (size) of the test.

There are other tests available for a binomial proportion. Among all tests, the exact test and the score test are popular and commonly used in applications. The exact test guarantees that the type I error rates never exceed the nominal level α. However, the exact test is too conservative (type I error rates are often much smaller than the nominal level), and so it is less powerful than the score test. The score test is not exact (its type I error rates sometime exceed the nominal level) but controls the type I error rates around the nominal level, and is more powerful than the exact test. For these reasons, the score test is recommended for practical applications. Finally, we note that the Wald test is not accurate even for large samples. On the basis of extensive numerical studies, Brown et al. (2001) advocated that the score test is preferable to the Wald test in terms of type I error rates and powers.

As the score test and the exact test are most popular, we shall consider the power and sample size calculation only for these two tests. *StatCalc* uses expression (4.5) to calculate the exact power. The dialog box [StatCalc→Discrete→Binomial →Tests for Proportions and Power Calculation] computes the p-value for testing binomial proportion. This also computes the power for a given (p, p_0, α, n), and computes the sample size for a given $(p, p_0, \alpha, \text{power})$.

Example 4.4. (*Calculation of p-values*) When $n = 20$, $k = 8$, and $p_0 = 0.2$, it is desired to test $H_0 : p \leq 0.2$ vs. $H_a : p > 0.2$ at the level of 0.05. To compute the p-value, based on the exact test, select [StatCalc→Discrete→Binomial→Tests for Proportions and Power Calculation], and enter the values of n, k, and p_0 in the dialog box; click [p-values] under "Exact Test" to get 0.0321427; the p-value based on the score test is 0.0126737. If the nominal level is 0.05, then both tests reject the null hypothesis, and they provide evidence to indicate that the true p is greater than 0.2.

Suppose the hypotheses are $H_0 : p = 0.2$ vs. $H_a : p \neq 0.2$. For these hypotheses, the p-value of the exact test is $2 \times 0.0321427 = 0.0642854$, and the p-value of the score test is $2 \times 0.0126737 = 0.0253474$. Note that, at the level of .05, the score test rejects H_0, whereas the exact test does not reject H_0. This is because the exact test is conservative.

Example 4.5. (*Calculation of p-values*) A pharmaceutical company claims that 75% of doctors prescribe one of its drugs to treat a particular disease. In a random sample of 40 doctors, 23 prescribed the drug to their patients. Does this information provide sufficient evidence to indicate that the actual percentage of doctors who prescribe the drug is less than 0.75?
Solution: Let p be the actual proportion of doctors who prescribe the drug to their patients. The hypotheses of interest are

$$H_0 : p \geq 0.75 \quad \text{vs.} \quad H_a : p < 0.75.$$

To compute the p-value for testing above hypotheses, select the dialog box [StatCalc→ Discrete→Binomial→Tests for Proportion ...], enter 40 for n, 23 for observed k, and 0.75 for [Value of p0]. Click on [p-values], to get the p-value of the exact test as 0.0115614, and the p-value of the score test as 0.00529357. Both p-values are less than 0.05, and so we can conclude, contrary to the manufacturer's claim, that less than 75% of doctors prescribe the drug. As in the preceding example, the score test produced smaller p-value than the exact test.

Power and Sample Size Calculation: For a given n, population proportion p, and a hypothesized value p_0, [StatCalc→ Discrete→Binomial→Tests for Proportion ...] computes the powers of the exact test and score test. Furthermore, for a given p, p_0, level α, and the power, *StatCalc* computes the sample size so that the type I error rate is no more than α, and the power is at least the specified value. According to this criterion, the sample sizes required for both tests (involving one-sided hypotheses) to attain a specified power are the same. This is because the exact one-sided test is uniformly the most powerful level α test; for more details, see Krishnamoorthy and Peng (2007).

Example 4.6. When $n = 35$, $p_0 = 0.2$, nominal level $= 0.05$, and $p = 0.4$, the power of the exact test for $H_0 : p \leq 0.2$ vs. $H_a : p > 0.2$, is 0.804825. To compute the power, select the dialog box [StatCalc→ Discrete→Binomial→Tests for Proportion ...] enter 1 to indicate right-tailed test, 0.05 for the level, 0.2 for [Null p0], 0.4 for [Guess p], and 35 for [S Size] and click on [Power] to get 0.804825. The power of the score test for the same sample size is 0.887746. Note that the power of the score test is considerably larger than that of the exact test. However, the type I error rate of the score test (this can be obtained by entering $p = p_0 = 0.2$) is 0.074709, which is not appreciably larger than the nominal level 0.05.

For testing $H_0 : p = 0.2$ vs. $H_a : p \neq 0.2$, the power of the exact test is 0.69427 when the true p is .4.; to compute this power, enter 3 to indicate two-tailed test, and click [Power]. The power of the score test is 0.80483. We once again notice that the power of the score test is larger than that of the exact test. For this two-sided hypothesis, the type I error rate of the score test (which can be obtained by entering 0.2 for [Guess p]) is 0.0533975, which is very close to the nominal level .05.

Example 4.7. (*Sample Size Calculation*) Suppose that a researcher believes that a new drug is 20% more effective than the existing drug, which has a success rate of 70%. The required sample size to test his belief (at the level 0.05 and power 0.90) can be computed as follows. Enter 1 in the dialog box [StatCalc→Discrete→Binomial→Tests for Proportions and Sample Size Calculation] to indicate right-tailed test, 0.05 for the level, 0.7 for [Null p0], 0.9 for [Guess p], 0.9 for [Power], and click on [S Size] to get 37; now click on [Power] to get 0.928915, which is the actual power when the sample size is 37. Note that when the sample size is 36, the power of the exact test is 0.854603.

To find the sample size required for the score test to attain the same power of .90, enter 37 for [S Size] under [Score Test], and click on [Power] to get 0.928915, which is the power of the exact test noted in the preceding paragraph. In order to understand the power calculation of the score test, let us try some values of sample size smaller than 37. For example, when [S Size] under [Score Test] is 35, the power is 0.944817, then why can't we use the sample size 35 instead of 37? The reason for using 37 is that the type I error rate of the score test when [S Size] is 35, [Guess p] = [Null p0] = .7, is 0.0649987, which is larger than the nominal level .05. On the other hand, the type I error rate of the score test at $n = 37$ is 0.0439672 $< .05$. As noted earlier, *StatCalc* determines the sample size for the score test so that the type I error rate is no more than the nominal level, and the power is at least the specified value. According to this criterion, for testing one-sided hypotheses, the score test and the exact test require the same sample size to attain a power no less than the specified value.

Remark 4.1. Regarding sample sizes for a two-tailed test with a specified power, score test requires smaller sample sizes in many cases. In some situations, the sample sizes for the score test (with type I error rates smaller than the nominal level) are smaller than those for the exact method. For a two-tailed test, the score test may be preferable to the exact test.

4.4.3 Confidence Intervals

Confidence intervals for a binomial proportion can be obtained by inverting the tests in the preceding section. In the following, we shall outline the Wald, score, exact, and fiducial confidence intervals.

The Wald Confidence Interval

The set of values of p_0 for which the null hypothesis in (4.3) is accepted is a confidence set. In particular, the set of values of p_0 for which

$$\left| \frac{\widehat{p} - p_0}{\sqrt{\widehat{p}(1 - \widehat{p})/n}} \right| < z_{1-\frac{\alpha}{2}},$$

where z_q is the qth quantile of the standard normal distribution, is a $1-\alpha$ confidence interval, and is given by

$$\widehat{p} \pm z_{1-\frac{\alpha}{2}} \sqrt{\frac{\widehat{p}(1-\widehat{p})}{n}}.$$

The Score Confidence Interval

The acceptance region of the score test is given by

$$\left\{ p : \left| \frac{\widehat{p} - p}{\sqrt{p(1-p)/n}} \right| < z_{1-\alpha/2} \right\}.$$

The above acceptance region is an interval with endpoints determined by the roots of the quadratic equation

$$\left(\frac{\widehat{p} - p}{\sqrt{p(1-p)/n}} \right)^2 - c^2 = 0,$$

where $c = z_{1-\alpha/2}$. The score confidence interval, determined by these roots, is given by

$$\left(\frac{\widehat{p} + \frac{c^2}{2n}}{1 + \frac{c^2}{n}} \right) \pm \frac{\frac{c}{\sqrt{n}} \sqrt{\widehat{p}(1-\widehat{p}) + c^2/(4n)}}{1 + \frac{c^2}{n}}. \tag{4.6}$$

An Exact Confidence Interval

For a given sample size n and an observed number of successes k, the lower limit p_L for p is the solution of the equation

$$\sum_{i=k}^{n} \binom{n}{i} p_L^i (1-p_L)^{n-i} = \frac{\alpha}{2}, \tag{4.7}$$

and the upper limit p_U is the solution of the equation

$$\sum_{i=0}^{k} \binom{n}{i} p_U^i (1-p_U)^{n-i} = \frac{\alpha}{2}. \tag{4.8}$$

Using a relation between the binomial and beta distributions (see Section 16.6.2), it can be shown that

$$p_L = B_{k,n-k+1;\frac{\alpha}{2}} \quad \text{and} \quad p_U = B_{k,n-k+1;1-\frac{\alpha}{2}}, \tag{4.9}$$

where $B_{a,b;q}$ denotes the qth quantile of a beta distribution with the shape parameters a and b. The interval (p_L, p_U) is an exact $1 - \alpha$ confidence interval for p, in the sense that the coverage probability is always greater than or equal to the specified confidence level $1 - \alpha$. One-sided $1 - \alpha$ lower confidence limit for p is $B_{k,n-k+1;\alpha}$ and one-sided $1 - \alpha$ upper confidence limit for p is $B_{k,n-k+1;1-\alpha}$. When $k = n$, the upper limit is 1 and the lower limit is $\alpha^{\frac{1}{n}}$; when $k = 0$, the lower limit is 0 and the upper limit is $1 - \alpha^{\frac{1}{n}}$.

The above exact confidence interval and one-sided confidence limits can be obtained by inverting the tests described in Section 4.4.1. For instance, the two-tailed test based on (4.4), and the one that rejects the null hypothesis in (4.3) whenever the exact confidence interval (p_L, p_U) in (4.9) does not include p_0, are the same. Clopper and Pearson (1934) have proposed the above confidence interval using a fiducial argument, and it is commonly referred to as the Clopper-Pearson confidence interval.

Fiducial Confidence Interval

To describe the fiducial approach, let $X \sim$ binomial(n, p), and let $B_{a,b}$ denote the beta random variable (Chapter 17) with shape parameters a and b. It is well-known that, for an observed value k of X,

$$P(X \geq k|n, p) = P(B_{k,n-k+1} \leq p) \quad \text{and} \quad P(X \leq k|n, p) = P(B_{k+1,n-k} \geq p).$$
(4.10)

On the basis of this relation, we see that there is a pair of fiducial distributions for p, namely, $B_{k,n-k+1}$ for setting lower limit for p and $B_{k+1,n-k}$ for setting upper limit for p. Indeed, the Clopper-Pearson exact confidence interval in (4.9) is based on the pair of fiducial distributions. Instead of having two fiducial variables, a random quantity that is "stochastically between" $B_{k,n-k+1}$ and $B_{k+1,n-k}$ can be used as a single approximate fiducial variable for p (Stevens, 1950). Hypothesis test or confidence interval for p can be obtained from the distribution of $B_{k+.5,n-k+.5}$. Specifically, the interval

$$\left(B_{k+\frac{1}{2},n-k+\frac{1}{2};\frac{\alpha}{2}}, \ B_{k+\frac{1}{2},n-k+\frac{1}{2};1-\frac{\alpha}{2}} \right),$$
(4.11)

where $B_{a,b;\alpha}$ denote the 100α percentile of $B_{a,b}$, is an approximate $1 - \alpha$ confidence interval for p. This is also the Bayesian confidence interval with the Jeffreys prior, and so it is also referred to as the Jeffreys confidence interval. The above confidence interval for the binomial proportion is quite comparable with the score confidence interval in (4.6). The fiducial approach can be extended in a straightforward manner to find confidence intervals for many summary indices involving binomial proportions (Krishnamoorthy and Lee, 2010, 2013), such as the relative risk and odds ratio as shown in the later sections.

Among the confidence intervals described above, the score and fiducial confidence intervals are satisfactory in terms of coverage probability and precision. These two confidence intervals are certainly preferable to the Wald confidence interval. The exact confidence interval is often very conservative, and is wider than the approximate score and fiducial confidence intervals. Most researchers now recommend the score and fiducial confidence intervals for practical applications. See Agresti and Coull (1998) and Brown et al. (2001).

4.4.4 Sample Size Calculation for a Given Precision

For a given sample size n and p, the expected length of a $1 - \alpha$ confidence interval (p_L, p_U) can be computed using the expression

$$\sum_{k=0}^{n} \binom{n}{k} p^k (1 - p)^{n-k} (p_U - p_L).$$
(4.12)

An approximation to the sample size required to construct a $1 - \alpha$ score confidence interval with a precision d is given by (see Krishnamoorthy and Peng, 2007)

$$
n \simeq \frac{z_{\alpha/2}^2 \left[(pq - 2d^2) + \sqrt{(pq - 2d^2)^2 - d^2(4d^2 - 1)} \right]}{2d^2}, \tag{4.13}
$$

where $q = 1 - p$.

The dialog box [StatCalc→Discrete→Binomial→Confidence intervals ...] calculates confidence intervals for p using the exact method and the score method. The fiducial confidence interval is not included in the dialog box because the required beta percentiles can be readily calculated using the dialog box [StatCalc→Continuous→Beta], and it is as good as the score confidence interval in terms of coverage probability and precision. Furthermore, this dialog box computes the required sample size for a given precision (margin of error); that is, one half of the expected width defined in (4.12).

Example 4.8. (*Confidence Intervals for p*) Suppose that a binomial experiment of 40 trials resulted in 5 successes. To find 95% confidence intervals, select the dialog box [StatCalc→Discrete→Binomial→Confidence intervals ...], enter 40 for n, 5 for the observed number of successes k, and 0.95 for the confidence level; click [2-sided] to get the exact confidence interval (0.0419, 0.2680), and the score confidence interval (0.0546, .2611). For one-sided limits, click [1-sided] to get 0.0506 and 0.2450 (exact); score limits are 0.0622 and .2353. That is, 95% exact one-sided lower limit for p is 0.0506, and 95% exact one-sided upper limit for p is 0.2450. Note that the score confidence intervals are shorter than the exact confidence intervals.

Example 4.9. (*Confidence Intervals for p*) The manufacturer of a product reports that at most 5 percent of his products could be defective. In a random sample of 25 such products, 4 of them were found to be defective. Find a 95% confidence interval for the true percentage of defective products.

Solution: To get a 95% confidence interval for the actual percentage of defective products, select the dialog box [StatCalc→Discrete→Binomial→Confidence Intervals ...] from *StatCalc*, enter 25 for n, 4 for k, and 0.95 for the confidence level, and click on [2-sided] to get (0.0454, 0.3608) (exact) and (0.0640, .3465) (score). If we use the exact confidence interval, then the actual percentage of defective items is somewhere between 4.5 and 36, with 95% confidence. Click [1-sided] to get the lower limit 0.05656; this means that the actual percentage of defective products is at least 5.66, with 95% confidence. The 95% one-sided lower score confidence limit is 0.0739. We note that the score confidence interval is shorter than the corresponding exact confidence interval.

Example 4.10. (*One-Sided Confidence Limits*) For the data in Example 4.3, let us find a 95% lower confidence limit for the true percentage of defective items in the shipment. Here, $X = 5$ and $n = 30$. To find a 95% lower confidence limit, select [StatCalc→Discrete→ Binomial→ Confidence Intervals ...] from *StatCalc*, enter 30 for n, 5 for k, and 0.95 for the confidence level. Click on [1-sided] to get 0.0681 (exact) and .0836 (score). If we use the score confidence limit, then we conclude that the true percentage of defective items in the shipment is at least 8.36 with confidence 0.95.

Example 4.11. (*Sample Size Calculation*) A researcher hypothesizes that the proportion of individuals with an attribute of interest in a population is 0.3, and he wants to estimate the true proportion within ±5% with 95% confidence. To compute the required sample size, select the dialog box [StatCalc→Discrete→ Binomial→Confidence Intervals ...] from *StatCalc*, enter .95 for [Conf Level], .3 for [Guess p], .05 for [Half-Width], and click on [exact] to get 340, and click on [score] to get 320. Thus, the sample size required to construct an exact confidence interval with the margin of error ±5% is 340. The required sample size for the score confidence interval is 320. The approximate formula for n in (4.13) also gives $319.58 \simeq 320$.

4.5 Prediction Intervals

The prediction problem concerns two independent binomial samples with the same "success probability" p. Given that X successes have been observed in n independent Bernoulli trials, we like to predict the number of successes Y in a future set of m independent Bernoulli trials. In particular, we like to find a $1 - \alpha$ prediction interval $[L(X; n, m, \alpha), U(X; m, n, \alpha)]$ so that

$$P_{X,Y}\left(L(X; n, m, \alpha) \leq Y \leq U(X; m, n, \alpha)\right) \geq 1 - \alpha.$$

Assume that $X \sim$ binomial(n, p) independently of $Y \sim$ binomial(m, p). The conditional distribution of X given the sum $X + Y = s$ is hypergeometric (Chapter 5) with the sample size s, number of "nondefects" n, and the lot size $n + m$. The conditional probability mass function is given by

$$P(X = x | X + Y = s, n, n + m) = \frac{\binom{n}{x}\binom{m}{y}}{\binom{m+n}{s}}, \quad \max\{0, s - m\} \leq x \leq \min\{n, s\}.$$

Let us denote the cumulative distribution function (cdf) of X given $X + Y = s$ by $H(t; s, n, n + m)$. That is,

$$H(t; s, n, n + m) = P(X \leq t | s, n, n + m) = \sum_{i=0}^{t} \frac{\binom{n}{i}\binom{m}{s-i}}{\binom{m+n}{s}}. \tag{4.14}$$

Note that the conditional cdf of Y given $X + Y = s$ is given by $H(t; s, m, n + m)$.

An Exact Prediction Interval

Thatcher (1964) developed the following exact prediction interval on the basis of the conditional distribution of X given $X + Y$. Let x be an observed value of X. The $1 - \alpha$ lower prediction limit L is the smallest integer for which

$$P(X \geq x | x + L, n, n + m) = 1 - H(x - 1; x + L, n, n + m) > \alpha. \tag{4.15}$$

The $1 - \alpha$ upper prediction limit U is the largest integer for which

$$H(x; x + U, n, n + m) > \alpha. \tag{4.16}$$

Furthermore, $[L, U]$ is a $1 - 2\alpha$ two-sided prediction interval for Y. Thatcher (1964) has noted that, for a fixed (x, n, m), the probability (4.16) is a decreasing function of U, and so a backward search, starting from m, can be used to find the largest integer U for which the probability in (4.16) is just greater than α. Similarly, we see that the probability in (4.15) is an increasing function of L, and so a forward search method, starting from a small value, can be used to find the smallest integer L for which this probability is just greater than α. The exact prediction intervals for extreme values of X are defined as follows. When $X = 0$, the lower prediction limit for Y is 0, and the upper one is determined by (4.16); when $X = n$, the upper prediction limit is m, and the lower prediction limit is determined by (4.15).

An Approximate Prediction Interval

Krishnamoorthy and Peng (2011) proposed the following prediction interval based on the complete sufficient statistic $X + Y$ for the binomial$(n+m, p)$ distribution. To describe their prediction interval, let $c_\alpha = z_{1-\alpha/2}$, where z_q is the $100q$ percentile of the standard normal distribution. Define

$$(L, U) = \frac{\left[\widehat{Y}\left(1 - \frac{c_\alpha^2}{m+n}\right) + \frac{mc_\alpha^2}{2n} \right] \pm c_\alpha \sqrt{\widehat{Y}(m - \widehat{Y})\left(\frac{1}{m} + \frac{1}{n}\right) + \frac{m^2 c_\alpha^2}{4n^2}}}{1 + \frac{mc_\alpha^2}{n(m+n)}}, \qquad (4.17)$$

where $\widehat{Y} = mX/n$ for $X = 0, 1, ..., n$. The $1 - \alpha$ prediction interval is given by $[\lceil L \rceil, \lfloor U \rfloor]$, where $\lceil x \rceil$ is the smallest integer greater than or equal to x, and $\lfloor x \rfloor$ is the largest integer less than or equal to x. As an example, if $(L, U) = (3.4, 6.7)$, then the prediction interval is $[4, 6]$.

   Krishnamoorthy and Peng's (2011) extensive comparison studies showed that the exact prediction intervals are too conservative, and the approximate prediction intervals control the coverage probabilities close to the nominal level and have shorter expected widths than those of exact ones. The dialog box [StatCalc→Discrete→Binomial→Confidence Intervals ...] computes both the exact and the approximate prediction intervals.

Example 4.12. (*Binomial Prediction Interval*) Suppose for a random sample of $n = 100$ devices tested, $x = 6$ devices are unacceptable. A 90% prediction interval is desired for the number of unacceptable devices in a future sample of $m = 50$ such devices. The sample proportion of unacceptable devices is $\widehat{p} = 6/100 = .06$ and $\widehat{Y} = m \times \widehat{p} = 3$. To compute the prediction intervals, select the dialog box [StatCalc→Discrete→Binomial→Confidence Intervals ...] from *StatCalc*, enter 100 for [No. of Trials, n], 6 for [No. Successes, k], 50 for [Future Sample Size, m], and .90 for [Conf Level]; click on [2-sided] to get approximate prediction interval $[1, 7]$, and the exact one $[0, 7]$.

Example 4.13. The manufacturer of an expensive piece of medical equipment has sold 40 units in the past year. Of these 40 units, two required repair/services in the past year. The manufacturer, who has currently received 60 orders for the equipment, is concerned on the number of service calls that he may receive from the hospitals that will use the equipment in the forthcoming year. Specifically, the manufacturer wants to predict the number of service calls in the forthcoming year. Formally, the problem is to find a prediction interval for $Y \sim \text{binomial}(60, p)$ based on $X \sim \text{binomial}(40, p)$, where p is the probability of receiving a service call for the equipment.

To find a 95% prediction interval in the above setup, select [StatCalc→Discrete→ Binomial→Confidence Intervals ...], enter 40 for [No. of Trials, n], 2 for [No. Successes,k], 60 for [Future Sample Size, m], and 0.95 for [Conf Level]; click on [2-sided] to get [0, 11] (approximate) and [0, 12] (exact).

Example 4.14. The data for this example are taken from the National Institute of Standards and Technology (NIST) webpage[1], and they represent fractions of defective chips in a sample of wafers. A chip in a wafer is considered to be defective whenever a misregistration, in terms of horizontal and/or vertical distances from the center, is recorded. On each wafer, locations of 50 chips were measured and the proportion of defective chips was recorded. As the original data (based on 30 wafers) was overdispersed (Wang and Tsung, 2009), we shall use a part of the data consisting of 21 wafers as given in Table 4.1. Krishnamoorthy, Xia, and Xie (2011) have tested the equality of the proportions of defective chips across the 21 wafers using a chi-square statistic

$$\sum_{i=1}^{21} \frac{n_i(\widehat{p}_i - \widehat{p})^2}{\widehat{p}(1 - \widehat{p})} = 19.58.$$

The p-value of the test is $P(\chi^2_{20} > 19.58) = 0.4842$, where χ^2_f denotes the chi-square random variable with degrees of freedom f. Here, the n_is are all equal to 50, $\widehat{p}_i$s are the sample fractions of defective given in Table 4.1, and the overall proportion of defective

$$\widehat{p} = \frac{\sum_{i=1}^{21} n_i \widehat{p}_i}{\sum_{i=1}^{21} n_i} = \frac{196}{1050} = 0.1867.$$

As the equality of proportions is tenable, we can use the combined estimate $\widehat{p}$ to estimate the true proportion of defective chips in a wafer.

TABLE 4.1: Fractions of Defective Chips in a Sample of 21 Wafers

Sample Number	Fraction of Defective, $\widehat{p}_i$	Sample Number	Fraction Defective, $\widehat{p}_i$	Sample Number	Fraction of Defective, $\widehat{p}_i$
1	.24	11	.10	21	.22
2	.16	12	.12	22	.18
3	.20	13	.24	23	.24
4	.14	14	.16	24	.14
5	.18	15	.20	25	.26
6	.28	16	.10	26	.18
7	.20	17	.26	27	.12

Suppose it is desired to find a 95% prediction interval for the number of defective chips in a wafer. To find the prediction interval, select the dialog box [StatCalc→Discrete →Binomial→Confidence Intervals ...], and enter 1050 for n, 196 for k, 50 for the future sample size m, .95 for the confidence level, and click [2-sided] to get [4, 14] (approximate PI) and [4, 15] (exact). Thus, if we decided to use the approximate prediction interval, the number of defective chips in a future wafer will be between 4 and 14.

---

[1]http://www.itl.nist.gov/div898/handbook/pmc/section3/pmc332.htm

4.6   Tolerance Intervals

As noted in Section 2.6.2, one-sided tolerance limits are one-sided confidence limits
of appropriate quantiles. As the binomial quantile involves only one unknown parameter (the success probability), a confidence limit for the quantile can be obtained
by replacing the parameter in the expression for quantile by a suitable confidence
limit. As a consequence, replacing the parameter by a better confidence limit in
the quantile expression, a better tolerance interval could be obtained. Hahn and
Chandra (1981) used the exact confidence interval (Section 4.4.3) for the binomial
parameter to find tolerance intervals for a binomial distribution. These tolerance
intervals are exact in the sense that the minimum coverage probability is at least
the nominal level $1 - \alpha$.

4.6.1   Equal-Tailed and Two-Sided Tolerance Intervals

Let X be a binomial(n, p) random variable, and let k be an observed value of
X. We shall describe the method of constructing tolerance intervals for a future
binomial(m, p) distribution based on k, n and m. For $0 < \beta < 1$, the 100β percentile
of the binomial(m, p) distribution is the smallest integer $k_\beta(p, m)$ for which

$$P(Y \leq k_\beta(p, m) | m, p) = \sum_{i=0}^{k_\beta(p,m)} \binom{m}{i} p^i (1 - p)^{m-i} \geq \beta.$$

The $(\beta, 1 - \alpha)$ upper tolerance limit for the binomial(m, p) distribution is $k_\beta(p_u, m)$,
where p_u is a $1 - \alpha$ upper confidence limit of p based on k and n. Specifically,
$k_\beta(p_u, m)$ is the 100β percentile of the binomial(m, p_u) distribution, defined as the
smallest integer that satisfies

$$P(Y \leq k_\beta(p_u, m) | m, p) = \sum_{i=0}^{k_\beta(p_u,m)} \binom{m}{i} p_u^i (1 - p_u)^{m-i} \geq p. \qquad (4.18)$$

The $(\beta, 1 - \alpha)$ lower tolerance limit for the binomial(m, p) distribution is $k_{1-\beta}(p_l, m)$,
where p_l is a $1 - \alpha$ lower confidence limit of p based on k and n. In particular,
$k_{1-\beta}(p_l, m)$ is the largest integer for which

$$P(Y \geq k_{1-\beta}(p_l, m) | m, p_l) = \sum_{i=k_{1-\beta}(p_l,m)}^{m} \binom{m}{i} p_l^i (1 - p_l)^{m-i} \geq \beta. \qquad (4.19)$$

The $(\beta, 1 - \alpha)$ equal-tailed tolerance interval is given by

$$\left[k_{\frac{1-\beta}{2}}(p_l, m), \quad k_{\frac{1+\beta}{2}}(p_u, m) \right], \qquad (4.20)$$

where (p_l, p_u) is a $1 - \alpha$ two-sided confidence interval for p based on a realization of
a binomial(n, p) random variable. In other words, $k_{\frac{1-\beta}{2}}(p_l, m)$ is the largest integer
that satisfies (4.19), and $k_{\frac{1+\beta}{2}}(p_u, m)$ is the smallest integer that satisfies (4.18).

The tolerance interval (4.20) based on the exact confidence interval for p is an exact equal-tailed tolerance interval, and we refer to the one based on the score confidence interval for p as the score equal-tailed tolerance interval.

4.6.2 Tolerance Intervals Based on Approximate Quantiles

Recall that one-sided tolerance limits are essentially confidence bounds on appropriate quantiles. On the basis of the normal approximation to the quantity

$$\frac{Y - mp}{\sqrt{mp(1-p)}},$$

the p quantile of a binomial(m, p) distribution is expressed as

$$k_\beta(p, m) \simeq mp + z_\beta \sqrt{mp(1-p)}.$$

Noting that the above quantile is an increasing function of p, an approximate $(\beta, 1 - \alpha)$ upper tolerance limit for the binomial(m, p) distribution can be obtained by replacing the p in the above expression by a $1 - \alpha$ upper confidence limit p_u. More specifically,

$$k_\beta(p_u, m) \simeq \left[mp_u + z_\beta \sqrt{mp_u(1-p_u)} \right]^*, \tag{4.21}$$

where $[x]^*$ is the integer nearest to x, is an approximate $(p, 1 - \alpha)$ upper tolerance limit. Similarly, an approximate $(p, 1 - \alpha)$ lower tolerance limit can be obtained as

$$k_\beta(p_l, m) \simeq \left[mp_l - z_\beta \sqrt{mp_l(1-p_l)} \right]^*. \tag{4.22}$$

If (p_l, p_u) is a $1 - \alpha$ confidence interval for p, then $\left[k_{\frac{1-\beta}{2}}(p_l, m), k_{\frac{1+\beta}{2}}(p_u, m) \right]$ is approximately equal to

$$\left[\left[mp_l - z_{\frac{1+\beta}{2}} \sqrt{mp_l(1-p_l)} \right]^*, \left[mp_u + z_{\frac{1+\beta}{2}} \sqrt{mp_u(1-p_u)} \right]^* \right]. \tag{4.23}$$

The above interval is an approximate $(\beta, 1-\alpha)$ equal-tailed tolerance interval. These tolerance intervals were proposed by Krishnamoorthy, Xia and Xie (2011).

Remark 4.2. In many applications, one needs two-sided tolerance intervals, not equal-tailed tolerance intervals (see Section 2.6.2). Note that $1-\alpha$ confidence interval for p is used in (4.20) and (4.23) to obtain an equal-tailed tolerance interval. Instead, if we use $1-2\alpha$ confidence interval for p, then the resulting tolerance interval includes at least a proportion β of the binomial distribution with coverage probability close to $1-\alpha$ (Krishnamoorthy, Xia and Xie, 2011), and so it can be used as an approximate $(\beta, 1 - \alpha)$ tolerance interval.

Remark 4.3. The tolerance intervals based on the exact confidence intervals are exact, in the sense that their coverage probabilities are at least the nominal confidence level $1 - \alpha$. However, the exact tolerance intervals are too conservative, as a result, unnecessarily wider. The tolerance intervals based on the score confidence interval are not exact, but they do have good coverage properties, and have shorter expected widths than those of the exact tolerance intervals. Finally, we note that the approximate tolerance intervals in (4.21), (4.22), and (4.23) are not only simple to compute, but they are also comparable with the score tolerance intervals based on (4.18) and (4.19).

The dialog box [StatCalc→Discrete →Binomial→Confidence Intervals ...] uses the above approach to compute the exact equal-tailed tolerance intervals (the ones in (4.18) and (4.19) with the exact confidence limits p_l and p_u), and the approximate equal-tailed tolerance intervals in (4.23) using the score confidence limits p_l and p_u. To find a $(p, 1 - \alpha)$ two-sided tolerance intervals, just enter $1 - 2\alpha$ for [Conf Level] in the dialog box.

Example 4.15. Let us use "wafers data" in Example 4.14 to illustrate the methods of finding tolerance intervals. To compute $(0.90, 0.95)$ one-sided as well as $(0.90, 0.95)$ two-sided tolerance intervals using the approaches given in the preceding sections, we note that $n = \sum_{i=1}^{21} n_i = 1050$, $k =$ the total number of defective chips, which is 196, and $m = 50$. To construct tolerance intervals for the binomial$(m = 50, p)$ distribution using *StatCalc*, select [StatCalc→Discrete →Binomial→Confidence Intervals ...], enter 1050 for [No. of Trials, n], 196 for [No. Successes, k], 50 for [Future Sample Size, m], .90 for [Content Level], .95 for [Conf Level], and click on [2-sided] to get [4, 15] (exact equal-tailed tolerance interval and the approximate one are the same). Thus, with 95% confidence, we can say that at least 90% of wafers have 4 to 15 defective chips. Notice that this tolerance interval [4, 15] is an equal-tailed $(.90, .95)$ tolerance interval. To find an approximate $(.90, .95)$ two-sided tolerance interval, just enter .90 (instead of .95) for confidence level, and click [2-sided] to get [4, 15], which is the same as the $(.90, .95)$ equal-tailed tolerance interval.

To find one-sided tolerance limits, click on [1-sided] to get 5 and 14. That is, a $(0.90, 0.95)$ lower tolerance limit is 5, and a $(0.90, 0.95)$ upper tolerance limit is 14. Thus, with 95% confidence, we can say that at least 90% of wafers have five or more defective chips. Furthermore, at least 90% of wafers have 14 or fewer defective chips, with confidence 95%. Finally, we note that both approaches produced the same results because of the large sample size 1,050.

Remark 4.4. Suppose that a binomial experiment is repeated m times, and let k_j denote the number of successes, and let n_j denote the number of trials at the jth time, $j = 1, 2, ..., m$. Then, all the inferential procedures in the preceding sections are valid with (n, k) replaced by $\left(\sum_{j=1}^{m} n_j, \sum_{j=1}^{m} k_j \right)$.

4.7 Tests for the Difference between Two Proportions

Let $X_1 \sim$ binomial(n_1, p_1) distribution independently of $X_2 \sim$ binomial(n_2, p_2) distribution. Let (k_1, k_2) be an observed value of (X_1, X_2). We shall describe some tests for comparing these two success probabilities p_1 and p_2. These tests are applicable to compare two proportions in large populations. Specifically, let X_i denote the number of individuals in a sample of size n_i from the population i with an attribute of interest, $i = 1, 2$. The following tests can be used to compare the population proportions on the basis of sample proportions $\widehat{p}_1 = \frac{X_1}{n_1}$ and $\widehat{p}_2 = \frac{X_2}{n_2}$.

4.7.1   Approximate Tests

The Wald Test

Let $\widehat{p}_i = X_i/n_i$, $i = 1, 2$. The Wald test is on the basis of the result that

$$Z = \frac{(\widehat{p}_1 - \widehat{p}_2) - (p_1 - p_2)}{\sqrt{\frac{\widehat{p}_1(1-\widehat{p}_1)}{n_1} + \frac{\widehat{p}_2(1-\widehat{p}_2)}{n_2}}} \sim N(0, 1), \text{ asymptotically.} \tag{4.24}$$

Let z_0 be an observed value of Z. For testing

$$H_0 : p_1 = p_2 \quad \text{vs.} \quad H_a : p_1 \neq p_2, \tag{4.25}$$

the Wald test rejects the null hypothesis when $|z_0| > z_{1-\frac{\alpha}{2}}$.

An Unconditional Test

Let $\widehat{p}_i = X_i/n_i$ and $\widehat{p}_{i0} = k_i/n_i$, where (k_1, k_2) is an observed value of (X_1, X_2). Consider the test statistic

$$Z(X_1, X_2, n_1, n_2) = \frac{X_1 - X_2}{\sqrt{\widehat{p}_X(1 - \widehat{p}_X)\left(\frac{1}{n_1} + \frac{1}{n_2}\right)}}, \text{ where } \widehat{p}_X = \frac{X_1 + X_2}{n_1 + n_2}.$$

An estimate of the p-value for testing hypotheses $H_0 : p_1 \leq p_2$ vs. $H_a : p_1 > p_2$ can be computed using the formula

$$\begin{aligned} P(k_1, k_2, n_1, n_2) &= \sum_{x_1=0}^{n_1} \sum_{x_2=0}^{n_2} f(x_1|n_1, \widehat{p}_k) f(x_2|n_2, \widehat{p}_k) \\ &\times \ I\left(Z(x_1, x_2, n_1, n_2) \geq Z(k_1, k_2, n_1, n_2)\right), \end{aligned} \tag{4.26}$$

where

$$\widehat{p}_k = \frac{k_1 + k_2}{n_1 + n_2},$$

$I(.)$ is the indicator function, and

$$f(x_i|n_i, \widehat{p}_k) = \binom{n_i}{x_i} \widehat{p}_k^{x_i - 1} (1 - \widehat{p}_k)^{n_i - x_i}, \ i = 1, 2.$$

The terms $Z(k_1, k_2, n_1, n_2)$ is equal to $Z(x_1, x_2, n_1, n_2)$ with x replaced by k. The null hypothesis will be rejected when the p-value in (4.26) is less than or equal to the nominal level α. This test is due to Storer and Kim (1990). Even though this test is approximate, its type I error rates seldom exceed the nominal level, and it is more powerful than Fisher's conditional test in the following section.

The p-values for a left-tailed test and for a two-tailed test can be computed similarly. Specifically, the p-value for a two-tailed test can be obtained by replacing $Z(x_1, x_2, n_1, n_2)$ and $Z(k_1, k_2, n_1, n_2)$ in (4.26) by their absolute values.

4.7.2   Fisher's Exact Test

Let $X \sim$ binomial(n_1, p_1) independently of $Y \sim$ binomial(n_2, p_2) random variable. When $p_1 = p_2$, the conditional probability of observing $X = k$, given that $X + Y = m$ is given by

$$P(X = k | X + Y = m) = \frac{\binom{n_1}{k}\binom{n_2}{m-k}}{\binom{n_1+n_2}{m}}, \quad \max\{0, m - n_2\} \le k \le \min\{n_1, m\}.$$

$$(4.27)$$

The probability mass function in (4.27) is known as the hypergeometric$(m, n_1, n_1 + n_2)$ pmf (Chapter 5). This conditional distribution can be used to test the hypotheses regarding $p_1 - p_2$. For example, when

$$H_0 : p_1 \le p_2 \quad \text{vs.} \quad H_a : p_1 > p_2,$$

the null hypothesis will be rejected if the p-value $P(X \ge k | X + Y = m)$ is less than or equal to the nominal level α. Similarly, the p-value for testing $H_0 : p_1 \ge p_2$ vs. $H_a : p_1 < p_2$ is given by $P(X \le k | X + Y = m)$.
In the form of 2×2 table we have

TABLE 4.2: 2×2 Table

Sample	Successes	Failures	Totals
1	k	$n_1 - k$	n_1
2	$m - k$	$n_2 - m + k$	n_2
Totals	m	$n_1 + n_2 - m$	$n_1 + n_2$

4.7.3   Powers and Sample Size Calculation

For a given $(n_1, n_2, p_1, p_2, \alpha)$, the exact power of the tests described in the preceding sections can be computed using the expression

$$\sum_{k_1=0}^{n_1} \sum_{k_2=0}^{n_2} f(k_1 | n_1, p_1) f(k_2 | n_2, p_2) I \, (\text{p} - \text{value} \le \alpha), \qquad (4.28)$$

where $f(x | n, p) = \binom{n}{x} p^x (1 - p)^{n-x}$ and $I(.)$ is the indicator function. The powers of a left-tailed test and a two-tailed test can be computed similarly. *StatCalc* uses the above formula for computing the powers of the unconditional test and Fisher's exact test.

For a given 2×2 table of "failures and successes," the dialog box [StatCalc→ Discrete→ Binomial→Two-Sample ...] computes the probability of observing k or more successes (as well as the probability of observing k or less number of successes) in the cell (1,1). If either probability is less than $\frac{\alpha}{2}$, then the null hypothesis of equal proportion will be rejected at the level α. This dialog box also calculates p-values of the unconditional test in Section 4.7.1, and sample sizes required for both tests to attain a specified power.

Example 4.16. (*P-values of the Conditional and the Unconditional Test*) Suppose a sample of 25 observations from population 1 yielded 20 successes, and a sample of 20 observations from population 2 yielded 10 successes. Let p_1 and p_2 denote the proportions of successes in populations 1 and 2, respectively. We want to test

$$H_0 : p_1 \leq p_2 \quad \text{vs.} \quad H_a : p_1 > p_2.$$

To compute the p-values, select the dialog box [StatCalc→Discrete→ Binomial→Two-Sample ...] from *StatCalc*, enter the numbers of successes and failures for each sample, and click on [Pr $<=$ (1,1) cell] to get the p-value of the conditional test as .0355347. The p-value for the two-tailed test ($H_0 : p_1 = p_2$ vs. $H_a : p_1 \neq p_2$) is $2 \times .0355347 = .0711$. Thus, at the level of 5%, we can conclude that $p_1 > p_2$, but we cannot conclude that p_1 is significantly different from p_2.

To find the p-values of the unconditional test, enter 0 for [H0: p1-p2=d], and click on [p-values for] to get .024635 (for $H_a : p_1 > p_2$) and .040065 (for $H_a : p_1 \neq p_2$). Note that the p-values of the unconditional test are smaller than the corresponding p-values of the conditional test.

Example 4.17. A physician believes that one of the causes of a particular disease is long-term exposure to a chemical. To test his belief, he examined a sample of adults and obtained the following 2 × 2 table:

Group	Symptoms Present	Symptoms Absent	Totals
Exposed	13	19	32
Unexposed	4	21	25
Totals	17	40	57

The hypotheses of interest are

$$H_0 : p_e \leq p_u \quad \text{vs.} \quad H_a : p_e > p_u,$$

where p_e and p_u denote, respectively, the actual proportions of exposed people and unexposed people who have the symptom. To find the p-value, select the dialog box [StatCalc→ Discrete→ Binomial→Two-Sample ...], enter the cell frequencies, and click [Prob $<=$ (1,1) cell]. The p-value is 0.04056. Thus, at the 5% level, the data provide sufficient evidence to indicate that there is a positive association between the prolonged exposure to the chemical and the disease.

To compute the p-value of the unconditional test, click [p-values - difference] to get 0.022654. We observe that the p-value of the unconditional test is smaller than the p-value of the Fisher exact test. This is because the Fisher test is conservative.

Example 4.18. (*Sample Size Calculation for Power*) Suppose the sample size for each group needs to be determined to carry out a two-tailed test at the level of significance $\alpha = 0.05$ and power $= 0.80$. Furthermore, the guess values of the proportions are given as $p_1 = 0.45$ and $p_2 = 0.15$. To determine the sample size, select [StatCalc→Discrete→ Binomial→Two-Sample ...] from *StatCalc*, enter 2 for two-tailed test, 0.05 for [Level], 0.45 for p_1, 0.15 for p_2, and 28 for each sample size. Click [Power] to get 0.697916. By trial-error, we can find the required sample size from both groups is 36, and the corresponding power is 0.81429. Also, note that the power at $n_1 = n_2 = 35$ is 0.799666. If we choose to use the Fisher test, then the required sample size from both groups is 41, and the corresponding power is .8062.

In general, the sample size required for the Fisher test is larger than that for the unconditional test, because the former test is conservative.

The power of the tests can also be computed for unequal sample sizes. For instance, when $n_1 = 30$, $n_2 = 42$, $p_1 = 0.45$, $p_2 = 0.15$, the power of a two-tailed unconditional test at the nominal level 0.05 is 0.799335. For the same values of p_1 and p_2, a power of 0.797573 can be attained if $n_1 = 36$ and $n_2 = 35$.

Example 4.19. (*Sample Size Calculation for Power*) Suppose that an experimenter wants to apply Fisher's exact test to compare two proportions. His guess on the proportion p_1 is around .45, and on p_2 is around .15, and he wants to compute the required sample sizes to have a power of 0.9 for testing

$$H_0 : p_1 \leq p_2 \text{ vs. } H_a : p_1 > p_2$$

at the level 0.05.

To determine the sample size required from each population, enter 28 (this is our initial guess) for both sample sizes, 0.45 for p_1, 0.15 for p_2, 0.05 for level, and click power to get 0.724359. This is less than 0.9. After trying a few values larger than 28 for each sample size, we find the required sample size is 45 from each population. In this case, the actual power is 0.90683.

4.8 Two-Sample Confidence Intervals for Proportions

Let $X_1 \sim \text{binomial}(n_1, p_1)$ independently of $X_2 \sim \text{binomial}(n_2, p_2)$. Let (k_1, k_2) be an observed value of (X_1, X_2). It should be noted that there are several interval estimation methods are available in the literature. Among them the score method appears to be popular, and comparison studies by several authors indicated that score confidence intervals are satisfactory in terms of coverage properties and precision. In the following, we shall describe the Wald, score, and fiducial confidence intervals for various problems.

4.8.1 Difference

Wald Confidence Interval

Let $\widehat{p}_i = \frac{X_i}{n_i}$ and $\widehat{q}_i = 1 - \widehat{p}_i$, $i = 1, 2$. The Wald confidence interval is based the result that

$$\frac{\widehat{p}_1 - \widehat{p}_2 - d}{\sqrt{\frac{\widehat{p}_1 \widehat{q}_1}{n_1} + \frac{\widehat{p}_2 \widehat{q}_2}{n_2}}} \sim N(0, 1), \text{ asymptotically.} \tag{4.29}$$

On the basis of the above distributional result, an approximate $1 - \alpha$ confidence interval for $p_1 - p_2$ is obtained as

$$\widehat{p}_1 - \widehat{p}_2 \pm z_{1-\frac{\alpha}{2}} \sqrt{\frac{\widehat{p}_1 \widehat{q}_1}{n_1} + \frac{\widehat{p}_2 \widehat{q}_2}{n_2}}, \tag{4.30}$$

where z_α denotes the α quantile of the standard normal distribution.

Newcombe Confidence Interval

Newcombe (1998) proposed a confidence interval for the difference $d = p_1 - p_2$ based on the Wilson score intervals for individual proportions. Let (L_i, U_i) be the $1 - \alpha$ score confidence interval for p_i given in (4.6), $i = 1, 2$. Newcombe's confidence interval for $p_1 - p_2$ is based on (L_i, U_i)'s, and is given by

$$\left(\hat{d} - z_{1-\frac{\alpha}{2}} \sqrt{\frac{L_1(1 - L_1)}{n_1} + \frac{U_2(1 - U_2)}{n_2}}, \ \hat{d} + z_{1-\frac{\alpha}{2}} \sqrt{\frac{U_1(1 - U_1)}{n_1} + \frac{L_2(1 - L_2)}{n_2}} \right),$$

(4.31)

where $\hat{d} = \hat{p}_1 - \hat{p}_2$, and z_α denotes the α quantile of the standard normal distribution.

Fiducial Confidence Interval

A fiducial quantity (see Section 4.4.3) for p_i is given by $B_{k_i + \frac{1}{2}, n_i - k_i + \frac{1}{2}}$, where (k_1, k_2) is an observed value of (X_1, X_2), and $B_{a,b}$ denotes the beta random variable with shape parameters a and b. A fiducial quantity for the difference $d = p_1 - p_2$ is given by

$$Q_d = Q_{p_1} - Q_{p_2} = B_{k_1 + \frac{1}{2}, n_1 - k_1 + \frac{1}{2}} - B_{k_2 + \frac{1}{2}, n_2 - k_2 + \frac{1}{2}}.$$

(4.32)

The $\frac{\alpha}{2}$ quantile and the $1 - \frac{\alpha}{2}$ quantile of Q_d form a $1 - \alpha$ confidence interval for $p_1 - p_2$. Note that for a given (k_1, k_2), the percentiles of Q_d can be estimated using Monte Carlo simulation or obtained by numerical method. The percentiles of Q_d in (4.32) can be approximated using the modified normal-based approximations (2.14) and (2.15). A lower 100α percentile of Q_d can be approximated by

$$Q_{d;\alpha} \simeq \tilde{p}_1 - \tilde{p}_2 - \left[(\tilde{p}_1 - B_{1;\alpha})^2 + (\tilde{p}_2 - B_{2;1-\alpha})^2 \right]^{\frac{1}{2}}, \text{ for } 0 < \alpha \leq .5,$$

(4.33)

where

$$\tilde{p}_i = \frac{k_i + \frac{1}{2}}{n_i + 1} \text{ and } B_i = B_{k_i + \frac{1}{2}, n_i - k_i + \frac{1}{2}}, \quad i = 1, 2,$$

and $B_{i;\alpha}$ denotes the 100α percentile of B_i, $i = 1, 2$. An approximate $100(1 - \alpha)$ percentile of Q_d is expressed as

$$Q_{d;1-\alpha} \simeq \tilde{p}_1 - \tilde{p}_2 + \left[(\tilde{p}_1 - B_{1;1-\alpha})^2 + (\tilde{p}_2 - B_{2;\alpha})^2 \right]^{\frac{1}{2}}, \text{ for } 0 < \alpha \leq .5.$$

(4.34)

The interval $(Q_{d;\alpha}, Q_{d;1-\alpha})$ is an approximate $1 - 2\alpha$ confidence interval for the difference $p_1 - p_2$.

Score Confidence Interval

Consider the Wald statistic in (4.24) for testing $H_0 : p_1 - p_2 = d$. The Wald statistic Z has an asymptotic standard normal distribution. The Wald confidence interval is obtained by inverting the test based on Z. Instead of using the usual estimate of variance of $(\hat{p}_1 - \hat{p}_2)$ in (4.24), Miettinen and Nurminen (1985) have used the variance estimate based on the maximum likelihood estimate under the constraint that $p_1 = p_2 + d$. Specifically, they proposed the following test statistic:

$$T_M = \frac{\hat{p}_1 - \hat{p}_2 - d}{[(\tilde{p}_2 + d)(1 - \tilde{p}_2 - d)/n_1 + \tilde{p}_2(1 - \tilde{p}_2)/n_2]^{\frac{1}{2}} \sqrt{R_n}},$$

(4.35)

where $R_n = (n_1 + n_2)/(n_1 + n_2 - 1)$, and $\tilde{p}_2$ is the maximum likelihood estimator under the constraint that $p_1 - p_2 = d$. Even though the constrained likelihood equation (a function of $\widehat{p}_2$) is a polynomial of order three, the likelihood equation has a unique closed-form solution; see Appendix I of Miettinen and Nurminen (1985). Since the constrained maximum likelihood estimate (MLE) is also a function of d, an approximate confidence interval for $p_1 - p_2$ can be obtained by solving the equation $|T_M| = z_{1-\alpha/2}$ for d numerically. The R package "gsDesign" or the recent one "PropCIs" can be used to compute the score confidence interval for $p_1 - p_2$.

Comparison of Confidence Intervals for the Difference Between Two Proportions

On the basis of comparison studies by Newcombe (1998), Krishnamoorthy and Zhang (2015) and Krishnamoorthy, Lee and Zhang (2014), the score confidence interval, the Newcombe confidence interval and the fiducial confidence intervals are quite comparable in terms of coverage probabilities and expected widths, and these three confidence intervals are superior to the Wald confidence interval. On computational ease, the fiducial confidence interval and the Newcombe confidence intervals are easy to calculate, whereas the score confidence interval requires a root fining method for a nonlinear equation. Furthermore, numerical studies by Krishnamoorthy and Zhang (2015) indicated some situations where the score confidence intervals are inferior to the Newcombe and fiducial confidence intervals.

Example 4.20. To assess the effectiveness of a diagnostic test for detecting a certain disease, it was administered on 30 diseased people, and 28 were correctly diagnosed by the test. The test was also used on 60 nondiseased persons, and 8 were incorrectly diagnosed. Let p_t and p_f denote, respectively, the true positive diagnoses and false positive diagnoses. We shall find 95% confidence intervals for the difference $p_t - p_f$. Noticing that $\widehat{p}_t = 0.9333$ and $\widehat{p}_f = 0.1333$, we get the point estimate for the difference $p_t - p_f$ as 0.8.

To find the Newcombe confidence interval in (4.31), we find 95% score confidence intervals for p_t and p_f as $(L_t, U_t) = (.7868, .9815)$ and $(L_f, U_f) = (.0691, .2417)$, respectively. Using these numbers along with $z_{.975} = 1.96$ in (4.31), we find the 95% Newcombe confidence interval as $(.6177, .8803)$. The score confidence interval using the function

$$\text{ciBinomial}(28,8,30,60,\text{alpha}=.05,\text{adj}=1,\text{scale}=\text{``Difference''})$$

in R package "GsDesign" is computed as $(.6297, .8913)$. The fiducial confidence interval can be computed using [StatCalc→Discrete→ Binomial→Two-Sample ...], and is $(.6288, .8797)$. Note that all the confidence intervals are in good agrement.

4.8.2   Relative Risk and Odds Ratio

Let p_e denotes the probability of an event (such as death or adverse symptoms) in the exposed group of individuals, and p_c denotes the same in the control group, then the ratio p_e/p_c is a measure of relative risk for the exposed group. The ratio of odds is defined by $[p_e/(1 - p_e)]/[p_c/(1 - p_c)]$, which represents the relative odds of an event in the exposed group compared to that in the control group. In this section, we shall see some methods for finding confidence intervals for the relative risk and odds ratio.

Fiducial Confidence Intervals for the Relative Risk

Let $X_1 \sim \text{binomial}(n_1, p_1)$ independently of $X_2 \sim \text{binomial}(n_2, p_2)$. Let (k_1, k_2) be an observed value of (X_1, X_2). The fiducial variable for p_i is $B_{k_i+1/2, n_i-k_i+1/2}$, where $B_{a,b}$ denotes the beta random variable with shape parameters a and b.

A fiducial variable for the relative risk $RR = p_1/p_2$ can be obtained as

$$Q_{RR} = \frac{B_{k_1+1/2, n_1-k_1+1/2}}{B_{k_2+1/2, n_2-k_2+1/2}}, \tag{4.36}$$

where the beta random variables in (4.36) are independent. For a given (k_1, k_2), appropriate percentiles of Q_{RR} form a confidence interval for RR. Using the modified normal approximation (MNA) in (2.16), we find the approximate $1 - \alpha$ lower confidence limit for RR as

$$Q_{RR_L} = \frac{\tilde{p}_1\tilde{p}_2 - \sqrt{(\tilde{p}_1\tilde{p}_2)^2 - [\tilde{p}_2{}^2 - (u_2 - \tilde{p}_2)^2][\tilde{p}_1{}^2 - (l_1 - \tilde{p}_1)^2]}}{\tilde{p}_2{}^2 - (u_2 - \tilde{p}_2)^2}, \tag{4.37}$$

the approximate $1 - \alpha$ upper confidence limit as

$$Q_{RR_U} = \frac{\tilde{p}_1\tilde{p}_2 + \sqrt{(\tilde{p}_1\tilde{p}_2)^2 - [\tilde{p}_1{}^2 - (u_1 - \tilde{p}_1)^2][\tilde{p}_2{}^2 - (l_2 - \tilde{p}_2)^2]}}{\tilde{p}_2{}^2 - (l_2 - \tilde{p}_2)^2}, \tag{4.38}$$

where $\tilde{p}_i = \frac{k_i+.5}{n_i+1}$ and

$$(l_i, u_i) = \left(B_{k_i+.5, n_i-k_i+.5; \frac{\alpha}{2}}, \; B_{k_i+.5, n_i-k_i+.5; 1-\frac{\alpha}{2}} \right), \quad i = 1, 2.$$

Score Confidence Interval for the Relative Risk

To express the score confidence interval (based on the likelihood approach), we shall use Miettinen and Nurminen's (1985) derivation. Let $\zeta = p_1/p_2$. The MLE $\tilde{p}_2$ of p_2 under the constraint $p_1 = \zeta p_2$, is the solution to the equation $a\tilde{p}_2^2 + b\tilde{p}_2 + c = 0$, where $a = (n_1 + n_2)\zeta$, $b = -[(X_2 + n_1)\zeta + X_1 + n_2]$, and $c = X_1 + X_2$. The approximate $1 - \alpha$ confidence limits are the roots to the equation (with respect to ζ)

$$\left| \frac{\hat{p}_1 - \zeta\hat{p}_2}{\sqrt{\frac{\tilde{p}_1\tilde{q}_1}{n_1} + \frac{\tilde{p}_2\tilde{q}_2}{n_2}}} \right| = z_{1-\frac{\alpha}{2}}. \tag{4.39}$$

Notice that the above confidence interval is based on the asymptotic normality of the term within the absolute signs. Special algorithm is necessary to find the roots, and the R package "GsDesign" can be used to find the score confidence interval for the relative risk.

There are several confidence intervals for the relative risk available in the literature, and among them the score confidence interval is very popular for its accuracy. Recent studies by Krishnamoorthy, Lee and Zhang (2014) indicated that the fiducial confidence intervals are quite comparable with the score confidence intervals for most cases, and are better than the score confidence intervals in some cases. Furthermore, the fiducial confidence intervals are easier to compute. The dialog box [StatCalc→Discrete→Binomial→ Two-Sample ...] uses the fiducial approach to find confidence intervals for the relative risk. For other methods of constructing confidence intervals for p_1/p_2, see Bedrick (1987).

Example 4.21. (*Relative Risk*) Let us find 95% confidence intervals for the relative risk based on the exposure data in Example 4.17. Here $k_1 = 13, n_1 - k_1 = 19, k_2 = 4$, and $n_2 - k_2 = 21$. To compute the 95% confidence interval based on the fiducial approach, select the dialog box [StatCalc→Discrete→Binomial→Two-Sample ...] from *StatCalc*, enter 13 for [Successes 1], 19 for [Failures 1], 4 for [Successes 2], and 21 for [Failures 2]. After entering .95 for [Conf Level], click [CI-Ratio] to get (1.0269, 7.5244). The score confidence interval for the relative risk is computed using the function

$$\text{ciBinomial}(13,4,32,25,\text{alpha}=.05,\text{adj}=1,\text{scale}=\text{``RR''})$$

in the R package "GsDesign" as (1.0163, 6.8812). Both confidence intervals indicate that the risk for the exposed groups is higher than that for the unexposed group.

Confidence Intervals for the Odds Ratio

Let $X_1 \sim \text{binomial}(n_1, p_1)$ independently of $X_2 \sim \text{binomial}(n_2, p_2)$. Let (k_1, k_2) be an observed value of (X_1, X_2). In the following, we shall describe some methods for estimating the odds ratio $\eta = [p_1/(1 - p_1)]/[p_2/(1 - p_2)]$.

An Exact Confidence Interval

The exact $1 - \alpha$ confidence interval for the odds ratio η is based on the conditional distribution of X_1 given the total $X_1 + X_2 = m$. The pmf of this conditional distribution is given by

$$P(X_1 = x | \eta, m, n_1, n_2) = \frac{\binom{n_1}{x}\binom{n_2}{m-x}\eta^x}{\sum_{y=l}^{u}\binom{n_1}{y}\binom{n_2}{m-y}\eta^y}, \quad x = l, ..., u,$$

where $l = \max\{0, m - n_2\}$ and $u = \min\{n_1, m\}$. Like the Clopper-Pearson confidence interval for a binomial proportion, the endpoints of the exact $1 - \alpha$ confidence interval (η_L, η_U) are determined by

$$P(X_1 \geq k | \eta_L, m, n_1, n_2) = \frac{\alpha}{2} \text{ and } P(X_1 \leq k | \eta_U, m, n_1, n_2) = \frac{\alpha}{2}, \quad (4.40)$$

where k_1 is an observed values of X_1. This exact method was proposed by Cornfield (1956). Thomas and Gart (1977) have provided table values to calculate confidence intervals for the odds ratio.

Logit Confidence Interval

This confidence interval is based on the asymptotic normality of

$$\ln\left(\frac{\widehat{p}_1}{1 - \widehat{p}_1}\right) - \ln\left(\frac{\widehat{p}_2}{1 - \widehat{p}_2}\right),$$

where $\widehat{p}_i = X_i/n_i, i = 1, 2$. In order to handle zero counts, the formula for the confidence interval is adjusted by adding one half to each cell in a 2×2 contingency table. Let $\eta = p_1(1 - p_2)/[p_2(1 - p_1)]$ and $\widehat{\eta}_{\frac{1}{2}} = \widehat{p}_{1,\frac{1}{2}}(1 - \widehat{p}_{2,\frac{1}{2}})/[\widehat{p}_{2,\frac{1}{2}}(1 - \widehat{p}_{1,\frac{1}{2}})]$, where $\widehat{p}_{i,\frac{1}{2}} = (X_i + 1/2)/(n_i + 1/2), i = 1, 2$. An estimate of the standard deviation of $\ln(\widehat{\eta}_{\frac{1}{2}})$ is given by

$$\widehat{\sigma}\left(\widehat{\eta}_{\frac{1}{2}}\right) = \left(\frac{1}{X_1 + \frac{1}{2}} + \frac{1}{n_1 - X_1 + \frac{1}{2}} + \frac{1}{X_2 + \frac{1}{2}} + \frac{1}{n_2 - X_2 + \frac{1}{2}}\right)^{\frac{1}{2}}.$$

Using the above quantities, the $1 - \alpha$ logit confidence interval for the odds ratio is given by

$$\widehat{\eta}_{\frac{1}{2}}\left(\exp\left[-z_{\alpha/2}\widehat{\sigma}\left(\widehat{\eta}_{\frac{1}{2}}\right)\right], \exp\left[z_{\alpha/2}\widehat{\sigma}\left(\widehat{\eta}_{\frac{1}{2}}\right)\right]\right). \qquad (4.41)$$

Fiducial Confidence Intervals

Following the lines of fiducial approach for the risk ratio, a fiducial variable for the odds ratio $\eta = p_1(1 - p_2)/(p_2(1 - p_1))$ can be obtained by replacing the parameters by their fiducial variables. Thus, the fiducial variable for the odds ratio is given by

$$Q_\eta = \frac{B_{k_1+1/2,n_1-k_1+1/2}/(1 - B_{k_1+1/2,n_1-k_1+1/2})}{B_{k_2+1/2,n_2-k_2+1/2}/(1 - B_{k_2+1/2,n_2-k_2+1/2})} = \frac{G_1}{G_2}, \qquad (4.42)$$

where all the beta random variables are mutually independent. Percentiles of Q_η, which can be estimated using Monte Carlo simulation, form a $1 - \alpha$ confidence interval for the odds ratio. The percentiles of Q_η can be obtained from those of $\ln Q_\eta = \ln G_1 - \ln G_2$, and these percentiles can be approximated by modified normal-based approximations in (2.14) and (2.15). To express the approximate percentiles of $\ln Q_\eta$, let

$$\mu_{g1} = E\ln(G_1) = \psi(k_1+.5) - \psi(n_1+1) \text{ and } \mu_{g2} = E\ln(G_2) = \psi(k_2+.5) - \psi(n_2+1),$$

where ψ is the digamma function. Furthermore, let $B_{1;\alpha} = B_{k_1+.5,n_1-k_1+.5;\alpha}$ and $B_{2;\alpha} = B_{k_2+.5,n_2-k_2+.5;\alpha}$. Define $G_{i;\alpha} = \ln[B_{i;\alpha}/(1-B_{i;\alpha})]$, $i = 1,2$. Using the above expectations and percentiles in (2.14), we find the approximate 100α percentile of $\ln Q_\eta$ as

$$\ln Q_{\eta;\alpha} \simeq \mu_{g1} - \mu_{g2} - \left\{(\mu_{g1} - G_{1;\alpha})^2 + (\mu_{g2} - G_{2;1-\alpha})^2\right\}^{1/2}, \quad 0 < \alpha < .5, \quad (4.43)$$

and using (2.15), we find the approximate $100(1 - \alpha)$ percentile as

$$\ln Q_{\eta;1-\alpha} \simeq \mu_{g1} - \mu_{g2} + \left\{(\mu_{g1} - G_{1;1-\alpha})^2 + (\mu_{g2} - G_{2;\alpha})^2\right\}^{1/2}, \quad 0 < \alpha \le .5. \quad (4.44)$$

Note that $(\exp(\ln Q_{\eta;\alpha}), \exp(\ln Q_{\eta;1-\alpha}))$ is an approximate $1 - 2\alpha$ confidence interval for the odds ratio.

Among these three confidence intervals for the odds ratio, the exact ones are too conservative, yielding confidence intervals that are too wide. The logit confidence intervals are unsatisfactory even for large samples; they could be too conservative or liberal depending on the values of the sample sizes and the parameters (see Krishnamoorthy and Lee, 2010). Numerical investigation by Krishnamoorthy, Lee and Zhang (2014) suggested that the above fiducial confidence interval is very satisfactory when $\min\{k_1, n_1 - k_1, k_2, n_2 - k_2\} \ge 2$, equivalently, all cell entries in a 2×2 table are two or more. This approximation may be used when this condition is met, and the Monte Carlo estimates of the percentiles of Q_η in (4.42) can be used otherwise.

Example 4.22. To illustrate the results for the odds ratio, let us use the stillbirth and miscarriages data reported in Bailey (1987). In a group of 220 women who were exposed to diethylstilbestrol (DES), eight suffered stillbirths (57 miscarriages), and in a group of 224 women who were not exposed to DES, three suffered a stillbirth (36 miscarriages). The data are summarized in the following table.

group	sample size	still birth	miscarriage
exposed	$n_1 = 220$	8	57
unexposed	$n_2 = 224$	3	36

Note that $k_1 = 8, n_1 - k_1 = 212, k_2 = 3$, and $n_2 - k_2 = 221$. We shall now compute 95% confidence intervals for the ratio of odds of a stillbirth in the exposed group to that in the nonexposed group. The point estimate of the odds ratio is 2.78. The 95% logit confidence interval is (0.73, 10.62). To compute the confidence intervals using *StatCalc*, select the dialog box [StatCalc→Discrete→Binomial→Two-Sample ...], enter the data in appropriate edit boxes, and click on [Fiducial] to get fiducial confidence interval (0.79, 11.51). Click on [Exact] to get (.65, 16.45). The exact confidence interval using the conditional approach is the widest among these three intervals.

We shall now compute 95% confidence intervals for the ratio of odds of a miscarriage in the exposed group to that in the nonexposed group. Now, $k_1 = 57, n_1 - k_1 = 163, k_2 = 36$, and $n_2 - k_2 = 188$. The point estimate of the odds ratio is 1.83. The 95% logit confidence interval is (1.15, 2.87). Using *StatCalc*, we found the fiducial confidence interval as (1.15, 2.92), and the exact confidence interval based on the conditional approach is (1.12, 3.01), which is again the widest among these three intervals.

4.9 Confidence Intervals for a Linear Combination of Proportions

We shall now describe some confidence intervals for a linear combination $\xi = \sum_{i=1}^{g} w_i p_i$, where $w_1, ..., w_g$ are specified values, based on independent random variables $X_1, ..., X_g$ with $X_i \sim$ binomial(n_i, p_i), $i = 1, ..., g$.

Score Confidence Intervals

Let $\widehat{p}_i = X_i/n_i$, $i = 1, ..., g$, and let $\widehat{\xi} = \sum_{i=1}^{g} w_i \widehat{p}_i$. Consider testing

$$H_0 : \xi = \xi_0 \text{ vs. } H_1 : \xi \neq \xi_0, \tag{4.45}$$

where ξ_0 is a specified value. Let $S = \sum_{i=1}^{g} w_i p_i$. Then

$$z_0 = \frac{\widehat{\xi} - \xi_0}{\sqrt{V_0}} \sim N(0, 1) \text{ asymptotically}, \tag{4.46}$$

where $V_0 = \sum_{i=1}^{g} w_i^2 \widetilde{p}_{i0}(1 - \widetilde{p}_{i0})/n_i$, and the $\widetilde{p}_{i0}$ is the MLE of p_i obtained under $H_0 : \sum_{i=1}^{g} w_i p_i = \xi_0$. To calculate z_0, let $S_w = \sum_{i=1}^{g} w_i$ and $N = \sum_{i=1}^{g} n_i$. The value of z_0^2 is determined by

$$y(z_0^2) = N + (S_w - 2\xi)C - \sum_{i=1}^{g} R_i = 0, \tag{4.47}$$

where $C = z_0^2/(\widehat{\xi} - \xi_0)$, $R_i^2 = n_i^2 + 2n_i w_i b_i C + w_i^2 C^2$ and $b_i = 1 - 2\widehat{p}_i$ with $\widehat{p}_i = k_i/n_i$,

$i = 1, ..., g$. The null hypothesis in (4.45) is rejected if $y(z_{1-\alpha/2}^2) \geq 0$, where z_q is the $100q$ percentile of the standard normal distribution.

The endpoints of the score confidence interval are determined by the roots (with respect to ξ) of the equation $y(z_{1-\alpha/2}^2) = 0$. Even though the score test is easy to apply, finding the confidence interval requires a numerical iterative method. For more details, see Andrés et al. (2011).

Fiducial Confidence Intervals

A fiducial confidence interval for $\xi = \sum_{i=1}^g w_i p_i$ is formed by the percentiles of the fiducial quantity $W = \sum_{i=1}^g w_i B_{k_i+.5, n_i-k_i+.5}$, where $(k_1, ..., k_g)$ is an observed value of $(X_1, ..., X_g)$, and $B_{a,b}$ denotes the beta random variable with parameters a and b. The percentiles of this fiducial quantity can be approximated by the MNA in Section 2.16. In particular, noting that

$$E(B_{k_i+.5, n_i-k_i+.5}) = \frac{k_i + .5}{n_i + 1},$$

one can find approximations to the percentiles of W. Krishnamoorthy, Lee and Zhang (2014) have noted that a better approximation for the percentiles of W can be obtained by using $E(B_i) \simeq \hat{p}_i = k_i/n_i$, $i = 1, ..., g$. Letting

$$(l_i, u_i) = \left(B_{k_i+.5, n_i-k_i+.5; \frac{\alpha}{2}}, B_{k_i+.5, n_i-k_i+.5; 1-\frac{\alpha}{2}} \right), \quad \text{for } 0 < \alpha \leq .5,$$

we find

$$W_\alpha \simeq \sum_{i=1}^g w_i \hat{p}_i - \sqrt{\sum_{i=1}^g w_i^2 (\hat{p}_i - l_i^*)^2}, \quad \text{with} \quad l_i^* = \begin{cases} l_i & \text{if } w_i > 0, \\ u_i & \text{if } w_i < 0, \end{cases} \tag{4.48}$$

and

$$W_{1-\alpha} \simeq \sum_{i=1}^g w_i \hat{p}_i + \sqrt{\sum_{i=1}^g w_i^2 (\hat{p}_i - u_i^*)^2}, \quad \text{with} \quad u_i^* = \begin{cases} u_i & \text{if } w_i > 0, \\ l_i & \text{if } w_i < 0. \end{cases} \tag{4.49}$$

The interval $(W_\alpha, W_{1-\alpha})$ is an approximate $100(1 - 2\alpha)\%$ two-sided confidence interval for $\xi = \sum_{i=1}^g w_i p_i$.

The score confidence interval and the fiducial confidence interval are quite comparable with respect to coverage probabilities and precision. The fiducial confidence intervals have an advantage over the score confidence intervals, as they are easy to compute.

Example 4.23. In a study by Cohen et al. (1991), 120 rats were randomly assigned to four diets, as shown in Table 4.3. The absence or presence of a tumor was recorded for each rat. The data and the contrast of interest $L_i = (l_{i1}, ..., l_{i4})$, $i = 1, 2, 3$ are taken from Andrés et al. (2011), and they are presented in Table 4.3. Our 95% fiducial confidence interval for $\sum_{j=1}^4 l_{ij} p_j$ is formed by the lower 2.5th percentile and the upper 2.5th percentile of $\sum_{j=1}^4 l_{ij} B_{k_j+.5, n_j-k_j+.5}$, and these percentiles can be obtained using (4.48) and (4.49).

To find the fiducial confidence intervals, we first find the percentiles of $B_{k_i+.5, n_i-k_i+.5}$, $i = 1, ..., 4$. The needed percentiles are $B_{20.5, 10.5; .025} = .4889$,

$B_{14.5,16.5;.025} = .2981$, $B_{27.5,3.5;.025}.7566$, $B_{19.5,11.5;.025} = .4551$, $B_{20.5,10.5;.975} = .8140$, $B_{14.5,16.5,.975} = .6413$, $B_{27.5,3.5;.975} = .9710$, and $B_{19.5,11.5;.975} = .7872$. The estimated proportions are $\widehat{p}_1 = .6667$, $\widehat{p}_2 = .4667$, $\widehat{p}_3 = .9000$, and $\widehat{p}_4 = .6333$. For the contrast L_1 in Table 4.3, we find $\sum_{i=1}^{4} l_{1i}\widehat{p}_i = -.06667$. Substituting these quantities in (4.48), we found $W_{.025} = -0.3812$. Similarly, we found $W_{.975} = .3072$. Thus, the 95% fiducial confidence interval for the contrast L_1 is $(-0.3812, .3072)$, and this interval indicates that there is no fiber × fat interaction effect. The 95% fiducial confidence intervals for other contrasts were computed similarly, as shown in Table 4.3. The score confidence intervals reported in Table 4.3 are taken from Andrés et al. (2011).

TABLE 4.3: Types of Diets in Tumor Study

	Fiber		No Fiber	
	High fat	Low fat	High fat	Low fat
Sample size, n_i	30	30	30	30
Rats with tumor, k_i	20	14	27	19
$L_1 =$ Fiber × Fat	1	−1	−1	1
$L_2 =$ Fiber	1	1	−1	−1
$L_3 =$ Fat	1	−1	1	−1

95% confidence intervals for the contrasts

Method	L_1	L_2	L_3
Score	$(-.3883, .2445)$	$(-.7096, -.0772)$	$(.1420, .7742)$
Fiducial	$(-.3812, .2405)$	$(-.6979, -.0767)$	$(.1405, .7615)$

R function 4.1 "ci.bin.wavr" can be used to compute fiducial confidence intervals for $\sum_{i=1}^{k} w_i p_i$.

4.10   Properties and Results

4.10.1   Properties

1. Let $X_1, \ldots, X_m$ be independent random variables with $X_i \sim \text{binomial}(n_i, p)$, $i = 1, 2, \ldots, m$. Then

$$\sum_{i=1}^{m} X_i \sim \text{binomial}\left(\sum_{i=1}^{m} n_i, p\right).$$

2. Let X be a binomial(n, p) random variable. For fixed k,

$$P(X \leq k | n, p)$$

   is a nonincreasing function of p.

3. Recurrence Relations:
   (i) $P(X = k + 1) = \frac{(n-k)p}{(k+1)(1-p)} P(X = k)$, $k = 0, 1, 2, \ldots, n - 1$.

(ii) $P(X = k - 1) = \frac{k(1-p)}{(n-k+1)p}P(X = k)$, $k = 1, 2, \ldots, n$.

4. (i) $P(X \geq k) = p^k \sum_{i=k}^{n} \binom{i-1}{k-1}(1-p)^{i-k}$.

(ii) $\sum_{i=k}^{n} i\binom{n}{i}p^i(1-p)^{n-i} = npP(X \geq k) + k(1-p)P(X = k)$.

[Patel et al. (1976), p. 201]

R function 4.1. R function to find fiducial confidence intervals for $\sum_{i=1}^{k} w_i p_i$ [a]

```
# w = vector consisting of weights wi
# x = vector of numbers of successes
# n = vector of sample sizes; cl = confidence level
ci.bin.wavr = function(w, x, n, cl){
k = length(n); xls = seq(1:k); xus = seq(1:k)
alp = (1-cl)/2; alc = 1-alp
a = x+.5; b = n-x+.5; xl = qbeta(alp,a,b); xu = qbeta(alc,a,b)
u = x/n; cent = sum(w*u)
for(i in 1:k){
if(w[i] > 0){xls[i] = xl[i]; xus[i] = xu[i]}
else{xls[i] = xu[i]; xus[i] = xl[i]}
}
sel = sqrt(sum(w**2*(u-xls)**2)); seu = sqrt(sum(w**2*(u-xus)**2))
low = cent-sel; upp = cent+seu
return(c(low,upp))
}
```

---

[a]Electronic version of this R function can be found in HBSDA.r located in *StatCalc* directory.

4.10.2   Relation to Other Distributions

1. Bernoulli: Let $X_1, \ldots, X_n$ be independent Bernoulli(p) random variables with success probability p. That is, $P(X_i = 1) = p$ and $P(X_i = 0) = 1 - p$, $i = 1, \ldots, n$. Then

$$\sum_{i=1}^{n} X_i \sim \text{binomial}(n, p).$$

2. Hypergeometric: See Section 5.8.

3. Negative Binomial: See Section 8.7.2.

4. F Distribution: See Section 13.4.2.

5. Beta: See Section 17.6.2.

4.10.3   Approximations

1. Let n be such that $np > 5$ and $n(1-p) > 5$. Then,

$$P(X \leq k|n,p) \simeq P\left(Z \leq \frac{k-np+0.5}{\sqrt{np(1-p)}}\right),$$

and

$$P(X \geq k|n,p) \simeq P\left(Z \geq \frac{k-np-0.5}{\sqrt{np(1-p)}}\right),$$

where Z is the standard normal random variable.

2. Let $\lambda = np$. Then, for large n and small p,

$$P(X \leq k|n,p) \simeq P(Y \leq k) = \sum_{i=0}^{k} \frac{e^{-\lambda}\lambda^i}{i!},$$

where Y is a Poisson random variable with mean λ.

4.11   Random Number Generation

Input:
        n = number of trials
        p = success probability
        ns = desired number of binomial random numbers
Output:
        x(1),...,x(ns) are random numbers from the binomial(n, p)
        distribution

The following algorithm, which generates binomial(n,p) random numbers as the sum of n Bernoulli(p) random numbers, is satisfactory and efficient for small n.

Algorithm 4.1. Binomial variate generator

```
        Set   k = 0
        For j = 1 to ns
           For i = 1 to n
              Generate u from uniform(0, 1)
              If u <= p,   k = k + 1
           [end i loop]
2          x(j)   = k
           k = 0
        [end j loop]
```

The following algorithm first computes the probability and the cumulative probability around the mode of the binomial distribution, and then searching for k sequentially so that $P(X \leq k-1) < u \leq P(X \leq k)$, where u is a uniform random variate. Depending on the value of the uniform variate, forward or backward search from the mode will be carried out. If n is too large, search for k may be restricted in the interval $np \pm c\sqrt{np(1-p)}$, where $c \geq 7$. Even though this algorithm requires the computation of the cumulative probability around the mode, it is accurate and stable.

Algorithm 4.2. Binomial variate generator

```
Set k =  int((n + 1)*p)
    s = p/(1 - p)
    pk = P(X = k)
    df = P(X <= k)
    rpk = pk; rk = k;
For j = 1 to ns
    Generate u from uniform(0, 1)
    If u > df, go to 2
1   u = u + pk
    If k = 0 or u > df, go to 3
    pk = pk*k/(s*(n - k + 1))
    k = k - 1
    go to 1
2   pk = (n - k)*s*pk/(k + 1)
    u = u - pk
    k = k + 1
    If k = n or u <= df, go to 3
    go to 2
3   x(j) = k
    k = rk
    pk = rpk
[end j loop]
```

For other algorithms, see Kachitvichyanukul and Schmeiser (1988).

4.12 Computation of Probabilities

For small n, the probabilities can be computed in a straightforward manner. For large values of n, logarithmic gamma function $\ln\Gamma(x)$ (see Section 1.8) can be used to compute the probabilities.

To Compute $P(X = k)$:

Set $\quad x = \ln\Gamma(n+1) - \ln\Gamma(k+1) - \ln\Gamma(n-k+1)$
$\qquad y = k * \ln(p) + (n-k) * \ln(1-p)$
$\qquad P(X = k) = \exp(x + y).$

To Compute $P(X \leq k)$:

Compute $P(X = k)$

Set $m = \text{int}(np)$

If $k \leq m$, compute $P(X = k - 1)$ using the backward recursion relation

$$P(X = k - 1) = \frac{k(1 - p)}{(n - k + 1)p} P(X = k),$$

for $k - 1, k - 2, \ldots, 0$ or until convergence. Sum of these probabilities plus $P(X = k)$ is $P(X \leq k)$;

else compute $P(X = k + 1)$ using the forward recursion relation

$$P(X = k + 1) = \frac{(n - k)p}{(k + 1)(1 - p)} P(X = k),$$

for $k + 1, \ldots, n$, or until convergence; sum these probabilities to get $P(X \geq k + 1)$. The cumulative probability

$$P(X \leq k) = 1.0 - P(X \geq k + 1).$$

The following algorithm for computing the binomial cdf is based on the above method.

Algorithm 4.3. Calculation of binomial cdf

```
Input:
      k = nonnegative integer (0 <= k <= n)
      p = success probability (0 < p < 1)
      n = number of trials, n >= 1

Output: bincdf = P(X <= k)

Set  mode = int(n*p)
     bincdf = 0.0d0
     pk = P(X = k)
     if(k .le. mode) then
         For i = k to 0
         bincdf = bincdf + pk
         pk = pk * i*(1.0d0-p)/(en-i+1.0d0)/p
         (end i loop)
     else
         For i = k to n
         pk = pk * (en-i)*p/(i+1.0d0)/(1.0d0-p)
         bincdf = bincdf + pk
         [end i loop]
         bincdf = 1.0d0-bincdf+pk
     end if
```

The following R function computes the pmf and the cdf of a binomial(n, p) distribution. To calculate the logarithmic gamma function "alng," use R function 2.1 in Section 2.11.

R function 4.2. Calculation of the binomial pmf

```
binpr <- function(k,n,p)
{
cft <- alng(n+1)-alng(k+1)-alng(n-k+1)+k*log(p)+(n-k)*log(1-p)
return(exp(cft))
}
```

R function 4.3. Calculation of the binomial cdf[a]

```
bincdf <- function(k, n, p){
pk <- binpr(k, n, p); sum <- 1.0; term <- 1.0
if (k > n*p)
  {i <- 0
  cons <- p/(1.0-p)
  repeat{
    term <- term*(n-k-i)/(k+1.0+i)*cons
    sum <- sum + term
    if(term <= 1.0e-7 || i == n-k-1) break
    i <- i+1
  }
    ans <- 1.0-sum*pk+pk
}
else{
  cons <- (1.0-p)/p
    for (i in 1:k-1){
    term <- term*((k-i)/(n-k+i+1.0)*cons)
    sum <- sum + term
    if (term <= 1.0e-7 || i >= k-1) ans <- sum*pk}
  }
return(c(pk, ans, 1.0-ans+pk))
}
```

---

[a]Electronic version of this R function can be found in HBSDA.r located in *StatCalc* directory.

5

Hypergeometric Distribution

5.1 Description

Consider a lot consisting of N items of which M of them are defective and the remaining $N - M$ of them are nondefective. A sample of n items is drawn randomly without replacement. (That is, an item sampled is not replaced before selecting another item.) Let X denote the number of defective items that is observed in the sample. The random variable X is referred to as the hypergeometric random variable with parameters N and M. For a given set $\{N, M, n, k\}$, the probability $P(X = k|n, M, N)$ is the ratio of the number of samples of size that include exactly k defective items to the possible number samples of size n from the lot. Noting that the number of ways one can select b different objects from a collection of a different objects is

$$\binom{a}{b} = \frac{a!}{b!(a-b)!},$$

we find that the number of ways of selecting k defective items from M defective items is $\binom{M}{k}$; the number of ways of selecting $n - k$ nondefective items from $N - M$ nondefective items is $\binom{N-M}{n-k}$. Therefore, total number of ways of selecting n items with k defective and $n - k$ nondefective items is $\binom{M}{k}\binom{N-M}{n-k}$. Finally, the number of ways one can select n different items from a collection of N different items is $\binom{N}{n}$. Thus, the probability of observing k defective items in a sample of n items is given by

$$f(k|n, M, N) = P(X = k|n, M, N) = \frac{\binom{M}{k}\binom{N-M}{n-k}}{\binom{N}{n}}, \quad L \leq k \leq U, \quad (5.1)$$

where $L = \max\{0, M - N + n\}$ and $U = \min\{n, M\}$.

The cumulative distribution function of X is given by

$$F(k|n, M, N) = \sum_{i=L}^{k} \frac{\binom{M}{i}\binom{N-M}{n-i}}{\binom{N}{n}}, \quad L = \max\{0, M - N + n\}. \quad (5.2)$$

We shall denote the distribution by hypergeometric(n, M, N). The plots of probability mass functions are given in Figure 5.1 for a small lot size of $N = 100$ (the first set of four plots) and for a large lot size of $N = 5000$ (the second set of eight plots). The parameter-sample size configurations are chosen so that the hypergeometric plots can be compared with the corresponding binomial plots in Figure 4.1. The binomial plots with $n = 20$ are not in good agreement with the corresponding hypergeometric plots in Figure 5.1 with $(N = 100, n = 20)$, whereas all binomial plots are almost identical with the hypergeometric plots with $(N = 5000, n = 20)$ and with $(N = 5000, n = 100)$; see Burstein (1975).

5.2   Moments

Mean: $n\left(\frac{M}{N}\right)$

Variance: $n\left(\frac{M}{N}\right)\left(1 - \frac{M}{N}\right)\left(\frac{N-n}{N-1}\right)$

Mode: The largest integer $\leq \frac{(n+1)(M+1)}{N+2}$

Mean Deviation: $\frac{2x(N-M-n+x)\binom{M}{x}\binom{N-M}{n-x}}{N\binom{N}{n}}$,
                                 where x is the smallest integer larger than the mean. [Kamat (1965)]

Coefficient of Variation: $\left(\frac{(N-M)(N-n)}{nM(N-1)}\right)^{1/2}$

Coefficient of Skewness: $\frac{(N-2M)(N-2n)\sqrt{(N-1)}}{(N-2)\sqrt{nM(N-M)(N-n)}}$

Coefficient of Kurtosis: $\left(\frac{N^2(N-1)}{nM(N-M)(N-2)(N-3)(N-n)}\right)$
                                 $\times \left(\frac{3nM(N-M)(6-n)}{N} + N(N+1-6n) + 6n^2\right.$
                                 $\left. + 3M(N-M)(n-2) - \frac{18n^2M(N-M)}{N^2}\right)$

5.3   Probabilities, Percentiles, and Moments

The dialog box [StatCalc→Discrete→Hypergeometric →Probabilities, Critical Values and Moments] computes the probabilities, moments, and other parameters of a hypergeometric distribution.

Calculation of Probabilities: When $N = 100$, $M = 36$, $n = 20$, and $k = 3$,

$$P(X \leq 3) = 0.023231, \quad P(X \geq 3) = 0.995144, \quad \text{and} \quad P(X = 3) = 0.018375.$$

To Compute other Parameters: For any given four values from the set

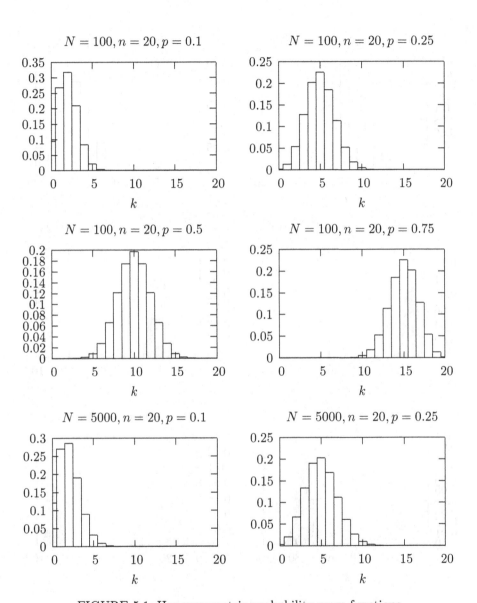

FIGURE 5.1: Hypergeometric probability mass functions

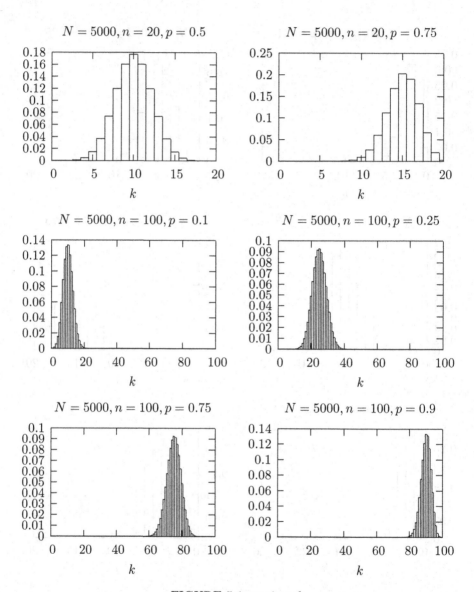

FIGURE 5.1 continued

$\{N, M, n, k, P(X \leq k)\}$, *StatCalc* computes the missing one. For example, to compute the value of M for a given N, n, k, and $P(X \leq k)$, enter the values of N, n, k, and $P(X \leq k)$, and then click on [E]. As an example, when $N = 300$, $n = 45$, $k = 12$, and $P(X \leq k) = 0.4321$, the value of M is 87. To carry out the computation, enter 300 for N, 45 for n, 12 for k, and 0.4321 for $P(X \leq k)$, and then click [E]. Note that $P(X \leq 12|45, 87, 300) = 0.429135$, which is close to the specified probability 0.4321. Because of the discrete nature of the distribution, there exists no integer value of M for which the probability is exactly equal to 0.4321. The values of N, n, and k can also be similarly calculated.

To compute moments: Enter the values of the N, M, and n; click [M].

Example 5.1. The following state lottery is well-known in the United States of America. A player selects 6 different numbers from $1, 2, \ldots, 44$ by buying a ticket for $1. Later in the week, the winning numbers will be drawn randomly by a device. If the player matches all winning numbers, then he or she will win the jackpot of the week. If the player matches 4 or 5 numbers, he or she will receive a lesser cash prize. If a player buys one ticket, what are the chances of matching (a) all numbers? (b) four numbers?

Solution: Let X denote the number of winning numbers in the ticket. If we regard winning numbers as defective, then X is a hypergeometric random variable with $N = 44$, $M = 6$, and $n = 6$. The probabilities can be computed using the dialog box [StatCalc→Discrete→Hypergeometric →Probabilities, Critical Values and Moments].

a.

$$P(X = 6) = \frac{\binom{6}{6}\binom{38}{0}}{\binom{44}{6}} = \frac{1}{\binom{44}{6}} = \frac{6! \, 38!}{44!} = \frac{1}{7059052}.$$

b.

$$P(X = 4) = \frac{\binom{6}{4}\binom{38}{2}}{\binom{44}{6}} = 0.0014938.$$

To find the probability in part (b) using *StatCalc*, enter 44 for N, 6 for M, 6 for sample size n, and 4 for observed k, and click [P] to get $P(X = 4) = 0.0014938$.

Example 5.2. A shipment of 200 items is under inspection. The shipment will be acceptable if it contains 10 or fewer defective items. The buyer of the shipment decided to buy the lot if he finds no more than one defective item in a random sample of n items from the shipment. Determine the sample size n so that the chances of accepting an unacceptable shipment is less than 10%.

Solution: Since we deal with a finite population, a hypergeometric model with $N = 200$ is appropriate for this problem. Let X denote the number of defective items in a sample of n items. The shipment is unacceptable if the number of defective items M is 11 or more. Furthermore, note that for M in $\{11, 12, \ldots, 200\}$, the chances of accepting an unacceptable shipment, that is, $P(X \leq 1|n, M, N)$, attains the maximum when $M = 11$. So we need to determine the value of n so that $P(X \leq 1|n, 11, 200) \leq 0.10$ and $P(X \leq 1|(n-1), 11, 200) > 0.10$. To compute the required sample size using *StatCalc*, select [StatCalc→Discrete→Hypergeometric→Probabilities, Critical Values and Moments], enter 200 for N, 11 for M, 1 for k, and 0.1 for $P(X \leq k)$; click [S] to get 61. Note that $P(X \leq 1|61, 11, 200) = 0.099901$.

Thus, the buyer has to inspect a sample of 61 items so that the chances of accepting an unacceptable shipment is less than 10%. Also, notice that when the sample size is 60, the probability of accepting an unacceptable lot is 0.106241, which is greater than 10%.

Example 5.3. Thirteen cards were drawn randomly from a deck of 52 cards. What are the chances of observing exactly 6 black cards and 7 red cards? At least one red card?

Solution: Note that there are 26 red cards and 26 black cards in a deck of 52 cards. So the probability of observing exactly 6 black and 7 red cards is

$$\frac{\binom{26}{6}\binom{26}{7}}{\binom{52}{7}} = 0.2384914.$$

Let X denote the number of red cards in a hand of 13 cards. The distribution of X is hypergeometric with sample size $n = 13$, number of red cards $M = 26$, and the lot size $N = 52$. Noting that $P(X \geq 1) = 1 - P(X = 0)$, we find the probability of observing at least one red card is

$$1 - P(X = 0) = 1 - \frac{\binom{26}{0}\binom{26}{13}}{\binom{52}{13}} = .99998.$$

5.4   Point Estimation

Let k denote the observed number of defective items in a sample of n items, selected from a lot of N items. Let M denote the number of defective items in the lot.

Point Estimation of M

The maximum likelihood estimator of M is given by

$$\widehat{M} = \left\lfloor \frac{k(N+1)}{n} \right\rfloor,$$

where $\lfloor x \rfloor$ denotes the largest integer less than or equal to x (floor function). If $k(N+1)/n$ is an integer, then both $k(N+1)/n$ and $k(N+1)/n - 1$ are the maximum likelihood estimators of M.

Estimation of the Lot Size

There are situations in which we want to estimate the lot size based on M, n, and k. For example, the capture–recapture technique is commonly used to estimate animal abundant in a given region [Thompson (1992), p.212]: A sample of n_1 animals was trapped, marked, and released in the first occurrence. After a while, another sample of n_2 animals was trapped. Let X denote the number of marked animals in the second trap. Then X follows a hypergeometric distribution with $M = n_1$ and $n = n_2$. For given $X = k$, we want to estimate N (lot size = total number of animals in the region). The maximum likelihood estimator of N is given by

$$\widehat{N} = \left\lceil \frac{n_1 n_2}{k} \right\rceil,$$

where $\lfloor x \rfloor$ denotes the largest integer less than or equal to x.

5.5 Test for the Proportion and Power Calculation

Suppose that we found k defective items in a sample of n items drawn from a lot of N items. Let M denote the true number of defective items in the population and let $p = M/N$.

An Exact Test

Let $M_0 = \text{int}(Np_0)$. For testing

$$H_0 : p \le p_0 \quad \text{vs.} \quad H_a : p > p_0, \tag{5.3}$$

the null hypothesis will be rejected if the p-value $P(X \ge k|n, M_0, N) \le \alpha$, for testing

$$H_0 : p \ge p_0 \quad \text{vs.} \quad H_a : p < p_0, \tag{5.4}$$

the null hypothesis will be rejected if the p-value $P(X \le k|n, M_0, N) \le \alpha$, and for testing

$$H_0 : p = p_0 \quad \text{vs.} \quad H_a : p \ne p_0, \tag{5.5}$$

the null hypothesis will be rejected if the p-value

$$2\min\{P(X \le k|n, M_0, N), P(X \ge k|n, M_0, N)\} \le \alpha.$$

The dialog box [StatCalc→Discrete→Hypergeometric→Test for Proportion ...] uses the above formulas for computing p-values of the test described above.

Example 5.4. (*Calculation of p-values*) When $N = 500$, $n = 20$, $k = 8$, and $p_0 = 0.2$, it is desired to test $H_0 : p \le p_0$ vs. $H_a : p > p_0$ at the level of 0.05. After entering these values in the aforementioned dialog box, click [p-values] to get 0.0293035. The null hypothesis H_0 will be rejected in favor of the alternative hypothesis H_a at the level of significance 0.05. However, if $H_0 : p = p_0$ vs. $H_a : p \ne p_0$ then, at the same nominal level, the H_0 in (5.5) can not be rejected because the p-value for this two-tailed test is 0.058607, which is not less than 0.05.

Example 5.5. (*Calculation of p-values*) A pharmaceutical company claims that 75% of doctors prescribe one of its drugs for a particular disease. In a random sample of 40 doctors from a population of 1000 doctors, 23 prescribed the drug to their patients. Does this information provide sufficient evidence to indicate that the true percentage of doctors who prescribe the drug is less than 75? Test at the level of significance $\alpha = 0.05$.

Solution: Let p denote the actual proportion of doctors who prescribe the drug to their patients. The hypotheses of interest are

$$H_0 : p \ge 0.75 \quad \text{vs.} \quad H_a : p < 0.75.$$

In the dialog box [StatCalc→Discrete→Hypergeometric→Test for p and Sample Size for Power], enter 1000 for N, 40 for n, 23 for observed k, and 0.75 for [Value of p0], and click on [p-values for]. The p-value for the above left-tailed test is 0.0101239, which is less than 0.05. Thus, we conclude, on the contrary to the manufacturer's claim, that less than 75% of doctors prescribe the drug.

Power of the Exact Test

For a given p, let $M = \text{int}(Np)$ and $M_0 = \text{int}(Np_0)$, where p_0 is the specified value of p under H_0 in (5.3). For a right-tailed test, the exact power at the level α can be computed using the expression

$$\sum_{k=L}^{U} \frac{\binom{M}{k}\binom{N-M}{n-k}}{\binom{N}{n}} I(P(X \geq k|n, M_0, N) \leq \alpha),$$

and for a two-tailed test the exact power can be expressed as

$$\sum_{k=L}^{U} \frac{\binom{M}{k}\binom{N-M}{n-k}}{\binom{N}{n}} I\left[P(X \geq k|n, M_0, N) \leq \frac{\alpha}{2} \text{ or } P(X \leq k|n, M_0, N) \leq \frac{\alpha}{2}\right],$$

where $I(.)$ is the indicator function.

Example 5.6. (*Power Calculation*) When lot size $N = 500$, sample size $n = 35$, $p_0 = 0.2$, nominal level $= 0.05$, and $p = 0.4$, the power of the test for the hypotheses in (5.3) is 0.813779. For hypotheses in (5.5), the power is 0.701371. The power can be computed using *StatCalc* as follows. Enter 500 for [Lot Size, N], 3 to indicate two-tailed test, .05 for [Nominal Level], 0.4 for [Guess p], 0.2 for [Null p0], and 35 for [S Size]; click on [Power].

Example 5.7. (*Sample Size Calculation*) Suppose that a researcher believes that a new drug is 20% more effective than the existing drug, which has a success rate of 70%. Assume that the size of the population of patients is 5000. The required sample size to test his belief (at the level 0.05 and power 0.90) can be computed using *Stat-Calc* as follows. Select the dialog box [StatCalc→Discrete→Hypergeometric→Test for p ...], enter 5000 for N, 1 to indicate the test is right-tailed, .05 for the nominal level, .9 for [Guess p] and 0.7 for [Null p0], and click on [S Size] to get 37. Note that the sample size is determined so that the power will be at least .90. To find the actual power at sample size 37, click on [Power] to get 0.929651.

Suppose we relax the requirement, and determine sample size so that the power of the test is close to the specified power of .90. By trial-error (in the dialog box [StatCalc→Discrete→Hypergeometric→Test for p ...]), we see that a sample of size 33 produces a power of .895. So a little compensation in power, lower the sample size considerably.

5.6 Confidence Interval and Sample Size Calculation

Suppose that we found k defective items in a sample of n items drawn from a finite population of N items. Let M denote the true number of defective items in the population.

Confidence Intervals

A lower confidence limit M_l for M is the largest integer such that

$$P(X \geq k|n, M_l, N) = \frac{\alpha}{2},$$

and an upper limit M_u for M is the smallest integer such that

$$P(X \leq k|n, M_u, N) = \frac{\alpha}{2}.$$

A $1 - \alpha$ confidence interval for the proportion of defective items in the lot is given by

$$(p_l, \ p_u) = \left(\frac{M_l}{N}, \ \frac{M_u}{N} \right). \tag{5.6}$$

A slightly better confidence interval can be obtained by choosing M_l as the smallest integer for which $P(X \geq k|n, M_l, N) > \alpha/2$ and M_u as the largest integer for which $P(X \leq k|n, M_u, N) > \alpha/2$.

The dialog box [StatCalc→Discrete→Hypergeometric→CI for p and Sample size for Precision] uses the better method in the preceding paragraph to compute $1 - \alpha$ confidence intervals for p. This dialog box also computes the sample size required to estimate the proportion of defective items within a given precision.

Example 5.8. (*Computing Confidence Interval*) Suppose that a sample of 40 units from a population of 500 items showed that 5 items are with an attribute of interest. To find a 95% confidence interval for the true proportion of the items with this attribute, enter 500 for N, 40 for n, 5 for the observed number of successes k, and 0.95 for the confidence level; click [2-sided] to get (0.044, 0.260). For one-sided limits, click [1-sided] to get 0.052 and 0.238. That is, 0.052 is 95% one-sided lower limit for p, and 0.238 is 95% one-sided upper limit for p.

Example 5.9. A shipment of 1000 items is submitted for inspection. Because of the cost of inspection, the purchaser decided to inspect only a sample of 30 items randomly selected from the shipment. The inspection showed that 2 out of 30 items are not acceptable (defective). Based on the result, find a 95% confidence interval for the true proportion of defective items in the shipment. The purchaser may accept the shipment if the proportion of defective items is no more that 10%. On the basis of the sampling result, do you recommend the purchaser to accept the shipment?

Solution: Let p denote the true proportion of defective items in the shipment. To find a 95% confidence interval for p, note that $N = 1000, n = 30$, and $k = 2$. Enter these data in the dialog box [StatCalc→Discrete→Hypergeometric→CI for p ...], and click [2-sided] to get (.008, .217).

To check if the true proportion of defective items is less than 10%, we need to show that a 95% (or 99%, depending on the risk) upper confidence limit for p is less than 10%. To calculate the 95% one-sided upper limit, click on [1-sided] in the dialog box [StatCalc→Discrete→Hypergeometric→CI for p ...] to get 0.192. That is, the true proportion of defective items is no more than 19.2% with confidence .95. Since the upper confidence limit is not less that 10%, we can not recommend the purchaser to buy the shipment.

Example 5.10. (*One-Sided Limit*) A highway patrol officer stopped a car for a minor traffic violation. Upon suspicion, the officer checked the trunk of the car, and found many bags. The officer arbitrarily checked 10 bags and found that all of them contained marijuana. A later count showed that there were 300 bags. Before the case went to trial, all the bags were destroyed without examining the remaining bags. Since the severity of the fine and punishment depends on the quantity of marijuana, it is desired to estimate the minimum number of marijuana bags. Based on the information, determine the minimum number marijuana bags in the trunk at the time of arrest.

Solution: The hypergeometric model, with lot size $N = 300$, sample size $n = 10$, and the observed number of defective items $k = 10$, is appropriate for this problem. So, we can use a 95% one-sided lower limit for M (total number of marijuana bags) as an estimate for the minimum number of marijuana bags out of these 300 bags. To get a 95% one-sided lower limit for M using the dialog box [StatCalc→Discrete→Hypergeometric→CI for p ...], enter the data in the appropriate edit boxes, and click [1-sided] to get 0.74667. That is, we estimate with 95% confidence that there were at least $300 \times 0.746667 = 224$ bags of marijuana at the time of arrest.

Sample Size for Precision

For a given N, n, and k, the dialog box [StatCalc→Discrete→Hypergeometric →CI for p and Sample Size for Precision] computes the sample size to find a confidence interval with a given precision.

Expected Length

For a given lot size N, p and a confidence level $1 - \alpha$, the expected length of the confidence interval in (5.6) can be computed as follows. For a given p, let $M = \text{int}(Np)$. Then the expected length is given by

$$\sum_{k=L}^{U} \frac{\binom{M}{k}\binom{N-M}{n-k}}{\binom{N}{n}}(p_u - p_l),$$

where L and U are as defined in (5.1). *StatCalc* computes the required sample size to have a confidence interval with a specified expected length.

Example 5.11. (*Sample Size for a Given Precision*) A researcher hypothesizes that the proportion of individuals with the attribute of interest in a population of size 1000 is 0.3, and he wants to estimate the true proportion within a margin of error of ±5% with 95% confidence. To compute the required sample size, enter 1000 for [Lot Size], .95 for [Conf Level], 0.3 for [Guess p] and .05 for [Margin of Error], and click [Appr S Size] to get 244. Click on [Actual ME] to get .0502386. By increasing the sample size to 246, we find the actual margin of error is .0499392. Thus, the required sample size is 246. See Example 4.11 to find out the required sample size if the population is infinite or the population size is unknown.

5.7 Test for Comparing Two Proportions in Finite Populations

Suppose that inspection of a sample of n_1 individuals from a population of N_1 units revealed k_1 units with a particular attribute, and a sample of n_2 individuals from another population of N_2 units revealed k_2 units with the same attribute. The problem of interest is to test the difference $p_1 - p_2$, where p_i denotes the proportion of individuals in the ith population with the attribute of interest, $i = 1, 2$.

The Test

Consider testing

$$H_0 : p_1 \leq p_2 \quad \text{vs.} \quad H_a : p_1 > p_2. \tag{5.7}$$

Define $\widehat{p}_k = \frac{k_1+k_2}{n_1+n_2}$ and $\widehat{M}_i = \text{int}(N_i\widehat{p}_k)$, $i = 1, 2$. Consider the test statistic

$$Z(k_1, k_2) = \frac{k_1 - k_2}{\sqrt{\widehat{p}_k(1 - \widehat{p}_k)\left(\frac{N_1-n_1}{n_1(N_1-1)} + \frac{N_2-n_2}{n_2(N_2-1)}\right)}}.$$

The p-value for testing the above hypotheses on the basis of $z(k_1, k_2)$ can be computed using the expression

$$
\begin{aligned}
P(k_1, k_2, n_1, n_2) &= \sum_{x_1=L_1}^{U_1} \sum_{x_2=L_2}^{U_2} f(x_1|n_1, \widehat{M}_1, N_1) f(x_2|n_2, \widehat{M}_2, N_2) \\
&\times I(Z(x_1, x_2) \geq Z(k_1, k_2)).
\end{aligned}
\tag{5.8}
$$

where $I(.)$ is the indicator function, $L_i = \max\{0, \widehat{M}_i - N_i + n_i\}$, $U_i = \min\{\widehat{M}_i, n_i\}$, $i = 1, 2$,

$$f(x_i|n_i, \widehat{M}_i, N_i) = \frac{\binom{\widehat{M}_i}{x_i}\binom{N_i-\widehat{M}_i}{n_i-x_i}}{\binom{N_i}{n_i}}, \quad i = 1, 2,$$

The term $Z(k_1, k_2)$ is equal to $Z(x_1, x_2)$ with x replaced by k.

The null hypothesis in (5.7) will be rejected when the p-value in (5.8) is less than or equal to α. For more details and properties of the test, see Krishnamoorthy and Thomson (2002). The p-value for a left-tailed test or for a two-tailed test can be computed similarly.

The dialog box [StatCalc→Discrete→Hypergeometric→Test for comparing ...] calculates the p-values of the above two-sample test.

Example 5.12. (*Calculation of p-values*) Suppose a sample of 25 observations from population 1 with size 300 yielded 20 successes, and a sample of 20 observations from population 2 with size 350 yielded 10 successes. Let p_1 and p_2 denote the proportions of successes in populations 1 and 2, respectively. Suppose we want to test

$$H_0 : p_1 \leq p_2 \quad \text{vs.} \quad H_a : p_1 > p_2.$$

To compute the p-value, select the aforementioned dialog box, enter 300 for [Lot Size 1], 350 for [Lot Size 2], 25 for [Sample Size 1], 20 for [Sample Size 2], 20 for [Obser defs 1], 10 for [Obser defs 2], 0 for [H0: p1-p2 = d], and click on [p-values] to get 0.0165608. The p-value for testing $H_0 : p_1 = p_2$ vs. $H_a : p_1 \neq p_2$ is 0.0298794.

Example 5.13. The quality control inspector of a manufacturing company has decided to check if the products produced during different shifts are similar. He inspected a sample of 30 products from a lot 500 products produced during day shifts, and a sample of 40 products from a lot of 450 products produced during night shifts. The inspection revealed 2 defective products in the sample of 30 products, and 4 defective products in the sample of 40. Let p_d and p_n denote the proportions of defective products produced during the day shift and night shift, respectively. Then the inspector may want to test

$$H_0 : p_d = p_n \quad \text{vs.} \quad H_a : p_d \neq p_n.$$

To compute the p-value, select the dialog box [StatCalc→Discrete→Hypergeometric→ Test for comparing ...], enter 500 for [Lot Size 1], 450 for [Lot Size 2], 30 for [Sample Size 1], 40 for [Sample Size 2], 2 for [Obser defs 1], 4 for [Obser defs 2], 0 for [H0: p1-p2 = d], and click on [p-values] to get 0.558236. This p-value indicates that there is no significant difference between the proportions of defective products among the products produced during different shifts.

Power Calculation

For a given p_1 and p_2, let $M_i = \text{int}(N_i p_i)$, $L_i = \max\{0, M_i - N_i + n_i\}$ and $U_i = \min\{n_i, M_i\}$, $i = 1, 2$. The exact power of the test described above can be computed using the expression

$$\sum_{k_1=0}^{n_1} \sum_{k_2=0}^{n_2} f(k_1|n_1, M_1, N_1) f(k_2|n_2, M_2, N_2) I(P(k_1, k_2, n_1, n_2) \leq \alpha), \qquad (5.9)$$

where $f(k|n, M, N)$ is the hypergeometric probability mass function, $M_i = \text{int}(N_i p_i)$, $i = 1, 2$, and the p-value $P(k_1, k_2, n_1, n_2)$ is given in (5.8). The powers of a left-tailed test and a two-tailed test can be computed similarly.

The dialog box [StatCalc→Discrete→Hypergeometric→Test for comparing ...] uses the above method for computing powers of the two-sample test.

Example 5.14. (*Sample Size Calculation for Power*) Suppose the sample size for each group needs to be determined to carry out a two-tailed test at the level of significance $\alpha = 0.05$ and power $= 0.80$. Assume that the lot sizes are 300 and 350. Furthermore, the guess values of the proportions are given as $p_1 = 0.45$ and $p_2 = 0.15$. To determine the sample size using *StatCalc*, enter 2 for two-tailed test, 0.05 for [Level], 0.45 for p_1, 0.15 for p_2, and 28 for each sample size. Click [Power] to get a power of 0.751881. Note that the sample size gives a power less than 0.80. This means, the sample size required to have a power of 0.80 is more than 28. Enter 31 (for example) for both sample sizes and click on [Power] radio button. Now the power is 0.807982. It can be verified that the power at 30 is 0.78988. Thus, the required sample size from each population to attain a power of at least 0.80 is 31.

Remark 5.1. Note that the power can also be computed for unequal sample sizes. For instance, when $n_1 = 30$, $n_2 = 34$, $p_1 = 0.45$, and $p_2 = 0.15$, the power for testing $H_0 : p_1 = p_2$ vs. $H_a : p_1 \neq p_2$ at the nominal 0.05 is 0.804974. For the same configuration, a power of 0.800876 can be attained if $n_1 = 29$ and $n_2 = 39$.

5.8   Properties and Results

Recurrence Relations

a. $P(X = k + 1 | n, M, N) = \frac{(n-k)(M-k)}{(k+1)(N-M-n+k+1)} P(X = k | n, M, N).$

b. $P(X = k - 1 | n, M, N) = \frac{k(N-M-n+k)}{(n-k+1)(M-k+1)} P(X = k | n, M, N).$

c. $P(X = k | n + 1, M, N) = \frac{(N-M-n+k)}{(M+1-k)(N-M)} P(X = k | n, M, N).$

Relation to Other Distributions

1. Binomial: Let X and Y be independent binomial random variables with common success probability p and numbers of trials m and n, respectively. Then

$$P(X = k | X + Y = s) = \frac{P(X = k) P(Y = s - k)}{P(X + Y = s)},$$

which simplifies to

$$P(X = k | X + Y = s) = \frac{\binom{m}{k}\binom{n}{s-k}}{\binom{m+n}{s}}, \quad \max\{0, s - n\} \le k \le \min\{m, s\}.$$

Thus, the conditional distribution of X given $X + Y = s$ is hypergeometric$(s, m, m + n)$.

Approximations

1. Let $p = \frac{M}{N}$. Then, for large N and M,

$$P(X = k) \simeq \binom{n}{k} p^k (1 - p)^{n-k}.$$

2. Let $\frac{M}{N}$ be small and n is large such that $n\left(\frac{M}{N}\right) = \lambda$.

$$P(X = k) \simeq \frac{e^{-\lambda}\lambda^k}{k!} \left\{ 1 + \left(\frac{1}{2M} + \frac{1}{2n}\right)\left[k - \left(k - \frac{Mn}{N}\right)^2\right] + O\left(\frac{1}{k^2} + \frac{1}{n^2}\right)\right\}.$$

[Burr, 1973]

5.9   Random Number Generation

Input:

```
        N = lot size; M = number of defective items in the lot
        n = sample size; ns = number of random variates to be
            generated
```

Output:
        x(1),..., x(ns) are random number from the
        hypergeometric(n, M, N) distribution

The following generating scheme is essentially based on the probability mechanism involved in simple random sampling without replacement, and is similar to Algorithm 3.9.1 for the binomial case.

Algorithm 5.1. Hypergeometric variate generator

```
Set  k = int((n + 1)*(M + 1)/(N + 2))
     pk = P(X = k)
     df = P(X <= k)
     Low = max{0, M - N + n}
     High = min{n, M}
     rpk = pk; rk = k
For  j = 1 to ns
     Generate u from uniform(0, 1)
     If u > df, go to 2
1    u = u + pk
     If k = Low or u > df, go to 3
     pk = pk*k*(N - M - n + k)/((M - k + 1)*(n - k + 1))
     k = k - 1
     go to 1
2    pk = pk*(n - k)*(M -k)/((k + 1)*(N - M + k + 1))
     u = u - pk
     k = k + 1
     If k = High or u <= df, go to 3
     go to 2
3    x(j) = k
     pk = rpk
     k = rk
[end j loop]
```

For other lengthy but more efficient algorithms, see Kachitvichyanukul and Schmeiser (1985).

5.10   Computation of Probabilities

To compute $P(X = k)$

Set $U = \min\{n, M\}$; $L = \max\{0, M - N + n\}$
If $k > U$ or $k < L$ then return $P(X = k) = 0$
Compute $S_1 = \ln \Gamma(M + 1) - \ln \Gamma(k + 1) - \ln \Gamma(M - k + 1)$
                $S_2 = \ln \Gamma(N - M + 1) - \ln \Gamma(n - k + 1) - \ln \Gamma(N - M - n + k + 1)$

$$S_3 = \ln \Gamma(N+1) - \ln \Gamma(n+1) - \ln \Gamma(N-n+1)$$
$$P(X = k) = \exp(S_1 + S_2 - S_3)$$

To compute $\ln\Gamma(x)$, see Section 1.8.

To compute $P(X \le k)$

Compute $P(X = k)$
Set mode $= \text{int}((n+1)(M+1)/(N+2))$
If $k \le$ mode, compute the probabilities using the backward recursion relation

$$P(X = k-1|n, M, N) = \frac{k(N-M-n+k)}{(n-k+1)(M-k+1)} P(X = k|n, M, N)$$

for $k-1, \ldots, L$ or until a specified accuracy; add these probabilities and $P(X = k)$
to get $P(X \le k)$;
else compute the probabilities using the forward recursion

$$P(X = k+1|n, M, N) = \frac{(n-k)(M-k)}{(k+1)(N-M-n+k+1)} P(X = k|n, M, N)$$

for $k+1, \ldots, U$ or until a specified accuracy; add these probabilities to get
$P(X \ge k+1)$. The cumulative probability is given by

$$P(X \le k) = 1 - P(X \ge k+1).$$

The following algorithm for computing a hypergeometric cdf is based on the
above computational method.

Algorithm 5.2. Calculation of hypergeometric distribution function

```
Input:
      k = the value at which the cdf is to be evaluated
      n = the sample size
      m = the number of defective items in the lot
      lot = size of the lot
Output:
      hypcdf = P(X <= k|n,m,lot)

Set   one = 1.0d0
      lup = min(n, m)
      low = max(0, m-lot+n)

      if(k .lt. low) return hypcdf = 0.0d0
      if(k .gt. lup) return hypcdf = one

      mode = int(n*m/lot)
      hypcdf = 0.0d0
      pk = hypprob(k, n, m, lot)

      if(k .le. mode) then
         For i = k to low
            hypcdf = hypcdf + pk
```

```
        pk = pk*i*(lot-m-n+i)/(n-i+one)/(m-i+one)
      [end i loop]
   else
      For i = k to lup
         pk = pk * (n-i)*(m-i)/(i+one)/(lot-m-n+i+one)
         hypcdf = hypcdf + pk
      [end i loop]
      hypcdf = 1.0d0-hypcdf
   end if
```

The following R functions compute the pmf and cdf of a hypergeometric(n, m, lot) distribution.

R function 5.1. Calculation of the hypergeometric pmf

```
hyppr = function(k, n, m, lot){
lup = min(n, m); low = max(0, m-lot+n); one = 1.0
if(k < low | k > lup) return(0)
# alng(x) = logarithmic function R function 1.1
term1 = alng(m+one)-alng(k+one)-alng(m-k+one)
term2 = alng(lot-m+one)-alng(n-k+one)-alng(lot-m-n+k+one)
term3 = alng(lot+one)-alng(n+one)-alng(lot-n+one)
ans = exp(term1+term2-term3)
return(ans)
}
```

R function 5.2. Calculation of the hypergeometric cdf

```
hypcdf = function(k, n, m, lot){
lup = min(n, m); low = max(0, m-lot+n); one = 1.0
if(k > lup){return(1)}
if(k < low){return(0)}
mod = floor(n*m/lot)+1
ans = 0.0; pk = hyppr(k, n, m, lot)
if(k <= mod){
for(i in rev(k:low)){
ans = ans + pk;
    pk = pk*i*(lot-m-n+i)/(n-i+one)/(m-i+one)}
}
else{
for(i in k:lup){
    pk = pk * (n-i)*(m-i)/(i+one)/(lot-m-n+i+one)
    ans = ans + pk
}
    ans = one-ans}
return(ans)
}
```

6

Poisson Distribution

6.1 Description

Suppose events that occur over a period of time or space satisfy the following:

1. The numbers of events occurring in disjoint intervals of time are independent.

2. The probability that exactly one event occurs in a small interval of time Δ is $\Delta\lambda$, where $\lambda > 0$.

3. It is almost unlikely that two or more events occur in a sufficiently small interval of time.

4. The probability of observing a certain number of events in a time interval Δ depends only on the length of Δ and not on the beginning of the time interval.

Let X denote the number of events in a unit interval of time or in a unit distance. Then, X is called the Poisson random variable with mean number of events λ in a unit interval of time. The probability mass function of a Poisson distribution with mean λ is given by

$$f(k|\lambda) = P(X = k|\lambda) = \frac{e^{-\lambda}\lambda^k}{k!}, \quad k = 0, 1, 2, \dots . \tag{6.1}$$

The cumulative distribution function of X is given by

$$F(k|\lambda) = P(X \le k|\lambda) = \sum_{i=0}^{k} \frac{e^{-\lambda}\lambda^i}{i!}, \quad k = 0, 1, 2, \dots . \tag{6.2}$$

The Poisson distribution can also be developed as a limiting distribution of the binomial, in which $n \to \infty$ and $p \to 0$ so that np remains a constant. In other words, for large n and small p, the binomial distribution can be approximated by the Poisson distribution with mean $\lambda = np$. Some examples of the Poisson random variable are:

1. the number of radioactive decays over a period of time;

2. the number of automobile accidents per day on a stretch of an interstate road;

3. the number of typographical errors per page in a book;

4. the number of α particles emitted by a radioactive source in a unit of time;

5. the number of still births per week in a large hospital.

Poisson distribution gives probability of observing k events in a given period of time, assuming that events occur independently at a constant rate. The Poisson distribution is widely used in quality control, reliability, and queuing theory. It can be used to model the distribution of the number of defects in a piece of material, customer arrivals at a train station, auto insurance claims, and incoming telephone calls per period of time.

As shown in the plots of probability mass functions in Figure 6.1, Poisson distribution is right-skewed, and the degree of skewness decreases as λ increases.

6.2   Moments

Mean:	λ
Variance:	λ
Mode:	The largest integer less than or equal to λ. If λ is an integer, λ and $\lambda - 1$ are modes.
Mean Deviation:	$\frac{2e^{-\lambda}\lambda^{[\lambda]+1}}{[\lambda]!}$, where $[x]$ denotes the largest integer less than or equal to x. [Johnson et al. (1992), p. 157]
Coefficient of Variation:	$\frac{1}{\sqrt{\lambda}}$
Coefficient of Skewness:	$\frac{1}{\sqrt{\lambda}}$
Coefficient of Kurtosis:	$3 + \frac{1}{\lambda}$
Factorial Moments:	$E\left(\prod_{i=1}^{k}(X - i + 1)\right) = \lambda^k$ $E\left(\prod_{i=1}^{k}(X + i)\right)^{-1} = \frac{1}{\lambda^k}\left(1 - e^{-\lambda}\sum_{i=0}^{k-1}\frac{\lambda^i}{i!}\right)$
Moments about the Mean:	$\mu_k = \lambda\sum_{i=0}^{k-2}\binom{k-1}{i}\mu_i, \quad k = 2, 3, 4, \cdots,$ where $\mu_0 = 1$ and $\mu_1 = 0$. [Kendall 1943]
Moment Generating Function:	$\exp[\lambda(e^t - 1)]$
Probability Generating Function:	$\exp[\lambda(t - 1)]$

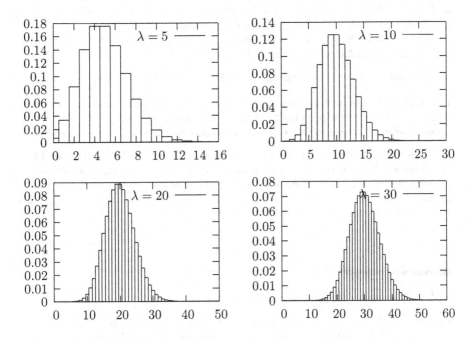

FIGURE 6.1: Poisson probability mass functions

6.3   Probabilities, Percentiles, and Moments

The dialog box [StatCalc→Discrete→Poisson→Probabilities, Critical Values, Moments] computes the tail probabilities, percentiles, and moments

To compute probabilities: Enter the values of the mean, and k at which the probability is to be computed; click [P]. As an example, when the mean = 6, $k = 5$, $P(X \leq 5) = 0.44568$, $P(X \geq 5) = 0.714944$, and $P(X = 5) = 0.160623$.

To compute other parameters: *StatCalc* also computes the mean or the value of k when other values are given. For example, to find the value of the mean when $k = 5$ and $P(X \leq k) = 0.25$, enter 5 for k, enter 0.25 for $P(X \leq k)$, and click [A] to get 7.4227. To find the value of k, when the mean = 4.5 and $P(X \leq k) = 0.34$, enter these values in appropriate edit boxes, and click [k] to get 3. Also, note that $P(X \leq 3) = 0.342296$ when the mean is 4.5.

To compute moments: Enter the value of the mean, and click [M].

Example 6.1. On average, four customers enter a fast food restaurant per every 3-min period during the peak hours 11 am - 1 pm. Assuming an approximate Poisson process, what is the probability of 26 or more customers arriving in a 15-min period?

Solution: Let X denote the number of customers entering in a 15-min period. Then, X follows a Poisson distribution with mean = $(4/3) \times 15 = 20$. To find the probability of observing 26 or more customers, select the dialog box referenced at the beginning of this section, enter 20 for the mean, 26 for the observed k, and click [P] to get $P(X \geq 26) = 0.1122$.

6.4   Model Fitting with Examples

Example 6.2. Rutherford and Geiger (1910) presented data on α particles emitted by a radioactive substance in 2608 periods, each of 7.5 sec. The data are given in Table 6.1.

a. Fit a Poisson model for the data.

b. Estimate the probability of observing 5 or fewer α particles in a period of 7.5 sec.

TABLE 6.1: Observed Frequency O_x of the Number of α Particles x in 7.5 Second Periods

x	0	1	2	3	4	5	6	7	8	9	10
O_x	57	203	383	525	532	408	273	139	45	27	16
E_x	54.6	211	408	526	508	393	253	140	67.7	29.1	17

Solution:

a. To fit a Poisson model, we estimate first the mean number λ of α particles emitted per 7.5 - sec period. Note that

$$\widehat{\lambda} = \frac{1}{2608} \sum_{x=0}^{10} x O_x = \frac{10086}{2608} = 3.867.$$

Using this estimated mean, we can compute the probabilities and the expected (theoretical) frequencies E_x under the Poisson($\widehat{\lambda}$) model. For example, E_0 is given by

$$E_0 = P(X = 0|\lambda = 3.867) \times 2608 = 0.020921 \times 2608 = 54.6.$$

Other expected frequencies are computed similarly. These expected frequencies are given in Table 5.1. We note that the observed and the expected frequencies are in good agreement. Furthermore, for this example, the chi-square statistic

$$\chi^2 = \sum_{x=0}^{10} \frac{(O_x - E_x)^2}{E_x} = 13.06,$$

and the df $= 11 - 1 - 1 = 9$ (see Section 2.4.2). The p-value for testing

H_0: The data fit Poisson(3.867) model vs. H_a: H_0 is not true

is given by $P(\chi_9^2 > 13.06) = 0.16$, which implies that the Poisson(3.867) model is tenable for the data.

b. Select the dialog box [StatCalc→Discrete→Poisson→Probabilities, Critical Values and Moments], enter 3.867 for the mean, and 5 for k; click [P(X <= k)] to get

$$P(X \le 5) = \sum_{k=0}^{5} \frac{e^{-3.867}(3.867)^k}{k!} = 0.805557.$$

Example 6.3. Data on the number of deaths due to kicks from horses, based on the observation of 10 Prussian cavalry corps for 20 years (equivalently, 200 corps-years), are given in Table 5.2. Prussian officials collected this data during the latter part of the 19th century in order to study the hazards that horses posed to soldiers (Bortkiewicz, 1898). In this situation, the chances of death due to a kick from a horse

TABLE 6.2: Horse Kick Data

Number of deaths k:	0	1	2	3	4	5
Number of corps-years in which k deaths occurred, O_x:	109	65	22	3	1	0
Expected number of corps-years, E_x:	108.7	66.3	20.2	4.1	0.6	0

is small, while the number of soldiers exposed to the risk is quite large. Therefore, a Poisson distribution may well fit the data. As in Example 6.2, the mean number of deaths per period can be estimated as

$$\widehat{\lambda} = \frac{0 \times 109 + 1 \times 65 + \ldots + 5 \times 0}{200} = 0.61.$$

Using this estimated mean, we can compute the expected frequencies as in Example 6.2. They are given in the third row of Table 5.2. For example, the expected frequency in the second column can be obtained as

$$P(X = 1|\lambda = 0.61) \times 200 = 0.331444 \times 200 = 66.3.$$

We note that the observed and the expected frequencies are in good agreement. Furthermore, for this example, the chi-square statistic

$$\chi^2 = \sum_{x=0}^{5} \frac{(O_x - E_x)^2}{E_x} = 0.7485,$$

and the df $= 4$. The p-value for testing

$$H_0: \text{The data fit Poisson}(0.61) \text{ model} \quad \text{vs.} \quad H_a: H_0 \text{ is not true}$$

is given by $P(\chi_4^2 > 0.7485) = 0.9452$, which is greater than any practical level of significance. Therefore, the Poisson(0.61) model is tenable.

6.5 One-Sample Inference

Let $X_1, \ldots, X_n$ be independent observations from a Poisson(λ) population. Then,

$$K = \sum_{i=1}^{n} X_i \sim \text{Poisson}(n\lambda).$$

The following inferences about λ are based on K. The maximum likelihood estimator of λ is given by

$$\widehat{\lambda} = \frac{1}{n} \sum_{i=1}^{n} X_i,$$

which is also the uniformly minimum variance unbiased estimator.

6.6 Test for the Mean

An Exact Test

Let K_0 be an observed value of K. Then, for testing

$$H_0 : \lambda \leq \lambda_0 \quad \text{vs.} \quad H_a : \lambda > \lambda_0, \tag{6.3}$$

the null hypothesis will be rejected if the p-value $P(K \geq K_0|n\lambda_0) \leq \alpha$, for testing

$$H_0 : \lambda \geq \lambda_0 \quad \text{vs.} \quad H_a : \lambda < \lambda_0, \tag{6.4}$$

the null hypothesis will be rejected if the p-value $P(K \leq K_0|n\lambda_0) \leq \alpha$, and for testing

$$H_0 : \lambda = \lambda_0 \quad \text{vs.} \quad H_a : \lambda \neq \lambda_0, \tag{6.5}$$

the null hypothesis will be rejected if the p-value

$$2 \min\{P(K \leq K_0|n\lambda_0), P(K \geq K_0|n\lambda_0)\} \leq \alpha. \qquad (6.6)$$

The Score Test

The score test statistic is given by

$$Z = \frac{\widehat{\lambda} - \lambda_0}{\sqrt{\lambda_0/n}},$$

which follows a standard normal distribution. The score test rejects the null hypothesis in (6.3) if the p-value

$$P(Z \geq z_0) = 1 - \Phi(z_0) \leq \alpha,$$

where Φ denotes the standard normal distribution and z_0 is an observed value of Z. P-values for testing other hypotheses are obtained similarly.

Example 6.4. It is desired to assess the average number defective spots per 100-ft of an electric cable. Inspection of a sample of 20 100 ft cables showed an average of 2.7 defective spots. Does this information indicate that the true mean number of defective spots per 100 ft is more than 2? Assuming a Poisson model, test at the level $\alpha = 0.05$.

Solution: Let X denote the number defective spots per 100-ft cable. Then, X follows a Poisson(λ) distribution, and we want to test

$$H_0 : \lambda \leq 2 \quad \text{vs.} \quad H_a : \lambda > 2.$$

In the dialog box [StatCalc→Discrete→Poisson→One-Sample: Test, CI and Power], enter 20 for the sample size, $20 \times 2.7 = 54$ for the total count, 2 for [Value of M0], and click the [p-values for] to get 0.0199946. Since the p-value is smaller than 0.05, we can conclude that true mean is greater than 2.

The score test statistic is

$$\frac{\widehat{\lambda} - \lambda_0}{\sqrt{\lambda_0/n}} = \frac{2.7 - 2}{\sqrt{2/20}} = 2.2136,$$

and the p-value is

$$1 - \Phi(2.2136) = .0134.$$

Powers of the Exact Test

The exact powers of the tests described in the preceding section can be computed using Poisson probabilities and an indicator function. For example, for a given λ and λ_0, the power of the test for hypotheses in (6.3) can be computed using the following expression.

$$\sum_{k=0}^{\infty} \frac{e^{-n\lambda}(n\lambda)^k}{k!} I(P(K \geq k|n\lambda_0) \leq \alpha), \qquad (6.7)$$

where $K \sim \text{Poisson}(n\lambda)$. Powers of the right-tailed test and two-tailed test can be expressed similarly.

The dialog box [StatCalc→Discrete→Poisson→One-Sample ...] uses the above exact method to compute the power.

Example 6.5. (*Sample Size Calculation*) Suppose that a researcher hypothesizes that the mean of a Poisson process has increased from 3 to 4. He likes to determine the required sample size to test his claim at the level 0.05 with power 0.80. To find the sample size, select [StatCalc→Discrete→Poisson→One-Sample ...], enter 1 to select right-tailed test, 0.05 for the level, 4 for [Guess M], 3 for [Null M0], and 0.80 for [Power]; click on [S Size] to get 23. To find the actual power at this sample size, click on [Power] to get 0.811302.

6.7 Confidence Intervals for the Mean

Let $X_1, \ldots, X_n$ be a sample from a Poisson(λ) population, and let $K = \sum_{i=1}^{n} X_i$. The following inferences about λ are based on K.

An Exact Confidence Interval

An exact $1 - \alpha$ confidence interval for λ is given by (λ_L, λ_U), where λ_L satisfies

$$P(K \geq k | n\lambda_L) = \exp(-n\lambda_L) \sum_{i=k}^{\infty} \frac{(n\lambda_L)^i}{i!} = \frac{\alpha}{2},$$

and λ_U satisfies

$$P(K \leq k | n\lambda_U) = \exp(-n\lambda_U) \sum_{i=0}^{k} \frac{(n\lambda_U)^i}{i!} = \frac{\alpha}{2},$$

where k is an observed value of K. Furthermore, using a relation between the Poisson and chi-square distributions, it can be shown that

$$\lambda_L = \frac{1}{2n} \chi^2_{2k;\alpha/2} \text{ and } \lambda_U = \frac{1}{2n} \chi^2_{2k+2;1-\alpha/2},$$

where $\chi^2_{m;p}$ denotes the pth quantile of a chi-square distribution with df $= m$. These formulas should be used with the convention that $\chi^2_{0;p} = 0$.

Score Confidence Intervals

The score confidence interval is on the basis of asymptotic normality of the score test statistic

$$T(\widehat{\lambda}, \lambda) = \frac{\widehat{\lambda} - \lambda}{\sqrt{\lambda/n}},$$

where $\widehat{\lambda} = \frac{K}{n}$. In particular, the endpoints of the $1 - \alpha$ score confidence interval are the roots of the quadratic equation $T^2(\widehat{\lambda}, \lambda) = c^2$, where $c = z_{1-\alpha/2}$, and are given by

$$(\lambda_l, \lambda_u) = \widehat{\lambda} + \frac{c^2}{2n} \pm \frac{c}{\sqrt{n}} \sqrt{\widehat{\lambda} + \frac{c^2}{4n}}. \tag{6.8}$$

The dialog box [StatCalc→Discrete→Poisson→CI for Mean and Sample Size for Width] computes the exact and the score confidence intervals for a Poisson mean.

Example 6.6. (*Confidence Interval for the Mean*) Let us compute a 95% confidence interval for the data given in Example 6.4. Recall that $n = 20$, sample mean $= 2.7$, and so the total count is 54. To find confidence intervals for the mean number of defective spots, select [StatCalc→Discrete→Poisson→One-Sample: Test, confidence interval and Power], enter 20 for [Sample Size], 54 for [Total], and 0.95 for [Conf Level]; click [2-sided] to get (2.02832, 3.52291) (exact) and (2.06952, 3.52255) (score). To find 95% one-sided confidence limits, enter .90 for confidence level, and click [2-sided] to get (2.12537, 3.387) (exact) and (2.15951, 3.37577) (score).

Example 6.7. The following data represent the number of serious earthquakes over a period of 75 years 1903–1977 (Blaesild and Granfeldt, 2003). An earthquake is considered serious if its magnitude is 7.5 or above on the Richter scale or at least 100 people were killed.

No. of serious earthquakes	0	1	2	3	4
Frequency		31	28	14	1	1

Let us find a 95% confidence intervals for the mean number of serious earthquakes per year. For this example, $N = 75$ years, and the total number of serious earthquakes is

$$0 \times 3 + 1 \times 28 + \ldots + 4 \times 1 = 63.$$

To find 95% confidence intervals, select the dialog box [StatCalc→Discrete→Poisson→One-Sample ...], enter 75 for [Sample Size], 63 for [Total Count], .95 for [Conf Level], and click on [2-sided] to get the exact confidence interval $(.645, 1.075)$, and the score confidence interval $(.657, 1.075)$. If we use the score confidence interval, then we estimate the mean number of serious earthquakes per year is between .657 and 1.075, or equivalently, 7 to 11 every decade. To find a 95% one-sided confidence limits, enter .90 for [Conf Level], and click on [2-sided]. For this example, 95% one-sided confidence limits are 0.673828 (exact lower), 1.03603 (exact upper), 0.68303 (score lower), and 1.03304 (score upper).

Sample Size Calculation for Precision

For a given n and λ, the expected length of the $1 - \alpha$ confidence interval (λ_L, λ_U) in Section 6.7 can be expressed as

$$\sum_{k=0}^{\infty} \frac{e^{-n\lambda}(n\lambda)^k}{k!} (\lambda_U - \lambda_L) = \frac{1}{2n} \sum_{k=0}^{\infty} \frac{e^{-n\lambda}(n\lambda)^k}{k!} (\chi^2_{2k+2,1-\alpha/2} - \chi^2_{2k,\alpha/2}).$$

The dialog box [StatCalc→Discrete→Poisson→One-Sample ...] also computes the sample size required to estimate the mean within a given precision.

Example 6.8. (*Sample Size Calculation*) Suppose that a researcher hypothesizes that the mean of a Poisson process is 3. He likes to determine the required sample size to estimate the mean within ±0.3 and with confidence 0.95. To find the sample size, select [StatCalc→Discrete→Poisson→One-Sample ...], enter .95 for [Conf Level], 3 for [Guess M], .3 for [Half-Width], and click [exact] to get 131. That is, if the experimenter decides to use the exact confidence interval, then the required sample size to estimate the mean within ±.3 is 131; for the score method, the required sample size is 129.

6.8   Prediction Intervals

Let X be the total counts in a sample of size n from a Poisson distribution with mean λ. Note that $X \sim$ Poisson($n\lambda$). Let Y denote the future total counts that can be observed in a sample of size m from the same Poisson distribution so that $Y \sim$ Poisson($m\lambda$). We shall describe some prediction intervals for Y based on an observed value of X.

The Exact Prediction Interval

The exact prediction interval is based on the conditional distribution of X given $X + Y$. The conditional distribution of X, conditionally given $X + Y = s$, is binomial($s, n/(m + n)$). Let us denote the cumulative distribution function of a binomial random variable with the number of trials N and success probability π by $B(x; N, \pi)$.

Let x be an observed value of X. The smallest integer L that satisfies

$$1 - B(x - 1; x + L, n/(n + m)) > \alpha \qquad (6.9)$$

is the $1 - \alpha$ lower prediction limit for Y. The $1 - \alpha$ upper prediction limit U is the largest integer for which

$$B(x; x + U, n/(n + m)) > \alpha. \qquad (6.10)$$

For $X = 0$, the lower prediction limit is defined to be zero, and the upper prediction limit is determined by (6.10). The interval (L, U) is a $1 - 2\alpha$ prediction interval for Y.

The Prediction Interval Based on the Joint Sampling Approach

Let $\widehat{\lambda}_{xy} = \frac{X+Y}{m+n}$, and let $\widehat{\text{var}}(m\widehat{\lambda}_{xy} - Y) = \frac{mn\widehat{\lambda}_{xy}}{m+n}$. The quantity

$$\frac{m\widehat{\lambda}_{xy} - Y}{\sqrt{\widehat{\text{var}}(m\widehat{\lambda}_{xy} - Y)}} = \frac{(mX - nY)}{\sqrt{mn(X + Y)}} \sim N(0, 1), \text{ asymptotically.}$$

In order to handle the zero count, we take X to be 0.5 when it is zero. The $1 - 2\alpha$ prediction interval is determined by the roots (with respect to Y) of the quadratic equation

$$\frac{(m\widehat{\lambda} - Y)^2}{\widehat{\lambda}_{xy}m(1 + m/n)} = z_{1-\alpha}^2.$$

Based on these roots, the $1 - 2\alpha$ prediction interval is given by $[\lceil L \rceil, \lfloor U \rfloor]$, where

$$[L, U] = \left(\widehat{Y} + \frac{mz_{1-\alpha}^2}{2n}\right) \pm z_{1-\alpha}\sqrt{m\widehat{Y}\left(\frac{1}{m} + \frac{1}{n}\right) + \frac{m^2 z_{1-\alpha}^2}{4n^2}}, \qquad (6.11)$$

where $\widehat{Y} = m\frac{X}{n}$, for $X = 1, 2, ...,$ and is $\frac{.5m}{n}$ for $X = 0$. Furthermore, $\lceil x \rceil$ denotes the smallest integer greater than or equal to x (ceiling function), and $\lfloor x \rfloor$ denotes the largest integer less than or equal to x (floor function). The prediction interval in (6.11), developed by Krishnamoorthy and Peng (2011), seems to be shorter than the exact ones determined by (6.9) and (6.10), and it controls the coverage probability very satisfactorily.

The dialog box [StatCalc→Discrete→Poisson→Tolerance Intervals and Prediction Intervals] uses the exact approach and the joint sampling approach to compute prediction intervals for a Poisson distribution.

Example 6.9. Consider the data on number of serious earthquakes in Example 6.7. Based on the data, we can predict $Y =$ the number of serious earthquakes in the next 10-year period as follows. Note that the observed value $X = 63$, the total number of serious earthquakes over a period of $n = 75$ years. To find the prediction interval, select the dialog box [StatCalc→Discrete→Poisson→Tolerance Intervals ...], enter 75 for [Sample Size, n], 63 for [# of events], 10 for [Future Sam Size], .95 for [Conf Level], and click on [2-sided] to get [3, 14] (approximate in (6.11)) and [3, 15] (exact). If we decide to use the exact prediction interval, then the number of serious earthquakes in a future 10-year period is between 3 and 15 with confidence 95%.

Example 6.10. This example is adapted from Bain and Patel (1993). For a random sample of 400 devices tested, 20 devices are unacceptable. A 90% prediction interval is desired for the number of unacceptable devices in a future sample of 100 such devices. The sample proportion of unacceptable devices is $\widehat{p} = 20/400 = .05$ and $\widehat{Y} = m \times \widehat{p} = 5$. To find 90% prediction intervals using *StatCalc*, enter 400 for [Sample Size n], 20 for [# of events], 100 for [Future Sam Size], and .90 for [Conf Level]; click [2-sided] to get approximate prediction interval based on the joint sampling approach as [2, 9], and the exact one as [1, 10]. Note that the approximate prediction interval is shorter than the exact one.

6.9 Tolerance Intervals

One-sided as well as equal-tailed tolerance intervals for a Poisson distribution can be obtained using the methods similar to the ones for the binomial case in Section 4.6. Let $X_1, ..., X_n$ be a sample from a Poisson(λ) distribution so that $S = \sum_{i=1}^n X_i \sim$ Poisson($n\lambda$).

Exact Tolerance Intervals

The $(p, 1 - \alpha)$ upper tolerance limit is the smallest integer $k_p(\lambda_u)$ so that

$$P(X \le k_p(\lambda_u)|\lambda_u) \ge p, \qquad (6.12)$$

where λ_u is the $1 - \alpha$ upper confidence limit for λ based on an observed value s of S. Similarly, a $(p, 1 - \alpha)$ lower tolerance limit is the largest integer $k_{1-p}(\lambda_l)$ so that

$$P(X \geq k_{1-p}(\lambda_l)|\lambda_l) \geq p, \qquad (6.13)$$

where λ_l is the $1 - \alpha$ lower confidence limit for λ. If (λ_l, λ_u) is a $1 - \alpha$ confidence interval λ, then

$$\left[k_{\frac{1-p}{2}}(\lambda_l), k_{\frac{1+p}{2}}(\lambda_u) \right] \qquad (6.14)$$

is a $(p, 1 - \alpha)$ equal-tailed tolerance interval.

An Approximate Method

The tolerance intervals based on the normal approximation to a Poisson quantile are as follows: Let λ_l and λ_u be a $1 - \alpha$ one-sided lower and upper confidence limits for λ, respectively. Then $\lambda_u + z_p \sqrt{\lambda_u}$ is a $(p, 1 - \alpha)$ upper tolerance limit, $\lambda_l - z_p \sqrt{\lambda_l}$ is a $(p, 1 - \alpha)$ lower tolerance limit. If (λ_l, λ_u) is a $1 - \alpha$ confidence interval for λ, then

$$\left[\lambda_l - z_{\frac{1+p}{2}} \sqrt{\lambda_l}, \ \lambda_u + z_{\frac{1+p}{2}} \sqrt{\lambda_u} \right]$$

is a $(p, 1 - \alpha)$ equal-tailed tolerance interval. If λ_l and λ_u are score confidence limits, then we refer to the corresponding tolerance intervals as the approximate-score tolerance intervals.

As in the binomial case (Section 4.6.2), we suggest to use $1 - 2\alpha$ confidence interval (λ_l, λ_u) for λ so that

$$\left[k_{\frac{1-p}{2}}(\lambda_l), \ k_{\frac{1+p}{2}}(\lambda_u) \right]$$

can be used as a two-sided tolerance interval with the minimum coverage probability close to the nominal level $1 - \alpha$.

In comparison among tolerance intervals, the exact ones are too conservative, producing tolerance intervals that are unnecessarily wide. The approximate ones based on the normal approximation are simple to compute and are satisfactory in terms of coverage probabilities.

The dialog box [StatCalc→Discrete→Poisson→Tolerance Intervals and Prediction Intervals] calculates the exact tolerance intervals based on the exact confidence intervals for λ in Section 6.7, and the approximate ones with the score confidence intervals for λ to compute tolerance intervals.

Example 6.11. This example concerns the number of surface defects in steel plates. The data are given in Montgomery (1996), and as in Wang and Tsung (2009), we use a part of the data for constructing Poisson tolerance intervals. The counts of surface defects on 21 steel plates are

$$1, \ 0, \ 4, \ 3, \ 1, \ 2, \ 0, \ 2, \ 1, \ 1, \ 0, \ 0, \ 2, \ 1, \ 3, \ 4, \ 3, \ 1, \ 0, \ 2, \ 4.$$

The maximum likelihood estimate $\widehat{\lambda} = \frac{35}{21} = 1.6667$. To compute $(0.90, 0.95)$ two-sided tolerance intervals, select the dialog box [StatCalc→Discrete→Poisson→

Tolerance Intervals and Prediction Intervals], enter 21 for [Sample Size, n], 3 for [# of events], 1 for [Future Sam Size], and click on [2-sided TI] to get [0, 5] (approximate), and [0, 5] (exact). This means that at leat 90% steel plates have 0 to 5 surface defects with confidence 95%.

6.10 Tests for Comparing Two Means and Power Calculation

Let $X_{i1}, \ldots, X_{in_i}$ be a sample from a Poisson(λ_i) population. Then,

$$K_i = \sum_{j=1}^{n_i} X_{ij} \sim \text{Poisson}(n_i \lambda_i), \ i = 1, 2.$$

The following tests about (λ_1/λ_2) are based on the conditional distribution of K_1, given $K_1 + K_2 = m$, is binomial($m, n_1\lambda_1/(n_1\lambda_1 + n_2\lambda_2)$).

A Conditional Test for the Ratio of Two Means

Consider testing

$$H_0 : \frac{\lambda_1}{\lambda_2} \leq c \quad \text{vs.} \quad H_a : \frac{\lambda_1}{\lambda_2} > c, \tag{6.15}$$

where c is a given positive number. The p-value based on the conditional distribution of K_1, given $K_1 + K_2 = m$, is given by

$$P(K_1 \geq k | m, p) = \sum_{x=k}^{m} \binom{m}{x} p^x (1 - p)^{m-x}, \ \text{where } p = \frac{n_1 c/n_2}{1 + n_1 c/n_2}. \tag{6.16}$$

The conditional test rejects the null hypothesis whenever the p-value is less than or equal to the specified nominal α (Chapman, 1952). The p-value of a left-tailed test or of a two-tailed test can be expressed similarly.

The dialog box [StatCalc→Discrete→Poisson→Two-Sample ...] uses the above exact approach to compute the p-values of the conditional test for the ratio of two Poisson means.

Example 6.12. (*Calculation of p-value*) Suppose that a sample of 20 observations from a Poisson(λ_1) distribution yielded a total of 40 counts, and a sample of 30 observations from a Poisson(λ_2) distribution yielded a total of 22 counts. We would like to test

$$H_0 : \frac{\lambda_1}{\lambda_2} \leq 1.5 \quad \text{vs.} \quad H_a : \frac{\lambda_1}{\lambda_2} > 1.5.$$

To compute the p-value using *StatCalc*, enter the sample sizes, total counts, and 1.5 for the value of c in [H0:M1/M2 = c], and click on [p-values-ratio] to get 0.01501. Thus, there is enough evidence to indicate that λ_1 is larger than one and a half times λ_2.

Example 6.13. (*Calculation of p-value*) Suppose that the number of work-related accidents over a period of 12 months in a manufacturing industry (say, A) is 14. In another manufacturing industry B, which is similar to A, the number of work-related accidents over a period of 9 months is 8. Assuming that the numbers of accidents in both industries follow Poisson distributions, it is desired to test if the mean number of accidents per month in industry A is greater than that in industry B. That is, we want to test

$$H_0 : \frac{\lambda_1}{\lambda_2} \leq 1 \quad \text{vs.} \quad H_a : \frac{\lambda_1}{\lambda_2} > 1,$$

where λ_1 and λ_2, respectively, denote the true mean numbers of accidents per month in A and B. To find the p-value using *StatCalc*, select [StatCalc→Discrete→ Poisson→Two-Sample ...], enter 12 for [Sam Size 1], 9 for [Sam Size 2], 14 for [No. Events 1], 8 for [No. Events 2], 1 for c in [H0:M1/M2 = c], and click [p-values for] to get 0.348343. Thus, there is not enough evidence to conclude that $\lambda_1 > \lambda_2$.

An Unconditional Test for the Difference between Two Means

This test is more powerful than the conditional test given in Section 6.10. However, this test is approximate, and in some situations, the type I error rates are slightly more than the nominal level. For more details, see Krishnamoorthy and Thomson (2004).

Consider testing

$$H_0 : \lambda_1 - \lambda_2 \leq d \quad \text{vs.} \quad H_a : \lambda_1 - \lambda_2 > d, \tag{6.17}$$

where d is a specified number. Let (k_1, k_2) be an observed value of (K_1, K_2), and let

$$\widehat{\lambda}_d = \frac{k_1 + k_2}{n_1 + n_2} - \frac{dn_1}{n_1 + n_2}.$$

The p-value for testing (6.17) is given by

$$P(k_1, k_2) = \sum_{x_1=0}^{\infty} \sum_{x_2=0}^{\infty} \frac{e^{-\eta}\eta^{x_1}}{x_1!} \frac{e^{-\delta}\delta^{x_2}}{x_2!} I(Z(x_1, x_2) \geq Z(k_1, k_2)), \tag{6.18}$$

where $\eta = n_1(\widehat{\lambda}_d + d)$, $\delta = n_2\widehat{\lambda}_d$,

$$Z(x_1, x_2) = \frac{\frac{x_1}{n_1} - \frac{x_2}{n_2} - d}{\sqrt{\frac{x_1}{n_1^2} + \frac{x_2}{n_2^2}}}$$

and $Z(k_1, k_2)$ is $Z(x_1, x_2)$ with x replaced by k. The null hypothesis will be rejected whenever the p-value is less than or equal to the nominal level α.

The dialog box [StatCalc→Discrete→Poisson→Two-Sample ...] in *StatCalc* uses the above formula to compute the p-values for testing the difference between two means.

Example 6.14. (*Unconditional Test*) Suppose that a sample of 20 observations from a Poisson(λ_1) distribution yielded a total of 40 counts, and a sample of 30 observations from a Poisson(λ_2) distribution yielded a total of 22 counts. We would like to test

$$H_0 : \lambda_1 - \lambda_2 \leq 0.7 \quad \text{vs.} \quad H_a : \lambda_1 - \lambda_2 > 0.7.$$

To compute the p-value, select the dialog box [StatCalc→Discrete→Poisson→Two-Sample ...], enter the sample sizes and the number of counts in appropriate edit boxes, 0.7 for [H0: M1-M2 = d], and click on [p-values-difference] to get 0.0459181. So, at the 5% level, we can conclude that there is enough evidence to indicate that λ_1 is 0.7 unit larger than λ_2.

Example 6.15. (*Unconditional Test*) Let us consider Example 6.13, where we used the conditional test for testing $\lambda_1 > \lambda_2$. We shall now apply the unconditional test for testing

$$H_0 : \lambda_1 - \lambda_2 \leq 0 \quad \text{vs.} \quad H_a : \lambda_1 - \lambda_2 > 0.$$

To find the p-value, enter 12 for the sample size 1, 9 for the sample size 2, 14 for [No. Events 1], 8 for [No. Events 2], 0 for d, and click [p-values for] to get 0.279551. Thus, we do not have enough evidence to conclude that $\lambda_1 > \lambda_2$.

As the unconditional test is more powerful than the conditional test, it produced a smaller p-value than that of the conditional test (see Example 6.13), which is 0.348343.

Power Study and Sample Size Calculation

For given sample sizes, guess values of the means and a level of significance, the exact power of the conditional test in (6.15) can be calculated using the following expression:

$$\sum_{k_1=0}^{\infty} \sum_{k_2=0}^{\infty} \frac{e^{-n_1\lambda_1}(n_1\lambda_1)^{k_1}}{k_1!} \frac{e^{-n_2\lambda_2}(n_2\lambda_2)^{k_2}}{k_2!} I(P(K_1 \geq k_1 | k_1 + k_2, p) \leq \alpha),$$

$$(6.19)$$

where $P(K_1 \geq k_1 | m, p)$ and p are as defined in (6.16). The powers of a two-tailed test and left-tailed test can be expressed similarly.

To compute the powers of the unconditional test, replace $P(K_1 \geq k_1 | k_1 + k_2, p)$ in (6.19) by the p-value $P(k_1, k_2)$ in (6.18).

The dialog box [StatCalc→Discrete→Poisson→Two-Sample ...] uses (6.19) to compute the power of the conditional test for the ratio of two Poisson means, and powers of the unconditional test for testing the difference between two Poisson means.

Example 6.16. (*Sample Size Calculation*) Suppose that a researcher hypothesizes that the mean $\lambda_1 = 3$ of a Poisson population is 1.5 times larger than the mean λ_2 of another population, and he would like to test

$$H_0 : \frac{\lambda_1}{\lambda_2} \leq 1.5 \quad \text{vs.} \quad H_a : \frac{\lambda_1}{\lambda_2} > 1.5.$$

To find the required sample size to get a power of 0.80 at the level 0.05, enter 30 for both sample sizes, 1 for one-tailed test, 0.05 for level, 3 for [Guess M1], 2 for [Guess M2], click [Cond. test] to get 0.76827, and click [Uncond. test] to get 0.791813. By trial–error, we see that the power of the conditional test is 0.804721 when both sample sizes are 33. The power of the unconditional test is 0.803148 when both sample sizes are 31.

We note that *StatCalc* also computes the power for unequal sample sizes. For example, when the first sample size is 27 and the second sample size is 41, the power is 0.803072 (conditional test). Furthermore, if it is desired to find sample sizes for testing the hypotheses

$$H_0 : \frac{\lambda_1}{\lambda_2} = 1.5 \quad \text{vs.} \quad H_a : \frac{\lambda_1}{\lambda_2} \neq 1.5,$$

then enter 2 for two-tailed test (while keep the other values as they are), and click [Power]. For example, when both sample sizes are 33, the power is 0.705986 (conditional test); when they are 40, the power is 0.791258, and when they are 41 the power is 0.801372. If we choose to use the unconditional test, then the required sample size from both populations is 39, and the power is 0.800053.

Example 6.17. (*Power Calculation*) Suppose a researcher hypothesizes that the mean $\lambda_1 = 3$ of a Poisson population is at least one unit larger than the mean λ_2 of another population, and he would like to test

$$H_0 : \lambda_1 - \lambda_2 \leq 0 \quad \text{vs.} \quad H_a : \lambda_1 - \lambda_2 > 0.$$

To find the required sample size to get a power of 0.80 at level of 0.05, enter 30 for both sample sizes, 0 for d in [H0: M1-M2 = d], 1 for one-tailed test, 0.05 for level, 3 for [Guess M1], 2 for [Guess M2], and click [Uncond. test] to get 0.791813. By raising the sample size to 31, we get a power of 0.803148. We also note that when the first sample size is 27 and the second sample size is 36, the power is 0.803128.

For the above example, if it is desired to find the sample sizes for testing the hypotheses

$$H_0 : \lambda_1 - \lambda_2 = 0 \quad \text{vs.} \quad H_a : \lambda_1 - \lambda_2 \neq 0,$$

then enter 2 for two-tailed test (while keeping the other values as they are), and click [Power]. For example, when both sample sizes are 33, the power is 0.730551; when they are 39, the power is 0.800053. (Note that if one choose to use the conditional test, then the required sample size from both populations is 41. See Example 6.16).

6.11   Confidence Intervals for the Ratio of Two Means

Let $K_1 \sim \text{Poisson}(\lambda_1)$ independently of $K_2 \sim \text{Poisson}(\lambda_2)$. In the following, we shall see a few methods of obtaining confidence intervals for the ratio $\frac{\lambda_1}{\lambda_2}$.

An Exact Confidence Interval

The exact confidence interval for (λ_1/λ_2) is based on the conditional distribution of K_1 given in (6.16). Let

$$p = \frac{\lambda_1}{\lambda_1 + \lambda_2} = \frac{(\lambda_1/\lambda_2)}{(\lambda_1/\lambda_2) + 1}. \tag{6.20}$$

For given $K_1 = k$ and $K_1 + K_2 = m$, a $1 - \alpha$ confidence interval for λ_1/λ_2 is

$$\left(\frac{p_L}{(1-p_L)}, \frac{p_U}{(1-p_U)} \right), \tag{6.21}$$

where (p_L, p_U) is a $1 - \alpha$ exact confidence interval for p based on k successes from a binomial(m, p) distribution (see Section 4.4.3).

Binomial-Score Confidence Interval

This confidence interval for λ_1/λ_2 is obtained from the score confidence interval (4.6) for p defined in (6.20). Let (p_{sl}, p_{su}) denote the score confidence interval (4.6) for p defined in (6.20) based on $K_1 = k$ successes out of $K_1 + K_2 = m$ trials. Let $a_1 = 2K_1 + z_{1-\alpha/2}^2$, $a_2 = 2K_2 + z_{1-\alpha/2}^2$, and $a_{12} = 4z_{1-\alpha/2}^2 K_1 K_2/(K_1 + K_2) + z_{1-\alpha/2}^4$. Then, the binomial-score confidence interval for λ_1/λ_2 is given by

$$\left(\frac{p_{sl}}{(1-p_{sl})}, \frac{p_{su}}{(1-p_{su})} \right) = \left(\frac{a_1 - \sqrt{a_{12}}}{a_2 + \sqrt{a_{12}}}, \frac{a_1 + \sqrt{a_{12}}}{a_2 - \sqrt{a_{12}}} \right). \tag{6.22}$$

Sato (1990) and Graham et al. (2003) developed likelihood-score confidence interval for λ_1/λ_2. Their likelihood-score confidence interval is the same as the one in (6.22) except that the likelihood-score confidence interval is not defined when $K_2 = 0$ and $K_1 \geq 0$.

Fiducial Confidence Interval for the Ratio of Poisson Means

A fiducial quantity for λ_i is given by $\frac{1}{2}\chi_{2k_i+1}^2$, where k_i is an observed value of K_i, $i = 1, 2$. A fiducial quantity for the ratio $\frac{\lambda_1}{\lambda_2}$ is obtained by substitution, and is given by $\frac{\chi_{2k_1+1}^2}{\chi_{2k_2+1}^2}$. The α and $1 - \alpha$ quantiles of this fiducial quantity form a $1 - 2\alpha$ confidence interval for $\frac{\lambda_1}{\lambda_2}$. Using the relation between the ratio of independent chi-square random variables and the F random variable, this confidence interval can be expressed as

$$\left(\frac{(2k_1+1)}{(2k_2+1)} F_{2k_1+1, 2k_2+1; \frac{\alpha}{2}}, \frac{(2k_1+1)}{(2k_2+1)} F_{2k_1+1, 2k_2+1; 1-\frac{\alpha}{2}} \right), \tag{6.23}$$

where $F_{m,n;q}$ denotes the $100q$ percentile of an $F_{m,n}$ distribution. Cox (1953) has proposed the confidence interval (6.23).

Remark 6.1. If K_i is the total count based on a sample of size n_i, then the above confidence intervals are for the ratio $\frac{n_1 \lambda_1}{n_2 \lambda_2}$. A confidence interval for the ratio $\frac{\lambda_1}{\lambda_2}$ can be obtained by multiplying the endpoints of the confidence interval for $\frac{n_1 \lambda_1}{n_2 \lambda_2}$ by $\frac{n_2}{n_1}$.

The exact confidence intervals are, in general, too conservative, that is, unnecessarily wider. Graham et al. (2003) have carried out extensive simulation studies comparing the binomial-score confidence interval (6.22)) with other asymptotic confidence intervals, and concluded that the likelihood-score confidence interval is the best. Krishnamoorthy and Lee's (2010) numerical comparison study indicates that the binomial-score confidence interval is conservative for small values of λ_1/λ_2. In general, the fiducial confidence interval maintains the coverage probability around the nominal level, except in a few cases it could be liberal. The coverage probabilities of both confidence intervals seldom fall below 0.94 when the nominal level is 0.95.

The dialog box [StatCalc→Discrete→Poisson→Two-Sample ...] calculates the exact and Cox's fiducial confidence intervals for the ratio of two Poisson means.

Example 6.18. (*Confidence Interval for the Ratio of Means*) Suppose that a sample of 20 observations from a Poisson(λ_1) distribution yielded a total of 40 counts, and a sample of 30 observations from a Poisson(λ_2) distribution yielded a total of 22 counts. To compute a 95% confidence interval for the ratio of means, select the above dialog box from *StatCalc*, enter 20 for [S Size 1], 30 for [S Size 2], 40 for [No. events 1], 22 for [No. events 2], .95 for [Conf Level], and click on [2-sided] to get the exact confidence interval (1.5824, 4.81807) and the Cox confidence interval (1.63566, 4.63471). To get one-sided confidence intervals click on [1-sided] to get 1.71496 (Cox, 1.77299) and 4.40773 (Cox, 4.24301). That is, 95% lower confidence limit for the ratio λ_1/λ_2 is 1.71496, and 95% upper confidence limit for the ratio λ_1/λ_2 is 4.40773. Note that fiducial confidence intervals (by Cox) are shorter than the corresponding exact confidence intervals.

Example 6.19. (*Confidence Intervals for the Ratio of Means*) This example is taken from Boice and Monson (1977), where two groups of women were compared to find whether those who had been examined using x-ray fluoroscopy during treatment for tuberculosis had a higher rate of breast cancer than those who had not been examined using the x-ray fluoroscopy. In the treatment group, 41 cases of breast cancer in 28,010 person-years at risk were reported, while in the control group of women not receiving x-ray fluoroscopy, 15 cases of breast cancer in 19,017 person-years at risk were reported. So, we have $K_1 = 41$, $n_1 = 28,010$, $K_2 = 15$, and $n_2 = 19,017$, and the problem of interest is to obtain a confidence interval for the ratio $\frac{\lambda_1}{\lambda_2}$, where λ_1 is the mean rate of breast cancers for the treatment group and λ_2 is that for the control group. To find 95% confidence intervals for $\frac{\lambda_1}{\lambda_2}$, enter the sample sizes and the numbers of cases in *StatCalc*, and click [2-sided] to get the exact confidence interval (1.01, 3.61), and the Cox fiducial confidence interval (1.05, 3.42). Note that both intervals indicate that λ_1 is significantly larger than λ_2. We also observe that the fiducial interval is shorter than the exact interval.

6.12   Confidence Intervals for the Difference between Two Means

Let K_i denote the total count based on a sample of size n_i from a Poisson(λ_i) distribution, $i = 1, 2$. The following confidence intervals for the difference $\lambda_1 - \lambda_2$ are based on $K_1 \sim$ Poisson($n_1\lambda_1$) and $K_2 \sim$ Poisson($n_2\lambda_2$).

The Wald Confidence Interval

Let $\widehat{\lambda}_i = \frac{K_i}{n_i}$, $i = 1, 2$. The $1 - \alpha$ confidence interval for $\lambda_1 - \lambda_2$ is given by

$$\widehat{\lambda}_1 - \widehat{\lambda}_2 \pm z_{1-\frac{\alpha}{2}} \sqrt{\frac{\widehat{\lambda}_1}{n_1} + \frac{\widehat{\lambda}_2}{n_2}}, \tag{6.24}$$

where z_p is the p quantile of the standard normal distribution.

The Score Confidence Interval

The score confidence interval is obtained by inverting the test statistic proposed in Krishnamoorthy and Thomson (2004) and is given by

$$\widehat{\lambda}_1 - \widehat{\lambda}_2 + \frac{z^2_{1-\alpha/2}}{2}\left(\frac{1}{n_1} - \frac{1}{n_2}\right) \pm z_{1-\alpha/2} \sqrt{\left(\frac{\widehat{\lambda}_1}{n_1} + \frac{\widehat{\lambda}_2}{n_2}\right) + \frac{z^2_{1-\alpha/2}}{4}\left(\frac{1}{n_1} - \frac{1}{n_2}\right)^2}, \tag{6.25}$$

where z_p denotes the p quantile of the standard normal distribution. Notice that the score confidence interval simplifies to the Wald confidence interval

$$\widehat{\lambda}_1 - \widehat{\lambda}_2 \pm z_{1-\alpha/2} \sqrt{\left(\frac{\widehat{\lambda}_1}{n_1} + \frac{\widehat{\lambda}_2}{n_2}\right)}, \tag{6.26}$$

when $n_1 = n_2$.

Fiducial Confidence Interval

The fiducial confidence interval is formed by the appropriate percentiles of the fiducial quantity

$$Q_{\lambda_1 - \lambda_2} = \frac{1}{2n_1}\chi^2_{2k_1+1} - \frac{1}{2n_2}\chi^2_{2k_2+1}, \tag{6.27}$$

where k_1 and k_2 are observed total counts. Let $Q_{\lambda_1 - \lambda_2;\alpha}$ denote the α quantile of $Q_{\lambda_1 - \lambda_2}$. Then $(Q_{\lambda_1 - \lambda_2;\alpha}, Q_{\lambda_1 - \lambda_2;1-\alpha})$ is a $1 - 2\alpha$ confidence interval for $\lambda_1 - \lambda_2$. Note that the percentiles of (6.27) can be estimated using Monte Carlo simulation. Approximate percentiles of $Q_{\lambda_1 - \lambda_2}$ based on the modified normal-based approximation in Section 2.9.1 are

$$Q_{\lambda_1 - \lambda_2;\alpha} \simeq M_1 - M_2 - \sqrt{\left(M_1 - \frac{\chi^2_{2k_1+1;\alpha}}{2n_1}\right)^2 + \left(M_2 - \frac{\chi^2_{2k_2+1;1-\alpha}}{2n_2}\right)^2}, \quad 0 < \alpha \le .5, \tag{6.28}$$

where $M_i = \frac{k_i+1/2}{n_i}$, $i = 1, 2$, and

$$Q_{\lambda_1 - \lambda_2;1-\alpha} \simeq M_1 - M_2 + \sqrt{\left(M_1 - \frac{\chi^2_{2k_1+1;1-\alpha}}{2n_1}\right)^2 + \left(M_2 - \frac{\chi^2_{2k_2+1;\alpha}}{2n_2}\right)^2}, \quad 0 < \alpha \le .5, \tag{6.29}$$

The score and the fiducial confidence intervals are proposed in Krishnamoorthy and Lee (2013). These authors have compared several confidence intervals, including the one by Li et al. (2011), and concluded that fiducial confidence intervals are preferable for smaller counts, and the score confidence interval is the best, provided $n_1\lambda_1$ and $n_2\lambda_2$ (expected total counts) are two or more.

Example 6.20. This example is taken from Jaech (1970). Three reactor fuel element failures were observed out of 310 process tubes for a given type of material. In a second type of material, 7 failures were observed out of 3,500 process tubes. Here binomial models are more appropriate to compare the failure rates. As the Poisson distribution is a limiting form of a binomial distribution, we can apply the interval estimation procedures given earlier. Let λ_1 be the failure rate of the first type material, and λ_2 be the failure rate of the second type material. To compute 90% confidence intervals for $\lambda_1 - \lambda_2$ using *StatCalc*, select the dialog box [StatCalc→Discrete→Poisson→Two-Sample ...], and enter the data as in the preceding example, .90 for [Conf Level], and click [2-sided] to get the fiducial confidence interval (.0012, .0206), and the approximate score confidence interval (.0016, .0217). We observe that the computed score and fiducial confidence intervals for the difference indicate that λ_1 is greater than λ_2.

Let us find a binomial-based fiducial confidence interval (Section 4.8.1) for the difference $\lambda_1 - \lambda_2$. Select the dialog box [StatCalc→Discrete→Binomial→Two-Sample ...], enter 3 for sample 1 successes, 307 for sample 1 failures, 7 for sample 2 successes, 3493 for sample 2 failures, .90 for [Conf Level] under [CI for p1 -p2], and click [CI-Diff] to get (.0012, .0204). Note that this confidence interval is very close to the one in the preceding paragraph.

6.13   Inference for a Weighted Sum of Poisson Means

Let $K_1, ..., K_g$ be independent random variables with $K_i \sim \text{Poisson}(n_i \lambda_i)$, $i = 1, ..., g$. We are interested in finding confidence intervals for $\sum_{i=1}^g c_i \lambda_i$, where c_i's are known positive constants. Without loss of generality, we can assume $c_i \in (0, 1)$, $i = 1, ..., g$ so that $\sum_{i=1}^g c_i = 1$ and $n_1 = ... = n_g = 1$. The case of unequal sample sizes can be handled by letting $w_i = c_i/n_i$ and $\xi_i = n_i \lambda_i$, $i = 1, ..., g$ so that $\sum_{i=1}^g w_i \xi_i = \sum_{i=1}^g c_i \lambda_i$.

A fiducial quantity for the weighted mean $\mu = \sum_{i=1}^g c_i \lambda_i$ is obtained by replacing the λ_is by their fiducial quantities. Letting $c_i^* = c_i/(2n_i)$, $i = 1, ..., g$, we write the fiducial quantity for μ as $Q_\mu = \sum_{i=1}^g c_i^* \chi_{2m_i+1}^2$, where m_i is an observed value of Y_i, $i = 1, ..., g$. For given sample sizes and observed counts, one can use Monte Carlo simulation to estimate the percentiles of Q_μ. As Q_μ is a linear combination of independent χ^2 variables, we can also approximate the distribution of Q_μ by $e\chi_f^2$, where e and f are to be determined by matching moments. By using this moment matching method, we find

$$e = \frac{\sum_{i=1}^g c_i^{*2}(2m_i + 1)}{\sum_{i=1}^g c_i^*(2m_i + 1)} \text{ and } f = \frac{\left(\sum_{i=1}^g c_i^*(2m_i + 1)\right)^2}{\sum_{i=1}^g c_i^{*2}(2m_i + 1)}.$$

Thus, an approximate $1 - \alpha$ CI for μ is given by

$$\left(e\chi_{f;\frac{\alpha}{2}}^2, e\chi_{f;1-\frac{\alpha}{2}}^2\right). \tag{6.30}$$

Lee's (2010) numerical studies indicate that the confidence interval based on the Monte Carlo method and the one in (6.30) are practically the same.

Example 6.21. We shall find confidence intervals for a weighted sum of Poisson means using the incidence rates given in Table III of Dobson et al. (1991). The data, reported in (WHO MONICA project, 1988), were collected from an urban area (reporting unit 1) and a rural area (reporting unit 2) in the Federal Republic of Germany. The 1986 incidence rates for nonfatal definite myocardial infarction in women aged 35–64 years stratified by 5-year age group and reporting unit. The incidence rates along with weights c_i for age groups (corresponding to the Segi World Standard Population) are reproduced here in Table 6.3.

We are interested in estimating the age-standardized incidence rates per 100,000 person-years, $\mu = \sum_{i=1}^{6} w_i \xi_i$ with $w_i = c_i/n_i$. The sample estimates for reporting units 1 and 2 are 2.75 and 1.41, respectively. The calculated 95% confidence intervals for μ based on the fiducial approach with simulation consisting of 10,000 runs, and the one in (6.30) are given in Table 6.3[1]. The confidence interval on the basis of the chi-square approximation and the one based on the simulation are practically the same for both reporting units.

TABLE 6.3: 95% Incidence Rates for Myocardial Infarction in Women by Age and Reporting Unit

Age (years)	c_i	Reporting unit 1		Reporting unit 2	
		Person-years, n_i	Events, Y_i	Person-years, n_i	Events, Y_i
35–39	6/31	7,971	0	10,276	0
40–44	6/31	7,084	0	9,365	1
45–49	6/31	9,291	1	11,623	0
50–54	5/31	7,743	2	8,684	4
55–59	4/31	7,798	4	7,926	0
60–64	4/31	8,809	10	8,375	3
Age standardized rate per 100,000		2.75		1.41	
95% fiducial CI		(2.05, 5.05)		(0.96, 3.26)	
95% CI based on χ^2 apprx.		(2.04, 5.04)		(0.97, 3.26)	

6.14 Properties and Results

6.14.1 Properties

1. For a fixed k, $P(X \leq k|\lambda)$ is a nonincreasing function of λ.

2. Let $X_1, \ldots, X_n$ be independent Poisson random variables with $E(X_i) = \lambda_i$, $i = 1, \ldots, n$. Then,

$$\sum_{i=1}^{n} X_i \sim \text{Poisson} \left(\sum_{i=1}^{n} \lambda_i \right).$$

---

[1]The first part of the table is reproduced with permission from Wiley.

3. Recurrence Relations:

$$P(X = k + 1|\lambda) = \frac{\lambda}{k+1}P(X = k|\lambda), \quad k = 0, 1, 2, \ldots$$
$$P(X = k - 1|\lambda) = \frac{k}{\lambda}P(X = k|\lambda), \quad k = 1, 2, \ldots$$

4. An identity: Let X be a Poisson random variable with mean λ and $|g(-1)| < \infty$. Then,

$$E[Xg(X - 1)] = \lambda E[g(X)]$$

provided the indicated expectations exist [Hwang, 1982].

6.14.2   Relation to Other Distributions

1. Binomial: Let X_1 and X_2 be independent Poisson random variables with means λ_1 and λ_2 respectively. Then, conditionally

$$X_1|(X_1 + X_2 = n) \sim \text{binomial}\left(n, \frac{\lambda_1}{\lambda_1 + \lambda_2}\right).$$

2. Multinomial: If $X_1, \ldots, X_m$ are independent Poisson(λ) random variables, then the conditional distribution of X_1 given $X_1 + \ldots + X_m = n$ is multinomial with n trials and cell probabilities $p_1 = \ldots = p_m = 1/m$.

3. Gamma: Let X be a Poisson(λ) random variable. Then

$$P(X \leq k|\lambda) = P(Y \geq \lambda),$$

where Y is Gamma($k+1, 1$) random variable. Furthermore, if W is a gamma(a, b) random variable, where a is an integer, then for $x > 0$,

$$P(W \leq x) = P(Q \geq a),$$

where Q is a Poisson(x/b) random variable.

6.14.3   Approximations

1. Normal:

$$P(X \leq k|\lambda) \simeq P\left(Z \leq \frac{k-\lambda+0.5}{\sqrt{\lambda}}\right),$$
$$P(X \geq k\lambda) \simeq P\left(Z \geq \frac{k-\lambda-0.5}{\sqrt{\lambda}}\right),$$

where X is the Poisson(λ) random variable and Z is the standard normal random variable.

6.15   Random Number Generation

Input:

```
    L = Poisson mean
    ns = desired number of random numbers
```

Output:
        x(1),..., x(ns) are random numbers from the
        Poisson(L) distribution

The following algorithm is based on the inverse method, and is similar to Algorithm 4.1 for the binomial random numbers generator.

Algorithm 6.1. Poisson variate generator

```
Set k = int(L); pk = P(X = k); df = P(X <= k)
    rpk = pk; rk = k
    max = L + 10*sqrt(L)
    If L > 100,  max = L + 6*sqrt(L)
    If L > 1000, max = L + 5*sqrt(L)

For j = 1 to ns
        Generate u from uniform(0, 1)
        If u > df, go to 2
1       u = u + pk
        If k = 0 or u > df, go to 3
        pk = pk*k/L
        k = k - 1
        go to 1
2       pk = L*pk/(k + 1)
        u = u - pk
        k = k + 1
        If k = max or u < df, go to 3
        go to 2
3       x(j) = k
        k = rk
        pk = rpk
[end j loop]
```

6.16   Computation of Probabilities

For a given k and small mean λ, $P(X = k)$ can be computed in a straightforward manner. For large values, the logarithmic gamma function can be used.
To Compute $P(X = k)$:

$$P(X = k) = \exp(-\lambda + k * \ln(\lambda) - \ln(\Gamma(k + 1)))$$

To compute $P(X \le k)$:
Compute $P(X = k)$
Set $m = \text{int}(\lambda)$

If $k \leq m$, compute the probabilities using the backward recursion relation

$$P(X = k - 1|\lambda) = \frac{k}{\lambda} P(X = k|\lambda),$$

for $k - 1, k - 2, \ldots, 0$ or until the desired accuracy; add these probabilities and $P(X = k)$ to get $P(X \leq k)$.
else compute the probabilities using the forward recursion relation

$$P(X = k + 1|\lambda) = \frac{\lambda}{k + 1} P(X = k|\lambda),$$

for $k + 1, k + 2, \ldots$ until the desired accuracy; sum these probabilities to get $P(X \geq k + 1)$; the cumulative probability $P(X \leq k) = 1 - P(X \geq k + 1)$.

R function 6.1. Calculation of the Poisson Cumulative Distribution Function

```
poicdf = function(k, lambda){
zero = 0.0; one = 1.0
if(k < 0) return(zero)
mod = floor(lambda)+1
pk = poipr(k, lambda)
i = k; ans = 0.0
if(k <= mod){
repeat{
ans = ans + pk
pk = pk*i/lambda
if(i == 0 | pk < 1.0e-14){break}
i = i - 1}
}
else{
repeat{
pk = pk*lambda/(i+one)
if(pk < 1.0e-14){break}
ans = ans + pk
i = i + 1}
ans = one - ans}
return(ans)
}
```

The following R functions based on the preceding method compute the cumulative distribution function (poicdf) and the probability mass function (poipr) of a Poisson distribution with mean λ. The logarithmic gamma function "alng" (R function 2.1) is required.

R function 6.2. Calculation of the Poisson pmf

```
poipr = function(k, lambda){
zero = 0.0; one = 1.0;
if(k < 0){return(zero)}
# alng(x) = logarithmic gamma function; R function 1.1
term = -alng(k+one)+k*log(lambda)-lambda
return(exp(term))
}
```

7

Geometric Distribution

7.1 Description

Consider a sequence of independent Bernoulli trials with success probability p. Let X denote the number of failures until the first success to occur. Then, the probability mass function of X is given by

$$
\begin{aligned}
P(X = k|p) &= P(\text{Observing } k \text{ failures}) \\
&\quad \times\ P(\text{Observing a success at the } (k+1)\text{st trial}) \\
&= (1-p)^k p, \quad k = 0, 1, 2, \ldots.
\end{aligned}
$$

This is the probability of observing exactly k failures until the first success to occur or the probability that exactly $(k+1)$ trials are required to get the first success. The cdf is given by

$$
F(k|p) = p \sum_{i=0}^{k} (1-p)^i = \frac{p[1 - (1-p)^{k+1}]}{1 - (1-p)} = 1 - (1-p)^{k+1}, \quad k = 0, 1, 2, \ldots
$$

Since the above cdf is a geometric series with finite terms, the distribution is called geometric distribution.

Memoryless Property: For nonnegative integers k and m,

$$
\begin{aligned}
P(X > m + k | X > m) &= \frac{P(X > m + k \text{ and } X > m)}{P(X > m)} \\
&= \frac{P(X > m + k)}{P(X > m)} \\
&= P(X > k).
\end{aligned}
$$

Thus, the probability of observing an additional k failures, given the fact that m failures have already been observed, is the same as the probability of observing k failures at the start of the sequence. That is, geometric distribution forgets what has occurred earlier.

7.2   Moments

Mean:	$\frac{1-p}{p}$
Variance:	$\frac{1-p}{p^2}$
Mode:	0
Mean Deviation:	$2u(1-p)^u$, where u is the smallest integer greater than the mean.
Coefficient of Variation:	$\frac{1}{\sqrt{(1-p)}}$
Coefficient of Skewness:	$\frac{2-p}{\sqrt{(1-p)}}$
Coefficient of Kurtosis:	$9 + \frac{p^2}{(1-p)}$
Moments about the Mean:	$\mu_{k+1} = (1-p)\left(\frac{\partial \mu_k}{\partial q} + \frac{k}{p^2}\mu_{k-1}\right)$, where $q = 1-p$, $\mu_0 = 1$ and $\mu_1 = 0$.
Moment Generating Function:	$p(1 - qe^t)^{-1}$
Probability Generating Function:	$p(1 - qt)^{-1}$

7.3   Probabilities, Percentiles, and Moments

The dialog box [StatCalc→Discrete→Geometric] computes the tail probabilities, critical points, parameters, and confidence intervals for a geometric distribution with parameter p.

To compute probabilities: Enter the number of failures k until the first success and the success probability p; click [P]. For example, the probability of observing the first success at the 12th trial, when the success probability is 0.1, can be computed as follows: Enter 11 for k and 0.1 for p; click on [P] to get

$$P(X \le 11) = 0.71757, \; P(X \ge 11) = 0.313811 \text{ and } P(X = 11) = 0.0313811.$$

To find the value of the success probability when $k = 11$ and P(X <= k) = 0.9, enter these numbers in the appropriate edit boxes, and click [s] to get 0.174596. For instance, to find the value of k when $p = 0.3$ and P(X <= k) = 0.8, enter these numbers in the white boxes, and click [k] to get 4.

To compute confidence intervals: Enter the observed number of failures k until the first success and the confidence level; click on [1-sided] to get one-sided limits, or click [2-sided] to get two-sided confidence intervals.

As an example, suppose that in an experiment consisting of sequence of Bernoulli trials, 12 trials were required to get the first success. To find a 95% confidence interval for the success probability p, enter 11 for k, 0.95 for confidence level; click [2-sided] to get (0.002, 0.285).

To compute moments: Enter a value for p in $(0, 1)$; click [M].

7.4   Properties and Results

1. $P(X \geq k+1) = (1-p)^{k+1}, \quad k = 0, 1, \dots$.

2. For fixed k, $P(X \leq k|p)$ is an increasing function of p.

3. If $X_1, \dots, X_r$ are independent geometric random variables with success probability p, then

$$\sum_{i=1}^{r} X_i \sim \text{negative binomial}(r, p).$$

7.5   Random Number Generation

Generate u from uniform$(0, 1)$
Set $k = $ integer part of $\frac{\ln(u)}{\ln(1-p)}$
k is a pseudo random number from the geometric(p) distribution.

8

Negative Binomial Distribution

8.1 Description

Consider a sequence of independent Bernoulli trials with success probability p. The distribution of the random variable that represents the number of failures until the first success is called geometric distribution. Now, let X denote the number of failures until the rth success. The random variable X is called the negative binomial random variable with parameters p and r, and its probability mass function is given by

$$
\begin{aligned}
P(X = k | r, p) \quad &= \quad P(\text{observing } k \text{ failures in the first } k + r - 1 \text{ trials}) \\
&\times \quad P(\text{observing a success at the } (k+r)\text{th trial}) \\
&= \quad \binom{r + k - 1}{k} p^{r-1}(1-p)^k \times p.
\end{aligned}
$$

Thus,

$$
f(k|r, p) = P(X = k | r, p) = \binom{r + k - 1}{k} p^r (1-p)^k, \quad k = 0, 1, 2, \ldots; \quad 0 < p < 1.
$$

This is the probability of observing k failures before the rth success or equivalently, probability that $k + r$ trials are required until the rth success to occur. In the binomial distribution, the number of successes out of a fixed number of trials is a random variable, whereas in the negative binomial, the number of trials required to have a given number of successes is a random variable.

The plots of the probability mass functions presented in Figure 8.1 show that the negative binomial distribution is always skewed to the right. The degree of skewness decreases as r increases. See the formula for coefficient of skewness in Section 8.2.

The following relation between the negative binomial and binomial distributions is worth noting.

$$
\begin{aligned}
P(X \le k) \quad &= \quad P(\text{observing } k \text{ or less failures before the } r\text{th success}) \\
&= \quad P((k + r) \text{ or less trials are required to have exactly } r \text{ successes}) \\
&= \quad P(\text{observing } r \text{ or more successes in } (k + r) \text{ trials}) \\
&= \quad \sum_{i=r}^{k+r} \binom{k + r}{i} p^i (1-p)^{k+r-i} \\
&= \quad P(Y \ge r),
\end{aligned}
$$

where Y is a binomial$(k + r, p)$ random variable.

8.2   Moments

Mean:	$\frac{r(1-p)}{p}$
Variance:	$\frac{r(1-p)}{p^2}$
Mode:	The largest integer $\leq \frac{(r-1)(1-p)}{p}$.
Mean Deviation:	$2u\binom{r+u-1}{u}(1-p)^u p^{r-1}$, where u is the smallest integer greater than the mean. [Kamat, 1965]
Coefficient of Variation:	$\frac{1}{\sqrt{r(1-p)}}$
Coefficient of Skewness:	$\frac{2-p}{\sqrt{r(1-p)}}$
Coefficient of Kurtosis:	$3 + \frac{6}{r} + \frac{p^2}{r(1-p)}$
Central Moments:	$\mu_{k+1} = q\left(\frac{\partial \mu_k}{\partial q} + \frac{kr}{p^2}\mu_{k-1}\right)$, where $q = 1-p$, $\mu_0 = 1$ and $\mu_1 = 0$.
Moment Generating Function:	$p^r(1-qe^t)^{-r}$
Probability Generating Function:	$p^r(1-qt)^{-r}$

8.3   Probabilities, Percentiles, and Moments

The dialog box [StatCalc→Discrete→Negative Binomial→Probabilities, Percentiles and Moments] computes the tail probabilities and percentiles.

To compute probabilities: Enter the number of successes r, number of failures until the rth success, and the success probability; click [P]. As an example, when $r = 20, k = 18$, and $p = 0.6$,

$$P(X \leq 18) = 0.862419, \ P(X \geq 18) = 0.181983, \ \text{and} \ P(X = 18) = 0.0444024.$$

To find the success probability when $k = 5$, P(X <= k) = 0.56, and $r = 4$, enter these values in appropriate edit boxes, and click [s] to get 0.417137.

To compute moments: Enter the values of r and the success probability p; click [M].

Example 8.1. A coin is to be flipped sequentially.

a. What are the chances that the 10th head will occur at the 12th flip?

b. Suppose that the 10th head had indeed occurred at the 12th flip. What can be said about the coin?

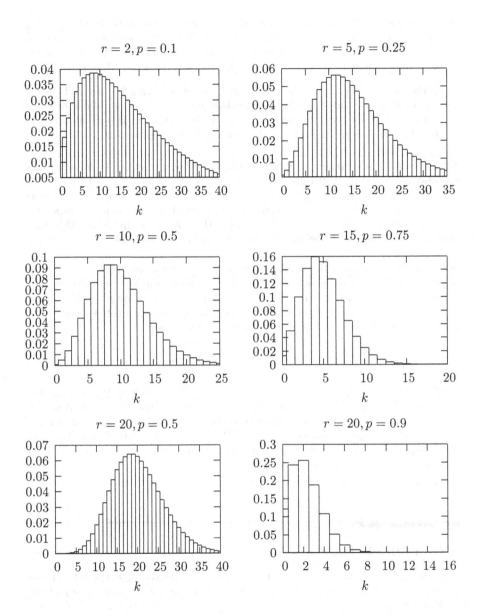

FIGURE 8.1: Negative binomial probability mass functions; k is the number of failures until the rth success

Solution:

a. Let us assume that the coin is balanced. To find the probability, select the dialog box [StatCalc→Discrete→Negative Binomial→Probabilities, Critical Values and Moments] from *StatCalc*, enter 10 for the number of successes, 2 for the number of failures, and 0.5 for the success probability; click $[P(X <= k)]$ to get 0.01343.

b. If the coin were balanced, then the probability of observing 2 or less tails before the 10th head is only 0.01929, which is less than 2%. Therefore, if one observes 10th head at the 12th flip, then it indicates that the coin is not balanced. To find this probability using *StatCalc*, just follow the steps in part (a).

Example 8.2. A shipment of items is submitted for inspection. In order to save the cost of inspection and time, the buyer of the shipment decided to adopt the following acceptance sampling plan: He decided to inspect a sample of not more than 30 items. Once the third defective item observed, he will stop the inspection and reject the lot; otherwise, he will continue the inspection up to the 30th item. What are the chances of rejecting the lot if it indeed contains 15% defective items?

Solution: Let X denote the number of nondefective items that must be examined in order to get 3 or more defective items. If we refer defective as "success" and nondefective as "failure," then X follows a negative binomial with $p = 0.15$ and $r = 3$. We need to find the probability of observing 27 or less nondefective items to get the third defective item. Thus, the required probability is

$$P(X \leq 27|3, \ 0.15) = 0.8486,$$

which can be computed using *StatCalc* as follows: Select the dialog box [StatCalc→Discrete→Negative Binomial→Probabilities, Percentiles and Moments] from *StatCalc*, enter 3 for the number of successes, 27 for the number of failures, and 0.15 for the success probability; click [P(X <= k)] to get 0.848599. Thus, for this acceptance sampling plan, the chances of rejecting the lot is about 85% if the lot actually contains 15% defective items.

8.4 Point Estimation

Suppose that a binomial experiment required $k + r$ trials to get the rth success. Then the uniformly minimum variance unbiased estimator of the success probability is given by

$$\widehat{p} = \frac{r - 1}{r + k - 1},$$

and its approximate variance is given by

$$\text{Var}(\widehat{p}) \simeq \frac{p^2(1 - p)}{2} \left(\frac{2k + 2 - p}{k(k - p + 2)} \right).$$

8.5   Test for the Proportion

Suppose that in a sequence of independent Bernoulli trials, each with success probability p, rth success was observed at the $(k+r)$th trial. Based on this information, we like to test about the true success probability p.

For testing

$$H_0 : p \le p_0 \quad \text{vs.} \quad H_a : p > p_0, \tag{8.1}$$

the null hypothesis will be rejected if the p-value $P(X \le k|r, p_0) \le \alpha$, for testing

$$H_0 : p \ge p_0 \quad \text{vs.} \quad H_a : p < p_0, \tag{8.2}$$

the null hypothesis will be rejected if the p-value $P(X \ge k|r, p_0) \le \alpha$, and for testing

$$H_0 : p = p_0 \quad \text{vs.} \quad H_a : p \ne p_0, \tag{8.3}$$

the null hypothesis will be rejected if the p-value

$$2\min\{P(X \le k|r, p_0), P(X \ge k|r, p_0)\} \le \alpha.$$

The dialog box [StatCalc→Discrete→Negative Binomial→Test and CI for Success Probability] computes the above p-values for testing the success probability.

Example 8.3.   A shipment of items is submitted for inspection. The buyer of the shipment inspected the items one-by-one randomly, and found the 5th defective item at the 30th inspection. Based on this information, can we conclude that the percentage of defective items in the shipment is less than 30? Find a point estimate of the percentage.

Solution: Let p denote the true proportion of defective items in the shipment. Then, we want to test

$$H_0 : p \ge 0.3 \quad \text{vs.} \quad H_a : p < 0.3.$$

To compute the p-value, select the dialog box [StatCalc→Discrete→ Negative Binomial→Test and CI ...], enter 5 for r, 25 for k, 0.3 for [p0], and click [p-values] to get 0.0378949. This is the p-value for the left-tailed test, and is less than 0.05. Therefore, we conclude that the true proportion of defective items in the shipment is less than 30%. A point estimate of the actual proportion of defective items is

$$\widehat{p} = \frac{r-1}{r+k-1} = \frac{5-1}{5+25-1} = 0.1379.$$

Suppose one inadvertently applies the binomial testing method described in Section 4.4.1 instead of negative binomial, then $n = 30$, and the number of successes is 5. Using [StatCalc→Discrete→Binomial →Tests for Proportion and Power Calculation], we get the p-value for testing above hypotheses as 0.0765948. Thus, contrary to the result based on the negative binomial, the result based on the binomial is not significant at 0.05 level.

8.6   Confidence Intervals for the Proportion

For a given r and k, an exact $1 - \alpha$ confidence interval for p can be computed using the Clopper–Pearson approach. The lower limit p_l satisfies

$$P(X \leq k | r, p_l) = \alpha/2,$$

and the upper limit p_u satisfies

$$P(X \geq k | r, p_u) = \alpha/2.$$

Using the relation between negative binomial and beta random variables (see Section 16.6.2), it can be shown that $p_l = B_{r,k+1;\frac{\alpha}{2}}$ and $p_u = B_{r,k;1-\frac{\alpha}{2}}$, where $B_{a,b;\alpha}$ denotes the αth quantile of a beta distribution with shape parameters a and b.

The dialog box [StatCalc→Discrete→Negative Binomial→Test and CI ...] uses the above methods to compute confidence intervals for p.

Example 8.4. A shipment of items is submitted for inspection. The buyer of the shipment inspected the items one by one randomly, and found the 6th defective item at the 30th inspection. Based on this information, find a 95% confidence interval for the true proportion of defective items in the shipment.

Solution: To find the 95% exact confidence interval for the true proportion of defective items, enter 6 for r, 24 for k, 0.95 for confidence level, and click [2-sided] to get (0.0771, 0.3577). That is, the true percentage of defective items in the shipment is between 7.7 and 36 with confidence 95%.

8.7   Properties and Results

In the following, X denotes the negative binomial(r, p) random variable.

8.7.1   Properties

1. For a given k and r, $P(X \leq k)$ is a nondecreasing function of p.

2. Let $X_1, \ldots, X_m$ be independent negative binomial random variables with

$$X_i \sim \text{negative binomial}(r_i, p), \quad i = 1, 2, \ldots, m.$$

Then,

$$\sum_{i=1}^{m} X_i \sim \text{negative binomial} \left(\sum_{i=1}^{m} r_i, p \right).$$

3. Recurrence Relations:

$$P(X = k + 1) = \frac{(r+k)(1-p)}{(k+1)} P(X = k)$$
$$P(X = k - 1) = \frac{k}{(r+k-1)(1-p)} P(X = k)$$

8.7.2   Relation to Other Distributions

1. Binomial: Let X be a negative binomial(r, p) random variable. Then

$$P(X \leq k | r, p) = P(Y \geq r), \; k = 1, 2, \ldots.$$

where Y is a binomial random variable with $k + r$ trials and success probability p.

2. Beta: See Section 17.6.2.

3. Geometric Distribution: Negative binomial distribution with $r = 1$ specializes to the geometric distribution described in Chapter 7

8.8   Random Number Generation

Algorithm 8.1. Negative binomial variate generator

```
Input:
        r = number of successes; p = success probability
        ns = desired number of random numbers

Output:
        k = random number from the negative binomial(r, p)
        distribution; the number of failures until the rth
        success

Set i = 1; k = 0

For j = 1 to ns
1       Generate u from uniform(0, 1)
        If u <= p, k = k + 1
        If k = r goto 2
        i = i + 1
        go to 1
2       x(j) = i - r
        k = 0
        i = 1
[end j loop]
```

The following algorithm is based on the inverse method, and is similar to Algorithm 4.1 for binomial variate generator.

Algorithm 8.2. Negative binomial variate generator

```
Set   k = int((r - 1.0)*(1 - p)/p)
      pk = P(X = k|r, p)
```

```
        df = P(X <= k|r, p)
        rpk = pk
        ik = k
        xb = r*(1 - p)/p
        s  = sqrt(xb/p)
        mu = xb + 10.0*s
        if(xb > 30.0) mu = xb + 6.0*s
        if(xb > 100.0) mu = xb + 5.0*s
        ml = max(0.0, mu - 10.0*s)
        if(xb > 30.0)  ml = max(0.0, xb - 6.0*s)
        if(xb > 100.0) ml = max(0.0, xb - 5.0*s)
For j = 1 to ns
        Generate u from uniform(0, 1)
        if(u > df) goto 2
1       u = u + pk
        if(k = ml or u > df) goto 3
        pk = pk*k/((r + k - 1.0)*(1.0 - p))
        k = k - 1
        goto 1
2       pk = (r + k)*(1 - p)*pk/(k + 1)
        u = u - pk
        k = k + 1
        If k = mu or u <= df, go to 3
        go to 2
3       x(j) = k
        k = rk
        pk = rpk
[end j loop]
```

8.9   Computation of Probabilities

For small values of k and r, $P(X = k)$ can be computed in a straightforward manner. For other values, logarithmic gamma function $\ln\Gamma(x)$ given in Section 1.8 can be used.

To compute $P(X = k)$:

Set $q = 1 - p$
$c = \ln\Gamma(r + k) - \ln\Gamma(k + 1) - \ln\Gamma(r)$
$b = k\ln(q) + r\ln(p)$
$P(X = k) = \exp(c + b)$

To compute $P(X \le k)$:

If an efficient algorithm for evaluating the cumulative distribution function of beta distribution is available, then the following relation between the beta and negative binomial distributions, $P(X \le k) = P(Y \le p)$, where Y is a beta variable with shape parameters r and $k + 1$, can be used to compute the cumulative probabilities.

The relation between the binomial and negative binomial distributions,

$$P(X \leq k) = 1.0 - P(W \leq r - 1),$$

where W is a binomial random variable with $k + r$ trials and success probability p, can also be used to compute the cumulative probabilities.

9

Logarithmic Series Distribution

9.1 Description

The probability mass function of a logarithmic series distribution with parameter θ is given by

$$P(X = k) = \frac{a\theta^k}{k}, \quad 0 < \theta < 1, \ k = 1, 2, \ldots,$$

where $a = -1/[\ln(1 - \theta)]$; the cumulative distribution function is given by

$$F(k|\theta) = P(X \leq k|\theta) = a \sum_{i=1}^{k} \frac{\theta^k}{k}, \quad 0 < \theta < 1, \ k = 1, 2, \ldots.$$

The logarithmic series distribution is useful to describe a variety of biological and ecological data. Specifically, the number of individuals per species can be modeled using a logarithmic series distribution. This distribution can also be used to fit the number of products requested per order from a retailer. Williamson and Bretherton (1964) used a logarithmic series distribution to fit the data that represent quantities of steel per order from a steel merchant; they also tabulated the cumulative probabilities for various values of the mean of the distribution. Furthermore, Chatfield et al. (1966) fitted the logarithmic series distribution to the distribution of purchases from a random sample of consumers.

The logarithmic series distribution is always right-skewed (see Figure 9.1).

9.2 Moments

Mean:	$\frac{a\theta}{1-\theta}, \quad a = -1/[\ln(1 - \theta)].$
Variance:	$\frac{a\theta(1-a\theta)}{(1-\theta)^2}$
Coefficient of Variation:	$\sqrt{\frac{(1-a\theta)}{a\theta}}$

Coefficient of Skewness:	$\frac{a\theta(1+\theta-3a\theta+2a^2\theta^2)}{[a\theta(1-a\theta)]^{3/2}}$
Coefficient of Kurtosis:	$\frac{1+4\theta+\theta^2-4a\theta(1+\theta)+6a^2\theta^2-3a^3\theta^{\,3}}{a\theta(1-a\theta)^2}$
Mean Deviation:	$\frac{2a\theta(\theta^m-P(X>m))}{1-\theta}$, where m denotes the largest integer $\leq$ the mean. [Kamat, 1965]
Factorial Moments:	$E\left(\prod_{i=1}^{k}(X-i+1)\right) = \frac{a\theta^k(k-1)!}{(1-\theta)^k}$
Moment Generating Function:	$\frac{\ln(1-\theta\exp(t))}{\ln(1-\theta)}$
Probability Generating Function:	$\frac{\ln(1-\theta t)}{\ln(1-\theta)}$

9.3   Probabilities, Percentiles, and Moments

The dialog box [StatCalc→Discrete→Logarithmic Series] in *StatCalc* computes the probabilities, moments, and the MLE of θ based on a given sample mean.

To compute probabilities: Enter the values of the parameter θ and the observed value k; click [P(X <= k)]. As an example, when $\theta = 0.3$ and $k = 3$, $P(X \leq 3) = 0.9925$, $P(X \geq 3) = 0.032733$, and $P(X = 3) = 0.025233$. *StatCalc* also computes the value of θ or the value of k. For example, when $k = 3$, $P(X \leq 3) = 0.6$, the value of θ is 0.935704. To get this value, enter 3 for k, 0.6 for [P(X <= k)], and click [T].

To compute the MLE of θ: Enter the sample mean, and click on [MLE]. For example, when the sample mean $= 2$, the MLE of θ is 0.715332.

To compute moments: Enter a value for θ in (0,1); click [M].

Example 9.1. A mail-order company recorded the number of items purchased per phone call or mail in form. The data are given in Table 8.1. We will fit a logarithmic series distribution for the number of item per order. To fit the model, we first need to estimate the parameter θ based on the sample mean, which is

$$\bar{x} = \frac{\sum x_i f_i}{\sum f_i} = \frac{2000}{824} = 2.4272.$$

To find the MLE of θ using *StatCalc*, enter 2.4272 for the sample mean, and click [MLE] to get 0.7923. Using this number for the value of θ, we can compute the probabilities $P(X = 1)$, $P(X = 2)$, etc. These probabilities are given in column 3 of Table 8.1. To find the expected frequencies, multiply the probability by the total frequency, which is 824 for this example. Comparison of the observed and expected frequencies indicates that the logarithmic series distribution is very well fitted for the data. The fitted distribution can be used to check whether the distribution of number of items demanded per order changes after a period of time.

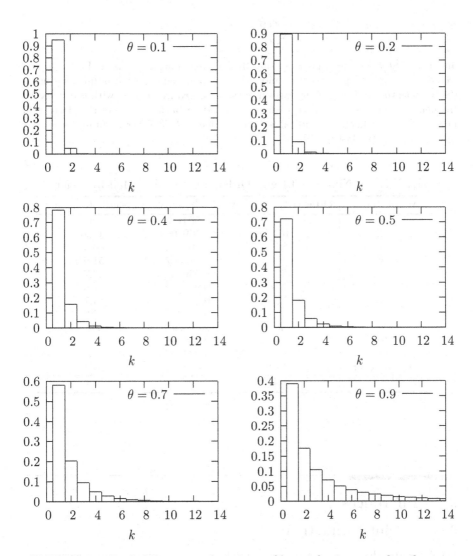

FIGURE 9.1: Probability mass functions of logarithmic series distribution

Example 9.2. Suppose that the mail-order company in the previous example collected new data a few months after the previous study, and recorded them as shown Table 9.2. First, we need to check whether a logarithmic series distribution still fits the data. The sample mean is

$$\bar{x} = \frac{\sum x_i f_i}{\sum f_i} = \frac{1596}{930} = 1.7161,$$

and using *StatCalc*, we find that the MLE of θ is 0.631316. As in Example 9.1, we can compute the probabilities and the corresponding expected frequencies using 0.631316 as the value of θ. Comparison of the observed frequencies with the expected frequencies indicate that a logarithmic series distribution still fits the data well; however, the smaller MLE indicates that the demand for fewer units per order has increased since the last study.

TABLE 9.1: Number of Items Ordered per Call or Mail-In Order

No. of item x_i	Observed frequency	Probability	Expected frequency
1	417	0.504116	415.4
2	161	0.199706	164.6
3	84	0.105485	86.9
4	50	0.062682	51.6
5	39	0.039730	32.7
6	22	0.026232	21.6
7	12	0.017814	14.7
8	8	0.012350	10.2
9	7	0.008698	7.2
10	6	0.006202	5.1
11	5	0.004467	3.7
12 and over	13	0.012518	10.3
Total	824	1.0	824

9.4 Inferences

9.4.1 Point Estimation

Let $\bar{X}$ denote the mean of a random sample of n observations from a logarithmic series distribution with parameter θ. The maximum likelihood estimate $\widehat{\theta}$ of θ is the solution of the equation

$$\widehat{\theta} = \frac{\bar{X}\ln(1-\widehat{\theta})}{\bar{X}\ln(1-\widehat{\theta}) - 1},$$

which can be solved numerically for a given sample mean. Williamson and Bretherton (1964) tabulated the values of $\widehat{\theta}$ for $\bar{x}$ ranging from 1 to 50.

Patil (1962) derived an asymptotic expression for the variance of the MLE, and it is given by

$$\mathrm{Var}(\widehat{\theta}) = \frac{\theta^2}{n\mu_2},$$

where μ_2 denotes the variance of the logarithmic series distribution with parameter θ (see Section 8.2).

Patil and Bildikar (1966) considered the problem of minimum variance unbiased estimation. Wani (1975) compared the MLE and the minimum variance unbiased estimator (MVUE) numerically and concluded that there is no clear-cut criterion to choose between these estimators. It should be noted that the MVUE also cannot be expressed in a closed form.

TABLE 9.2: Number of Items Ordered per Call or Mail-In Order after a Few Months

No. of item x_i	Observed frequency	Probability	Expected frequency
1	599	0.632698	588.4
2	180	0.199716	185.7
3	75	0.084056	78.2
4	30	0.039799	37
5	20	0.020101	18.7
6	11	0.010575	9.9
7	5	0.005722	5.3
8	4	0.003161	2.9
9	3	0.001774	1.6
10	2	0.001010	0.9
11	0	0.000578	0
12	1	0.000811	0.8
Total	930	1.0	929.4

9.4.2   Interval Estimation

Let $X_1, \ldots, X_n$ be a random sample from a logarithmic series distribution with parameter θ. Let Z denote the sum of the X_is, and let $f(z \mid n, \theta)$ denote the probability mass function of Z (see Section 8.5). For an observed value z_0 of Z, a $(1 - \alpha)$ confidence interval is (θ_L, θ_U), where θ_L and θ_U satisfy

$$\sum_{k=z_0}^{\infty} f(k \mid n, \theta_L) = \frac{\alpha}{2}$$

and

$$\sum_{k=1}^{z_0} f(k \mid n, \theta_U) = \frac{\alpha}{2},$$

respectively. Wani (1975) tabulated the values of (θ_L, θ_U) for $n = 10, 15,$ and $20,$ and the confidence level 0.95.

9.5   Properties and Results

1. Recurrence Relations:

$$P(X = k + 1) = \frac{k\theta}{k+1}P(X = k), \quad k = 1, 2, \ldots$$
$$P(X = k - 1) = \frac{k}{(k-1)\theta}P(X = k), \quad k = 2, 3, \ldots.$$

2. Let $X_1, \ldots, X_n$ be independent random variables, each having a logarithmic series distribution with parameter θ. The probability mass function of the $Z = \sum_{i=1}^{n} X_i$ is given by

$$P(Z = k) = \frac{n!|S_k^{(n)}|\theta^k}{k![-\ln(1 - \theta)]^n}, \quad k = n, n + 1, \ldots.$$

where $S_k^{(n)}$ denotes the Stirling number of the first kind (Abramowitz and Stegun, 1965, p. 824).

9.6   Random Number Generation

The following algorithm is based on the inverse method. That is, for a random uniform$(0, 1)$ number u, the algorithm searches for k such that

$$P(X \leq k - 1) < u \leq P(X \leq k).$$

Input:

> θ = parameter
> ns = desired number of random numbers

Output:

> $x(1), \ldots, x(ns)$ are random numbers from the
> Logarithmic Series(θ) distribution

Algorithm 9.1. Logarithmic series variate generator

```
Set pk = −θ/ ln(1 − θ)
    rpk = pk
For j = 1 to ns
        Generate u from uniform(0, 1)
        k = 1
1       If u ≤ pk, go to 2
        u = u − pk
        pk = pk *θ*k/(k + 1)
        k = k + 1
        goto 1
2       x(j) = k
        pk = rpk
[end j loop]
```

9.7 Computation of Probabilities

For a given θ and k, $P(X = k)$ can be computed in a straightforward manner. To compute $P(X \leq k)$, compute first $P(X = 1)$, compute other probabilities recursively using the recurrence relation

$$P(X = i + 1) = \frac{i\theta}{i+1} P(X = i), \quad i = 1, 2, ..., k-1....$$

and then compute $P(X \leq k) = P(X = 1) + \sum_{i=2}^{k} P(X = i)$.

The above method is used to obtain the following algorithm.

Algorithm 9.2. Calculation of logarithmic series tail probabilities

```
Input:
      k = the positive integer at which the cdf is to be evaluated
      t = the value of the parameter 'theta'
      a = -1/ln(1-t)
Output:
      cdf = P(X <= k| t)

Set p1 = t
      cdf = p1
For i = 1 to k
      p1 = p1*i*t/(i+1)
      cdf = cdf + p1
(end i loop)
cdf = cdf*a
```

Continuous Distributions

10

Continuous Uniform Distribution

10.1 Description

The probability density function of a continuous uniform random variable over an interval [a, b] is given

$$f(x; a, b) = \frac{1}{b-a}, \quad a \leq x \leq b.$$

The cumulative distribution function is given by

$$F(x|a, b) = \frac{x-a}{b-a}, \quad a \leq x \leq b.$$

The uniform distribution with support [a, b] is denoted by uniform(a, b). This distribution is also called the *rectangular* distribution because of the shape of its probability density function (see Figure 10.1).

10.2 Moments

Mean:	$\frac{b+a}{2}$
Variance:	$\frac{(b-a)^2}{12}$
Median:	$\frac{b+a}{2}$
Coefficient of Variation:	$\frac{(b-a)}{\sqrt{3}(b+a)}$
Mean Deviation:	$\frac{b-a}{4}$
Coefficient of Skewness:	0

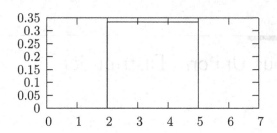

FIGURE 10.1: The probability density function of uniform(2,5) distribution

Central Moments:
$$\begin{cases} 0, & k = 1, 3, 5, \ldots, \\ \frac{(b-a)^k}{2^k(k+1)}, & k = 2, 4, 6, \ldots \end{cases}$$

Moments about the Origin: $E(X^k) = \frac{b^{k+1} - a^{k+1}}{(b-a)(k+1)}, \quad k = 1, 2, \cdots$

Moment Generating Function: $\frac{e^{tb} - e^{ta}}{(b-a)t}$

10.3   Inferences

Let $X_1, \ldots, X_n$ be a random sample from a uniform(a, b) distribution. Let $X_{(1)}$ denote the smallest order statistic and $X_{(n)}$ denote the largest order statistic.

1. When b is known,

$$\widehat{a}_u = \frac{(n+1)X_{(1)} - b}{n}$$

is the uniformly minimum variance unbiased estimator (UMVUE) of a; if a is known, then

$$\widehat{b}_u = \frac{(n+1)X_{(n)} - a}{n}$$

is the UMVUE of b.

2. When both a and b are unknown,

$$\widehat{a} = \frac{nX_{(1)} - X_{(n)}}{n-1} \quad \text{and} \quad \widehat{b} = \frac{nX_{(n)} - X_{(1)}}{n-1}$$

are the UMVUEs of a and b, respectively.

10.4   Properties and Results

1. *Probability Integral Transformation:* Let X be a continuous random variable with cumulative distribution function $F(x)$. Then,

$$U = F(X) \sim \text{uniform}(0, 1).$$

2. Let X be a uniform(0, 1) random variable. Then,

$$-2\ln(X) \sim \chi_2^2.$$

3. Let $X_1, \ldots, X_n$ be independent uniform(0,1) random variables, and let $X_{(k)}$ denote the kth order statistic. Then $X_{(k)}$ follows a beta($k, n-k+1$) distribution.

4. Relation to Normal: See Section 11.9.

10.5   Random Number Generation

Uniform(0, 1) random variates generator is usually available as a built-in intrinsic function in many commonly used programming languages such as Fortran and C. To generate random numbers from uniform(a, b), use the result that if $U \sim$ uniform(0,1), then $X = a + U * (b - a) \sim$ uniform(a, b).

11

Normal Distribution

11.1 Description

The probability density function of a normal random variable X with mean μ and standard deviation σ is given by

$$f(x|\mu,\sigma) = \frac{1}{\sigma\sqrt{2\pi}}\exp\left[-\frac{(x-\mu)^2}{2\sigma^2}\right], -\infty < x < \infty, \ -\infty < \mu < \infty, \sigma > 0.$$

This distribution is commonly denoted by $N(\mu,\sigma^2)$. The cumulative distribution function (cdf) is given by

$$F(x|\mu,\sigma) = \int_{-\infty}^{x} f(t|\mu,\sigma)dt.$$

The normal random variable with mean $\mu = 0$ and standard deviation $\sigma = 1$ is called the standard normal random variable, and its cdf is denoted by $\Phi(z)$.

If X is a normal random variable with mean μ and standard deviation σ, then

$$F(X|\mu,\sigma) = P\left(Z \le \frac{x-\mu}{\sigma}\right) = \int_{-\infty}^{(x-\mu)/\sigma} \exp(-t^2/2)dt = \Phi\left(\frac{x-\mu}{\sigma}\right).$$

The mean μ is the location parameter, and the standard deviation σ is the scale parameter. See the plots of the pdfs in Figures 11.1 and 11.2.

The normal distribution is the most commonly used distribution to model univariate data from a population or from an experiment.

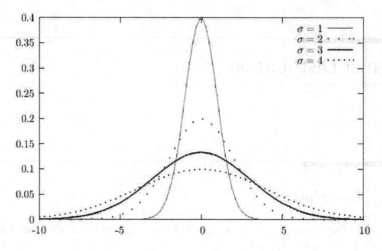

FIGURE 11.1: Normal probability density functions with $\mu = 0$

11.2 Moments

Mean:	μ
Variance:	σ^2
Coefficient of Variation:	σ/μ
Median:	μ
Mean Deviation:	$\sqrt{\frac{2\sigma^2}{\pi}}$
Coefficient Skewness:	0
Coefficient of Kurtosis:	3
Moments about the Origin:	$\begin{cases} \sigma^k \sum\limits_{i=1}^{(k+1)/2} \dfrac{k!\mu^{2i-1}}{(2i-1)![(k+1)/2-i]!2^{(k+1)/2-i}\sigma^{2i-1}}, \\ \qquad k = 1,3,5,\ldots \\ \sigma^k \sum\limits_{i=0}^{k/2} \dfrac{k!\mu^{2i}}{(2i)!(k/2-i)!2^{k/2-i}\sigma^{2i}}, \\ \qquad k = 2,4,6,\ldots \end{cases}$
Moments about the Mean:	$\begin{cases} 0, & k = 1,3,5,\ldots, \\ \dfrac{k!}{2^{k/2}(k/2)!}\sigma^k, & k = 2,4,6,\ldots \end{cases}$
Moment Generating Function:	$E(e^{tx}) = \exp\left(t\mu + t^2\sigma^2/2\right)$
A Recurrence Relation:	$E(X^k) = (k-1)\sigma^2 E(X^{k-2}) + \mu E(X^{k-1}),$ $\qquad k = 3,4,\cdots$

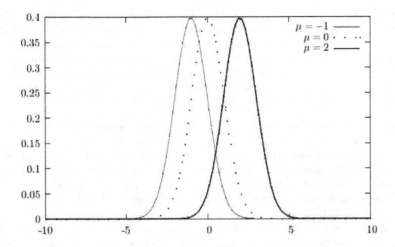

FIGURE 11.2: Normal probability density functions with $\sigma = 1$

11.3 Probabilities, Percentiles, and Moments

The dialog box [StatCalc→Continuous→Normal→Probabilities, Percentiles and Moments] computes the tail probabilities, critical points, parameters, and moments.

To compute probabilities: Enter the values of the mean, standard deviation, and the value of x at which the cdf is to be computed; click the [P(X <= x)] radio button. For example, when mean = 1.0, standard deviation = 2.0, and the value $x = 3.5$, $P(X \leq 3.5) = 0.89435$, and $P(X > 3.5) = 0.10565$.

To compute percentiles: Enter the values of the mean, standard deviation, and the cumulative probability P(X <= x); click on [x] radio button. For example, when mean = 1.0, standard deviation = 2.0, and the cumulative probability P(X <= x) = 0.95, the 95th percentile is 4.28971. That is,

$$P(X \leq 4.28971) = 0.95.$$

To compute the mean: Enter the values of the standard deviation, x, and P(X <= x). Click [Mean]. For example, when standard deviation = 3, $x = 3.5$, and P(X <= x) = 0.97, the value of the mean is -2.14238.

To compute the standard deviation: Enter the values of the mean, x, and P(X <= x). Click [Std Dev]. For example, when mean = 3, $x = 3.5$, and P(X <= x) = 0.97, the standard deviation is 0.265845.

To compute moments: Enter the values of the mean and standard deviation; click [M] button.

In the following example, we illustrate a method of checking whether a sample is from a normal population.

Example 11.1. (Assessing normality) An important problem in industrial hygiene is the exposure level of employees who are constantly exposed to workplace contaminants. In order to assess the exposure levels, hygienists monitor employees periodically. The following data represent the exposure measurements from a sample of 15 employees who were exposed to a chemical over a period of three months.

$$x: \quad 69 \ \ 75 \ \ 82 \ \ 93 \ \ 98 \ \ 102 \ \ 54 \ \ 59 \ \ 104 \ \ 63 \ \ 67 \ \ 66 \ \ 89 \ \ 79 \ \ 77$$

We want to test whether the exposure data are from a normal distribution. Following the steps of Section 2.4.2, we first order the data. The ordered data $x_{(j)}$s are given in the second column of Table 10.1. The cumulative probability level of $x_{(j)}$ is approximately equal to $(j - 0.5)/n$, where n is the number of data points. For these cumulative probabilities, standard normal quantiles are computed, and they are given in the fourth column of Table 10.1. For example, when $j = 4$, the observed 0.233333th quantile is 66, and the corresponding standard normal quantile $z_{(j)}$ is -0.7279. To compute the standard normal quantile for the 4th observation, select [StatCalc→Continuous→Normal→Probabilities, Percentiles and Moments], enter 0 for [Mean], 1 for [Std Dev], and 0.233333 for P(X<=x); click on [x] to get -0.7279. If the data are from a normal population, then the pairs $(x_{(j)}, z_{(j)})$ will be approximately linearly related. The plot of the pairs (Q–Q plot) is given in Figure 10.3. The Q–Q plot is nearly a line suggesting that the data are from a normal population.

  If a graphical technique does not give a clear-cut result, a rigorous test, such as Shapiro–Wilk test and the correlation test, can be used. We shall use the test based on the correlation coefficient to test the normality of the exposure data. The correlation coefficient between the $x_{(j)}$s and the $z_{(j)}$s is given by

$$r = \frac{\sum_{i=1}^{n}(X_{(i)} - \bar{x})(z_{(i)} - \bar{z})}{\sqrt{\sum_{i=1}^{n}(x_{(i)} - \bar{x})^2}\sqrt{\sum_{i=1}^{n}(z_{(i)} - \bar{z})^2}} = 0.984.$$

At the level 0.05, the critical value for $n = 15$ is 0.937 (see Looney and Gulledge, 1985). Since the observed correlation coefficient r is larger than the critical value, we have further evidence for our earlier conclusion that the data are from a normal population.

Example 11.2. An electric bulb manufacturer reports that the average lifespan of 100W bulbs is 1100 h with a standard deviation of 100 h. Assume that the life hours distribution is normal.

a. Find the percentage of bulbs that will last at least 1000 h.

b. Find the percentage of bulbs with lifetime between 900 and 1200 h.

c. Find the 90th percentile of the life hours.

Solution: Select the dialog box [StatCalc→Continuous→Normal→Probabilities, Percentiles and Moments].

a. To find the percentage, enter 1100 for the mean, 100 for the standard deviation,

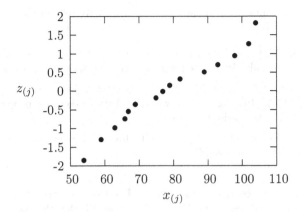

FIGURE 11.3: Q–Q plot of the exposure data

TABLE 11.1: Observed and Normal Quantiles for Exposure Data

j	Observed Quantiles $x_{(j)}$	Cumulative Probability Levels $(j-0.5)/15$	Standard Normal Quantile $z_{(j)}$
1	54	0.0333	-1.8339
2	59	0.1000	-1.2816
3	63	0.1667	-0.9674
4	66	0.2333	-0.7279
5	67	0.3000	-0.5244
6	69	0.3667	-0.3407
7	75	0.4333	-0.1679
8	77	0.5000	0.0000
9	79	0.5667	0.1679
10	82	0.6333	0.3407
11	89	0.7000	0.5244
12	93	0.7667	0.7279
13	98	0.8333	0.9674
14	102	0.9000	1.2816
15	104	0.9667	1.8339

and 1000 for the observed x; click [P(X <= x)] radio button to get $P(X \le 1000)$ = 0.1587 and $P(X > 1000)$ = 0.8413. This means that about 84% of the bulbs will last more than 1000 h.

b.
$$P(900 \le X \le 1200) = P(X \le 1200) - P(X \le 900)$$
$$= 0.841345 - 0.022750$$
$$= 0.818595.$$

That is, about 82% of the bulbs will last between 900 and 1200 h.

c. To find the 90th percentile, enter 1100 for the mean, 100 for the standard deviation, and 0.90 for the cumulative probability; click on [x] to get 1228.16. That is, 90% of the bulbs will last less than 1228 h; and 10% of the bulbs will last more than 1228 h.

Example 11.3. Suppose that the weekly demand for 5-lb sacks of onions at a grocery store is normally distributed with mean 140 sacks and standard deviation 10.

a. If the store stocks 160 sacks every week, find the percentage of weeks that the store has overstocked onions.

b. How many sacks should the store keep in stock each week in order to meet the demand for 95% of the weeks?

Solution:

a. Let X denote the weekly demand. We need to find the percentage of the weeks that the demand is less than the stock. Enter 140 for the mean, 10 for the standard deviation, and click [P(X <= x)] radio button to get $P(X \leq 160) = 0.97725$. This probability means that about 98% of the weeks the demand will be less than the supply.

b. Here, we need to find the 95th percentile of the normal distribution; that is, the value of x such that $P(X \leq x) = 0.95$. Using *StatCalc*, we get the value of $x = 156.449$. This means that the store has to stock 157 sacks each week in order to meet the demand for 95% of the weeks.

Example 11.4. A machine is set to pack 3 lb of ground beef per package. Over a long period of time, it was found that the average amount packed was 3 lb with a standard deviation of 0.1 lb. Assume that the weights of the packages are normally distributed.

a. Find the percentage of packages weighing more than 3.1 lb.

b. At what level should the machine be set, so that no more than 5% of the packages weigh less than 2.9 lb?

Solution: Let X be the actual weight of a randomly selected package. Then, X is normally distributed with mean 3 lb and standard deviation 0.1 lb.

a. To find the percentage, enter 3 for the mean, 0.1 for the standard deviation, and 3.1 for the x; click [P(X <= x)] radio button to get $P(X > 3.1) = 0.158655$. That is, about 16% of the packages will weigh more than 3.1 lb.

b. We are looking for the value of the mean μ such that $P(X < 2.9) = 0.05$. To get the value of the mean, enter 0.1 for the standard deviation, 2.9 for x, 0.05 for $P(X \leq x)$, and then click on [Mean] to get 3.06449. That is, the machine needs to be set at about 3.07 pounds so that fewer than 5% of the packages weigh less than 2.9 lb.

Example 11.5. A manufacturing company received a large quantity of bolts from one of its suppliers. A bolt is useable if it is 3.9 to 4.1 in long. Inspection of a sample of 50 bolts revealed that the average length is 3.95 in., with standard deviation 0.1 in. Assume that the distribution of lengths is normal.

a. Find an approximate proportion of bolts that are useable.

b. Find an approximate proportion of bolts that are longer than 4.1 in.

c. Find an approximate 95th percentile of the lengths of all bolts.

Solution: Assume that the lengths of bolts form a normal population with the mean μ and standard deviation σ. If μ and σ are known, then exact proportions and percentile can be computed using *StatCalc*. Since they are unknown, we can use the sample mean and standard deviation to find approximate solutions to the problem.

a. The proportion of bolts useable is given by $P(3.9 \le X \le 4.1) = P(X \le 4.1) - P(X \le 3.9)$, where X is a normal random variable with mean 3.95 and standard deviation 0.1. Using *StatCalc*, we get $P(X \le 4.1) - P(X \le 3.9) = 0.933193 - 0.308538 = 0.624655$. Thus, about 62% of bolts are useable.

b. This proportion is given by $P(X \ge 4.1) = 1 - P(X \le 4.1) = 1 - 0.933193 = 0.0668074$. That is, about 7% of bolts are longer than 4.1 inch.

c. To find an approximate 95th percentile, enter 3.95 for the mean, 0.1 for the standard deviation, 0.95 for the probability $[P(X <= x)]$, and click $[x]$ to get 4.11449. This means that approximately 95% of the bolts are shorter than 4.11449 inch.

11.4   One-Sample Inference

Let $X_1, \ldots, X_n$ be a random sample from a normal population with mean μ and standard deviation σ. The sample mean $\bar{X}$ and the variance S^2 are defined as

$$\bar{X} = \frac{1}{n} \sum_{i=1}^{n} X_i, \quad \text{and} \quad S^2 = \frac{1}{n-1} \sum_{i=1}^{n} (X_i - \bar{X})^2$$

are the uniformly minimum variance unbiased estimators of μ and σ^2, respectively. The sample mean is the maximum likelihood estimator of μ; however, the maximum likelihood estimator of σ^2 is $(n-1)S^2/n$, which is a biased estimate of σ^2.

11.4.1   Test for the Mean and Power Computation

t-test

The test statistic for testing null hypothesis $H_0 : \mu = \mu_0$ is given by

$$t = \frac{\bar{X} - \mu_0}{S/\sqrt{n}}, \tag{11.1}$$

which follows a t distribution with df $= n - 1$. Let $(\bar{x}, s^2)$ be an observed value of $(\bar{X}, S^2)$. Then $t_0 = \frac{\bar{x} - \mu_0}{s/\sqrt{n}}$ is the observed value of t in (11.1). For a given level α, the null hypothesis $H_0 : \mu = \mu_0$ will be rejected in favor of

$$H_a : \mu \ne \mu_0 \text{ if the p-value } P(|t| > |t_0|) < \alpha,$$

for testing $H_0 : \mu \ge \mu_0$ vs. $H_a : \mu < \mu_0$, the H_0 will be rejected if the p-value $P(t \le t_0) < \alpha$, and for testing $H_0 : \mu \le \mu_0$ vs. $H_a : \mu > \mu_0$, the H_0 will be rejected if the p-value $= P(t \ge t_0) < \alpha$.

Power Computation

Consider the hypotheses

$$H_0 : \mu \leq \mu_0 \quad \text{vs.} \quad H_a : \mu > \mu_0.$$

For a given nominal level α, the power of the t-test is the probability of rejecting the null hypothesis when the true mean μ is indeed greater than μ_0, and is given by

$$P(t > t_{n-1;1-\alpha}|H_a) = P(t_{n-1}(\delta) > t_{n-1;1-\alpha}), \qquad (11.2)$$

where t is given in (11.1), $t_{n-1,1-\alpha}$ denotes the $(1 - \alpha)$th quantile of the t-distribution (see Chapter 20) with degrees of freedom $n - 1$, and $t_{n-1}(\delta)$ denotes the noncentral t random variable with degrees of freedom $n-1$ and the noncentrality parameter

$$\delta = \frac{\sqrt{n}(\mu - \mu_0)}{\sigma}.$$

The power of a two-tailed test is similarly calculated. For a given μ, μ_0, σ and level α, StatCalc computes the power using (11.2).

To compute p-values for hypothesis testing about μ: Select [StatCalc→Continuous →Normal→One-Sample ...] from *StatCalc*, enter the values of the sample mean, sample standard deviation, sample size, and the value of the mean under the null hypothesis. Click [p-values (mean)] to get the p-values for various alternative hypotheses.

Example 11.6. *(Hypothesis testing)* Suppose that a sample of 20 observations from a normal population produced a mean of 3.4 and a standard deviation of 2.1. Consider testing

$$H_0 : \mu \leq 2.5 \quad \text{vs.} \quad H_a : \mu > 2.5.$$

To compute the p-value for testing above hypotheses, select [StatCalc→Continuous →Normal→One-Sample ...] from *StatCalc*, enter 20 for [Sample Size, n], 3.4 for [Sample Mean], 2.1 for [Sample SD], 2.5 for [H0: M = M0], and click [p-values (mean)] to get 0.0352254. That is, the p-value for testing the above hypotheses is 0.0352254. Thus, at 5% level, we have enough evidence to conclude that the true population mean is greater than 2.5.

Furthermore, note that for the two-sided hypothesis, that is,

$$H_0 : \mu = 2.5 \quad \text{vs.} \quad H_a : \mu \neq 2.5,$$

the p-value is $2 \times 0.0352254 = 0.0704508$. Now, the null hypothesis cannot be rejected at the level of significance 0.05. The value of the t-test statistic for this problem is 1.91663.

Sample Size for One-Sample t-test: For a given level of significance, the true mean and hypothesized mean of a normal population, and the standard deviation, the dialog box [StatCalc→Continuous → Normal → One-Sample ...] computes the sample size that is required to have a specified power. To compute the sample size, enter the hypothesized value of the population mean in [H0: M = M0], the population mean in [Population M], population standard deviation, level of the test, and power. Click [sample size for].

Example 11.7. *(Sample Size Calculation)* An experimenter believes that the actual mean of the population under study is 1 unit more than the hypothesized value $\mu_0 = 3$. From the past study, he learned that the population standard deviation is 1.3. He decides to use one-sample t-test, and wants to determine the sample size to attain a power of 0.90 at the level 0.05. The hypotheses for his study will be

$$H_0 : \mu \leq 3 \quad \text{vs.} \quad H_a : \mu > 3.$$

To compute the required sample size using the dialog box [StatCalc→Continuous → Normal → One-Sample ...], enter 3 for [H0: M = M0], 4 for [Population M], 1.3 for [Population SD], .05 for [Level], .9 for [Power], and click [Sample Size for] to get 16. Thus, a sample of 16 observations will be sufficient to test if the true mean is greater than 3 with a power of 90%.

11.4.2 Confidence Interval for the Mean

A $1 - \alpha$ confidence interval for the mean μ is given by

$$\bar{X} \pm t_{n-1,\, 1-\frac{\alpha}{2}} \frac{S}{\sqrt{n}},$$

where $t_{n-1, 1-\frac{\alpha}{2}}$ is the $(1 - \frac{\alpha}{2})$ quantile of a t distribution with df $= n - 1$. The above interval commonly referred to as the t-interval.

For a given sample mean, sample standard deviation, and sample size, the dialog box [StatCalc→Continuous→Normal→One-Sample ...] computes the p-value of the t-test and confidence interval for the mean.

To compute a confidence interval for μ: Select the dialog box [StatCalc→ Continuous→Normal→One-Sample ...], enter the values of the sample mean, sample standard deviation, sample size, and the confidence level. Click [1-sided] to get one-sided lower and upper limits for μ. Click [2-sided] to get confidence interval for the mean μ.

Example 11.8. Let us compute a 95% confidence interval for the true population mean based on summary statistics given in Example 11.6. In the dialog box [StatCalc→ Continuous →Normal→t-test and confidence interval for Mean], enter 3.4 for the sample mean, 2.1 for the sample standard deviation, 20 for the sample size, and 0.95 for the confidence level. Click [1-sided] to get 2.58804 and 4.21196. These are the one-sided limits for μ. That is, the interval (2.58804, ∞) would contain the population mean μ with 95% confidence. The interval (-∞, 4.21196) would contain the population mean μ with 95% confidence. To get a two-sided confidence interval for μ, click on [2-sided] to get 2.41717 and 4.38283. This means that the interval (2.41717, 4.38283) would contain the true mean with 95% confidence.

The following examples illustrate the one-sample inferential procedures for a normal mean.

Example 11.9. *(Confidence Interval for the Mean)* A marketing agency wants to estimate the average annual income of all households in a suburban area of a large city. A random sample of 40 households from the area yielded a sample mean of $65,000 with standard deviation approximately $8,000. Assume that the incomes follow a normal distribution.

a. Construct a 95% confidence interval for the true mean income of all the households in the suburb community.

b. Do these summary statistics indicate that the true mean income is greater than $63,000?

Solution:

a. To construct a 95% confidence interval for the true mean income, enter 65000 for the sample mean, 8000 for the sample standard deviation, and 40 for the sample size, and 0.95 for the confidence level. Click [2-sided] to get 62697.3 and 67302.7. That is, the actual mean income is somewhere between $62,697 and $67,303 with 95% confidence.

b. It is clear from part **a** that the mean income is greater than $62,697. However, to illustrate the *t*-test, we formulate the following hypothesis testing problem. Let μ denote the true mean income. We want to test

$$H_0 : \mu \leq 63000 \quad \text{vs.} \quad H_a : \mu > 63000.$$

To compute the p-value for the above test, enter the sample mean, standard deviation, and the sample size as in part **a**, and 63000 for [Ha: M = M0]. Click [p-values for] to get 0.0609618. Since this p-value is greater than .05, the data do not provide enough evidence to indicate that the mean income is greater than $63,000.

Example 11.10. *(Sample Size for the t-test)* A light bulb manufacturer considering a new method that is supposed to increase the average lifetime of bulbs by at least 100 h. The mean and standard deviation of the life hours of bulbs produced by the existing method are 1200 and 140 h, respectively. The manufacturer decides to test if the new method really increases the mean life hour of the bulbs. How many new bulbs should he test so that the test will have a power of 0.90 at the level of significance 0.05?

Solution: Let μ denote the actual mean life hours of the bulbs manufactured using the new method. The hypotheses of interest here are

$$H_0 : \mu \leq 1200 \quad \text{vs.} \quad H_a : \mu > 1200.$$

Enter 1200 for [H0: M = M0], 1300 for the population mean, 140 for the population standard deviation (it is assumed that the standard deviations of the existing method and old method are the same), 0.05 for the level and 0.9 for the power. Click [Sample Sizes for] to get 19. Thus, 19 new bulbs should be tested to check if the new method would increase the average life hours of the bulbs.

11.4.3   Confidence Interval for the Coefficient of Variation and Survival Probability

The coefficient of variation is a commonly used measure of variation, because it is not affected by the units of measurement. It is defined as the ratio of the standard deviation to the mean. In practical situations where the coefficient of variation is an appropriate measure of variability, the variable is usually positive. So for a $N(\mu, \sigma^2)$ population this measure is appropriate provided $\mu - 3\sigma > 0$, or equivalently, $\sigma/\mu <$

1/3; that is, the coefficient of variation must be at most .33 in practical situations where the coefficient of variation is a suitable measure of variability. Johnson and Welch (1940) have proposed an exact method of finding confidence intervals for the normal coefficient of variation, and it is described as follows.

An Exact Confidence Interval for the Coefficient of Variation

Let $\tau = \sigma/\mu$, $\nu = n - 1$ and $\hat{\tau} = s/\bar{x}$. An exact $1 - 2\alpha$ confidence interval (τ_L, τ_U) is determined by the roots of the following equations:

$$t_{\nu;\alpha}(\sqrt{n}/\tau_L) = \frac{\sqrt{n}}{\hat{\tau}} \quad \text{and} \quad t_{\nu;1-\alpha}(\sqrt{n}/\tau_U) = \frac{\sqrt{n}}{\hat{\tau}}.$$

For fixed ν and α, the percentile $t_{\nu;\alpha}(\delta)$ is an increasing function of δ, and so the roots of the above equations are unique and they can be found numerically.

Approximate Confidence Intervals for the Coefficient of Variation

The modified McKay's confidence interval for τ proposed by Vangel (1996) is given by

$$\left\{ \frac{s}{\bar{x}} \left[(u_1 - 1)\frac{s^2}{\bar{x}^2} + \frac{u_1}{\nu} \right]^{-1/2}, \ \frac{s}{\bar{x}} \left[(u_2 - 1)\frac{s^2}{\bar{x}^2} + \frac{u_2}{\nu} \right]^{-1/2} \right\}, \tag{11.3}$$

where $u_1 = \left(\chi^2_{\nu;1-\alpha/2} + 2\right)/n$ and $u_2 = \left(\chi^2_{\nu;\alpha/2} + 2\right)/n$ and $\nu = n - 1$.

Krishnamoorthy (2014) proposed the following approximate confidence interval based on the modified normal-based approximations in (2.14) and (2.15). Let

$$c_m = \frac{1}{\sqrt{e}}\left(1 + \frac{1}{m}\right)^{m/2}, \quad \text{with } m = n - 1. \tag{11.4}$$

Define

$$v_u^* = \begin{cases} \sqrt{\frac{\chi^2_{m;1-\alpha}}{m}} & \text{if } \hat{\tau} > 0, \\ \sqrt{\frac{\chi^2_{m;\alpha}}{m}} & \text{if } \hat{\tau} \le 0. \end{cases} \quad \text{and} \quad v_l^* = \begin{cases} \sqrt{\frac{\chi^2_{m;\alpha}}{m}} & \text{if } \hat{\tau} > 0, \\ \sqrt{\frac{\chi^2_{m;1-\alpha}}{m}} & \text{if } \hat{\tau} \le 0, \end{cases} \tag{11.5}$$

where $\hat{\tau} = \bar{x}/s$. In terms of these quantities, an approximate $1 - 2\alpha$ confidence interval for the coefficient of variation is given by

$$\left(\left[\frac{c_m}{\hat{\tau}} + \sqrt{\frac{(c_m - v_u^*)^2}{\hat{\tau}^2} + \frac{z_{1-\alpha}^2}{n}} \right]^{-1}, \ \left[\frac{c_m}{\hat{\tau}} - \sqrt{\frac{(c_m - v_l^*)^2}{\hat{\tau}^2} + \frac{z_{1-\alpha}^2}{n}} \right]^{-1} \right). \tag{11.6}$$

An Exact Confidence Interval for the Survival Probability

For $X \sim N(\mu, \sigma^2)$,

$$P(X > t) = P\left(Z > \frac{t - \mu}{\sigma}\right) = \Phi\left(\frac{t - \mu}{\sigma}\right),$$

where t is a specified value. If X represents the lifetime of a piece of equipment, then $P(X > t)$ is the probability that the equipment works for at least the specified time t. Because $\Phi(x)$ is a one-to-one function, it is enough to find confidence intervals for

$$\eta_t = \frac{t - \mu}{\sigma}.$$

If U is an upper confidence limit for $(t - \mu)/\sigma$, then $1 - \Phi(U)$ is a lower confidence limit for $P(X > t)$. A confidence limit for η_t can also be obtained from the noncentral t distribution. The exact $1 - 2\alpha$ confidence interval (η_{tL}, η_{tU}) due to Johnson and Welch (1940) is determined by

$$t_{n-1;1-\alpha}(\sqrt{n}\eta_{tL}) = \frac{\sqrt{n}(t - \bar{x})}{s} \text{ and } t_{n-1;\alpha}(\sqrt{n}\eta_{tU}) = \frac{\sqrt{n}(t - \bar{x})}{s}. \quad (11.7)$$

The $1 - 2\alpha$ exact confidence interval for the survival probability $P(X > t) = 1 - \Phi\left(\frac{t-\mu}{\sigma}\right)$ is given by

$$(1 - \Phi(\eta_{tU}), \ 1 - \Phi(\eta_{tL})).$$

An Approximate Confidence Interval for the Survival Probability

Krishnamoorthy (2014) proposed the following approximate confidence interval based on the modified normal-based approximations in (2.14) and (2.15). Let $\widehat{\eta}_t = (t - \bar{x})/s$. An approximate $1 - 2\alpha$ confidence interval for $\eta_t = (t - \mu)/\sigma$ is given by

$$(L_\eta, U_\eta) = \left(\widehat{\eta}_t c_m - \sqrt{\widehat{\eta}_t^2 (c_m - v_l^*)^2 + \frac{z_\alpha^2}{n}}, \ \widehat{\eta}_t c_m + \sqrt{\widehat{\eta}_t^2 (c_m - v_u^*)^2 + \frac{z_\alpha^2}{n}}\right), \quad (11.8)$$

where c_m is given in (11.4) and (v_l^*, v_u^*) is given in (11.5). An approximate $1 - 2\alpha$ confidence interval for the survival probability $\bar{\Phi}\left(\frac{t-\mu}{\sigma}\right)$ is

$$(1 - \Phi(U_\eta), 1 - \Phi(L_\eta)).$$

Remark 11.1. The dialog box [StatCalc→ Continuous→Normal→Coefficients of Variation ...] uses the exact methods to find confidence intervals for the coefficient of variation and for the survival probability for a given $(n, \bar{x}, s, t)$. The approximate confidence intervals are very accurate, and they can be easily calculated. In rare cases where the exact method has a convergence problem, these approximate confidence intervals can be used.

Example 11.11. *(Coefficient of Variation)* Consider the following simulated data

$$2.45,\ 1.28,\ \ 1.44,\ 1.91,\ 2.36,\ 1.78,\ 2.54,\ 2.15,$$
$$2.17,\ 1.72,\ \ 2.17,\ 1.68,\ 1.55,\ 1.49,\ 1.77$$

from a normal distribution. For the above data, $n = 15$, the mean $\bar{x} = 1.897$, the standard deviation $s = 0.3901$, and the sample coefficient of variation is $.3901/1.897 = 0.2056$. To compute the exact 95% confidence interval for the population coefficient of variation, select [StatCalc→ Continuous→Normal→Coefficients of Variation ...], enter 15 for [Sample Size], .95 for [Conf Level], 0.2056 for [Sample CV], and click on [confidence interval for CV] to get (.149,.333). That is, the population coefficient of variation is between .149 and .333 with 95% confidence.

To compute the approximate confidence interval by Vangel (1996), the chi-square percentiles are $u_2 = \chi^2_{14;.025} = 5.6287$ and $u_1 = \chi^2_{14;.975} = 26.119$. Also, note that $\bar{x}/s = 4.8629$. Using these numbers in (11.3), we find $(.149,.333)$, which is the same as the exact one. To compute the approximate confidence interval by Krishnamoorthy (2014), we first compute $r = \bar{x}/s = 4.8629$, $v_l^* = \sqrt{\chi^2_{14;.025}/14} = .6341$, $v_u^* = \sqrt{\chi^2_{14;.975}/14} = 1.3659$, and $c_m = .9831$. Substituting these numbers in (11.6), we find $(.149,.332)$, which is also practically the same as the exact one.

Example 11.12. To illustrate the estimation method for the survival probability, let us consider the exposure data of 15 employees in Example 11.1. In this situation, it is desired to find an upper confidence limit for $P(X > OEL)$, where OEL means the occupational exposure limit. This probability is referred to as the exceedance probability. Suppose for this chemical, the OEL is 105. For this example, the sample size is 15, the sample mean is 78.5 and the standard deviation is 15.9. To find the 95% exact upper confidence limit, select the dialog box [StatCalc→ Continuous→Normal→Coefficients of Variation ...], enter 15 for [Sample Size], 78.5 for [Sample Mean], 15.9 for [Sample SD], .9 for [Conf Level], 105 for [Value of t], and click on [confidence interval] to get .163. Note that we used .90 for confidence level to find 95% one-sided limits. Based on this upper confidence limit, we cannot conclude that the exceedance probability is less than .05 (compliance guidelines by federal agency such as the Occupational Safety and Health Administration (OSHA)).

To compute the approximate 95% upper confidence limit based on (11.8), we find the following quantities:

$$v_l^* = \sqrt{\frac{\chi^2_{14;.05}}{14}} = .6851,\ v_u^* = \sqrt{\frac{\chi^2_{14;.95}}{14}} = 1.3007,\ c_m = .9831, \widehat{\eta}_t = 1.6667.$$

Using these numbers, and noting that $z^2_{.95} = 2.7055$, we calculated the left endpoint L_η of the interval in (11.8) as .9850. Thus, the approximate 95% upper confidence limit for the exceedance probability is

$$1 - \Phi(L_\eta) = 1 - \Phi(.9850) = .162.$$

Note that this approximate limit is very close to the exact one, .163, reported in the preceding paragraph.

11.4.4   Prediction Intervals

Let $(\bar{X}, S)$ denote the (mean, SD) based on a sample of size n from a $N(\mu, \sigma^2)$ distribution. A $1-\alpha$ prediction interval for an individual (from the normal population from which the sample was drawn) is given by

$$\bar{X} \pm t_{n-1,\,1-\alpha/2} S \sqrt{1 + \frac{1}{n}}. \tag{11.9}$$

Upper Prediction Limits for at least l of m Observations from Each of r Locations

Construction of an upper prediction limit for at least l of m observations from a normal population at each of r locations is needed in ground water quality detection monitoring in the vicinity of hazardous waste management facilities (HWMF), and in process monitoring. For example, in groundwater quality monitoring near waste disposal facilities, a series of m sample observations from each of r monitoring wells located hydraulically downgradient of the HWMF are often compared with statistical prediction limits based on n measurements obtained from one or more upgradient sampling locations. The statistical problem is to construct upper prediction limit based on the n measurements so that it includes at least l of m samples at each of r downgradient monitoring wells. For a detailed discussion of this problem and strategies for monitoring ground-water quality, see Davis and McNichols (1987) and Bhaumik and Gibbons (2006). Davis and McNichols developed an exact method of constructing the upper prediction limit (UPL) assuming normality.

We shall outline Davis and McNichols' (1987) approach. The upper prediction limit is of the form

$$\bar{X} + k_u S,$$

where k_u is chosen so that at least l of m future observations are below $\bar{X} + k_u S$ on each of r locations, with probability $1 - \alpha$. Davis and McNichols (1987) showed that the factor k_u for constructing a $1 - \alpha$ level upper prediction limit can be obtained as the solution of the equation

$$\int_0^1 F\left(\sqrt{n}k_u \Big| n-1, \sqrt{\pi}\Phi^{-1}(x)\right) r\left(I(x;l,m+1-l)\right)^{r-1} \frac{x^{l-1}(1-x)^{m-l}}{B(l,m+1-l)} dx = 1 - \alpha, \tag{11.10}$$

where $F(x|\nu,\delta)$ denotes the cdf of the noncentral t random variable with df $= \nu$ and the noncentrality parameter δ, $B(a,b)$ denotes the usual beta function, and $I(x;a,b)$ denotes the cdf of a beta random variable with parameters a and b. Davis and McNichols tabulated values of k_u for some selected values of (n,r,l,m) and for $\gamma = 0.95$. It follows from the prediction interval in (11.9), the factor k_u should be $t_{n-1;1-\alpha}\sqrt{1+1/n}$ when $r = m = l = 1$.

Example 11.13. This example, along with the simulated data, are taken from Davis and McNichols (1987). Suppose in the vicinity of a hazardous waste management facility (HWMF), a single quantity, say, total organic compounds (TOC), has been specified for monitoring. A background sample size of $n = 20$ is chosen, and log(TOC) measurements from a single downgradient well and upgradient well were obtained. The differences are

.769	.093	-.669	-.284
.005	-.022	.036	1.028

.311	.233	.025	1.568
.276	.958	.369	.091
.590	1.000	-.402	.065

The mean and standard deviation for the background sample are $\bar{X} = .302$ and $S = .544$. Suppose we select the plan $(r, m, l) = (16, 4, 2)$, then the factor for constructing 95% upper prediction limit is 1.542. The upper prediction limit is $.302 + 1.542 \times .544 = 1.141$. Thus, at each of the semiannual sampling occasions, up to four measurements will be taken until two observations less than 1.141 are obtained. None or only one of such observations is less than 1.141 indicate that a change in the mean level of the downgradient-upgradient difference in log(TOC) has occurred.

The factor in the preceding paragraph can be obtained as follows. Select the dialog box [StatCalc→Continuous→Normal→Tolerance ...], enter 20 for n, 16 for r, 4 for m, 2 for l, .95 for confidence level, and click on [Factor k] to get 1.54238.

11.4.5   Test and Interval Estimation for the Variance

Let S^2 denote the variance of a sample of n observations from a normal population with mean μ and variance σ^2. The pivotal quantity for testing and interval estimation of a normal variance is given by

$$Q = \frac{(n-1)S^2}{\sigma^2},$$

which follows a chi-square distribution with df $= n - 1$. Let Q_0 be an observed value of Q. For testing

$$H_0 : \sigma^2 = \sigma_0^2 \text{ vs. } H_a : \sigma^2 \neq \sigma_0^2,$$

a size α test rejects H_0 if $2 \min\{P(\chi_{n-1}^2 > Q_0), P(\chi_{n-1}^2 < Q_0)\} < \alpha$. For testing $H_0 : \sigma^2 \leq \sigma_0^2$ vs. $H_a : \sigma^2 > \sigma_0^2$, the null hypothesis H_0 will be rejected if $P(\chi_{n-1}^2 > Q_0) < \alpha$, and for testing $H_0 : \sigma^2 \geq \sigma_0^2$ vs. $H_a : \sigma^2 < \sigma_0^2$, the null hypothesis H_0 will be rejected if $P(\chi_{n-1}^2 < Q_0) < \alpha$.

A $1 - \alpha$ confidence interval for the variance σ^2 is given by

$$\left(\frac{(n-1)S^2}{\chi_{n-1,1-\alpha/2}^2}, \frac{(n-1)S^2}{\chi_{n-1,\alpha/2}^2} \right),$$

where $\chi_{m,p}^2$ denotes the pth quantile of a chi-square distribution with df $= m$.

For a given sample variance and sample size, the dialog box [StatCalc →Continuous → Normal →One-Sample ...] computes the confidence interval for the population variance σ^2, and p-values for hypothesis testing about σ^2.

To compute a confidence interval for σ^2: Enter the value of the sample size, sample variance, and the confidence level. Click [1-sided] to get one-sided lower and upper confidence limits for σ^2. Click [2-sided] to get confidence interval for σ^2.

Example 11.14. Suppose that a sample of 20 observations from a normal population produced a variance of 12. To compute a 90% confidence interval for σ^2, select the dialog box [StatCalc →Continuous → Normal →One-Sample ...], enter 20 for [Sample Size], $\sqrt{12} = 3.464$ for [Sample SD], and 0.90 for [Confidence Level]. Click

[1-sided] to get 8.381 and 19.57. These are the one-sided limits for σ^2. That is, the interval (8.381, ∞) would contain the population variance σ^2 with 90% confidence. The interval (0, 19.57) would contain the population variance σ^2 with 90% confidence. To get a two-sided confidence interval for σ^2, click on [2-sided] to get 7.56381 and 22.5363. This means that the interval (7.564, 22.535) would contain σ^2 with 90% confidence.

To compute p-values for hypothesis testing about σ^2: Enter the summary statistics as in the above example, and the specified value of σ^2 under the null hypothesis. Click [p-values for] to get the p-values for various alternative hypotheses.

Example 11.15. Suppose we want to test

$$H_0 : \sigma^2 \leq 9 \quad \text{vs.} \quad H_a : \sigma^2 > 9 \tag{11.11}$$

at the level of 0.05 using the summary statistics given in Example 11.14. After entering the summary statistics, enter 9 for [H0: V = V0]. Click [p-values for] to get 0.1499. Since this p-value is not less than 0.05, the null hypothesis in (11.11) cannot be rejected at the level of significance 0.05. We conclude that the summary statistics do not provide sufficient evidence to indicate that the true population variance is greater than 9.

Example 11.16. *(Test about σ^2)* A hardware manufacturer was asked to produce a batch of 3-in screws with a specification that the standard deviation of the lengths of all the screws should not exceed 0.1 in. At the end of a day's production, a sample of 27 screws was measured, and the sample standard deviation was calculated as 0.09. Does this sample standard deviation indicate that the actual standard deviation of all the screws produced during that day is less than 0.1 inch?

Solution: Let σ denote the standard deviation of all the screws produced during that day. The appropriate hypotheses for the problem are

$$H_0 : \sigma \geq 0.1 \quad \text{vs.} \quad H_a : \sigma < 0.1 \iff H_0 : \sigma^2 \geq 0.01 \quad \text{vs.} \quad H_a : \sigma^2 < 0.01.$$

Note that the sample variance is $(0.09)^2 = 0.0081$. To compute the p-value for the above test, enter 27 for the sample size, 0.09 for the sample SD, 0.01 for [H0: V = V0], and click on [p-values (var)]. The computed p-value is 0.26114. Since this p-value is not smaller than any practical level of significance, we can not conclude that the standard deviation of all the screws made during that day is less than 0.1 in.

Example 11.17. *(Confidence interval for σ^2)* An agricultural student wants to estimate the variance of the yields of tomato plants. He selected a sample of 18 plants for the study. After the harvest, he found that the mean yield was 38.5 tomatoes with standard deviation 3.4. Assuming a normal model for the yields of tomato, construct a 90% confidence interval for the standard deviation of the yields of all tomato plants.

Solution: To construct a 90% confidence interval for the variance, enter 18 for the sample size, 3.4 for [Sample SD], and 0.90 for the confidence level. Click [2-sided] to get 7.12362 and 22.6621. Thus, the standard deviation of tomato yields is somewhere between $\sqrt{7.12362} = 2.669$ and $\sqrt{22.6621} = 4.760$ with 90% confidence.

11.5   Two-Sample Inference

Let S_i^2 denote the variance of a random sample of n_i observations from $N(\mu_i, \sigma_i^2)$, $i = 1, 2$. The following inferential procedures for the ratio σ_1^2/σ_2^2 are based on the F statistic given by

$$F = \frac{S_1^2}{S_2^2}. \tag{11.12}$$

11.5.1   Inference for the Ratio of Variances

Hypothesis Test for the Ratio of Variances

Consider testing $H_0 : \sigma_1^2/\sigma_2^2 = 1$. When H_0 is true, then the F statistic in (11.12) follows an F_{n_1-1,n_2-1} distribution. Let F_0 be an observed value of the F in (11.12). A size α test rejects the null hypothesis in favor of

$$H_a : \sigma_1^2 > \sigma_2^2 \text{ if the p-value } P(F_{n_1-1,n_2-1} > F_0) < \alpha.$$

For testing $H_0 : \sigma_1^2 \geq \sigma_2^2$ vs. $H_a : \sigma_1^2 < \sigma_2^2$, the null hypothesis will be rejected if the p-value $P(F_{n_1-1,n_2-1} < F_0) < \alpha$. For a two-tailed test, the null hypothesis $H_0 : \sigma_1^2 = \sigma_2^2$ will be rejected if either tail p-value is less than $\alpha/2$.

Interval Estimation for the Ratio of Variances

A $1 - \alpha$ confidence interval for σ_1^2/σ_2^2 is given by

$$\left(\frac{S_1^2}{S_2^2} F_{n_2-1,n_1-1,\frac{\alpha}{2}}, \ \frac{S_1^2}{S_2^2} F_{n_2-1,n_1-1,1-\frac{\alpha}{2}} \right),$$

where $F_{m,n,p}$ denotes the pth quantile of an F distribution with the numerator df $= m$ and the denominator df $= n$. The above confidence interval can be obtained from the distributional result that

$$\left(\frac{S_2^2/\sigma_2^2}{S_1^2/\sigma_1^2} \right) = \frac{\sigma_1^2}{\sigma_2^2} \frac{S_2^2}{S_1^2} \sim F_{n_2-1,n_1-1}.$$

The dialog box [StatCalc→Continuous → Normal → Two-Sample ...] computes confidence intervals as well as the p-values for testing the ratio of two variances.

To compute a confidence interval for σ_1^2/σ_2^2: Enter the values of the sample sizes and sample variances, and the confidence level. Click [1-sided] to get one-sided lower and upper limits for σ_1^2/σ_2^2. Click [2-sided] to get confidence interval for σ_1^2/σ_2^2.

Example 11.18. *(Confidence interval for σ_1^2/σ_2^2)* A sample of 8 observations from a normal population produced a variance of 4.41. A sample of 11 observations from another normal population yielded a variance of 2.89. To compute a 95% confidence interval for σ_1^2/σ_2^2, select the dialog box [StatCalc→Continuous → Normal → Two-Sample ...], enter the sample sizes, sample variances, and 0.95 for the confidence level. Click [1-sided] to get 0.4196 and 4.7846. This means that the interval $(0.4196, \infty)$

would contain the ratio σ_1^2/σ_2^2 with 95% confidence. Furthermore, we can conclude that the interval $(0, 4.7846)$ would contain the variance ratio with 95% confidence. Click [2-sided] to get 0.3863 and 7.2652. This means that the interval $(0.3863, 7.2652)$ would contain the variance ratio with 95% confidence.

Example 11.19. *(Hypothesis tests for σ_1^2/σ_2^2)* Suppose we want to test

$$H_0 : \sigma_1^2 = \sigma_2^2 \text{ vs. } H_a : \sigma_1^2 \neq \sigma_2^2,$$

at the level of 0.05 using the summary statistics given in Example 11.18. To compute the p-value, enter the summary statistics in the dialog box, and click on [p-values for] to get 0.5251. Since the p-value is greater than 0.05, we cannot conclude that the population variances are significantly different.

11.5.2   Inference for the Difference between Two Means

There are two procedures available to make inference about the mean difference $\mu_1 - \mu_2$. One is based on the assumption that $\sigma_1^2 = \sigma_2^2$ and another is valid for any variances. In practice, the equality of the variances is tested first using the F test in the previous section. If the assumption of equality of variances is tenable, then the two-sample t procedures given below are used to make inference about $\mu_1 - \mu_2$. otherwise, an approximate procedure (see Section 10.5.3) known as Welch's *approximate degrees of freedom* method can be used. This approach of selecting a two-sample test is criticized by many authors (see Moser and Stevens, 1992, and Zimmerman, 2004). In general, many authors suggested using the Welch test when the variances are unknown. Nevertheless, for the sake of completeness and illustrative purpose, we describe both approaches in the sequel.

Let $\bar{X}_i$ and S_i^2 denote, respectively, the mean and variance of a random sample of n_i observations from $N(\mu_i, \sigma_i^2)$, $i = 1, 2$. Let

$$S_p^2 = \frac{(n_1 - 1)S_1^2 + (n_2 - 1)S_2^2}{n_1 + n_2 - 2}. \tag{11.13}$$

The following inferential procedures for $\mu_1 - \mu_2$ are based on the sample means and the pooled variance S_p^2, and are valid only when $\sigma_1^2 = \sigma_2^2$.

Two-Sample t Test

The test statistic for testing $H_0 : \mu_1 = \mu_2$ is given by

$$t_2 = \frac{(\bar{X}_1 - \bar{X}_2)}{\sqrt{S_p^2 \left(\frac{1}{n_1} + \frac{1}{n_2} \right)}}, \tag{11.14}$$

which follows a t-distribution with degrees of freedom $n_1 + n_2 - 2$, provided $\sigma_1^2 = \sigma_2^2$. Let t_{20} be an observed value of t_2. For a given level α, the null hypothesis will be rejected in favor of

$$H_a : \mu_1 \neq \mu_2 \text{ if the p-value } P(|t_2| > |t_{20}|) < \alpha,$$

in favor of $H_a : \mu_1 < \mu_2$ if the p-value $P(t_2 < t_{20}) < \alpha$, and in favor of $H_a : \mu_1 > \mu_2$ if the p-value $P(t_2 > t_{20}) < \alpha$.

Power of the Two-Sample t-test

Consider the hypotheses

$$H_0 : \mu_1 \leq \mu_2 \quad \text{vs.} \quad H_a : \mu_1 > \mu_2.$$

For a given level α, the power of the two-sample t-test is the probability of rejecting the null hypothesis when μ_1 is indeed greater than μ_2, and is given by

$$P(t_2 > t_{n_1+n_2-2,1-\alpha}) = P(t_{n_1+n_2-2}(\delta) > t_{n_1+n_2-2,1-\alpha}), \qquad (11.15)$$

where t_2 is given in (11.14), $t_{n_1+n_2-2,1-\alpha}$ denotes the $(1-\alpha)$th quantile of a t-distribution with degrees of freedom $n_1 + n_2 - 2$, $t_{n_1+n_2-2}(\delta)$ denotes the noncentral t random variable with the degrees of freedom $n_1 + n_2 - 2$ and noncentrality parameter

$$\delta = \frac{(\mu_1 - \mu_2)}{\sigma\sqrt{\frac{1}{n_1} + \frac{1}{n_2}}}.$$

The power of a two-tailed test is similarly calculated. For given sample sizes, $\mu_1 - \mu_2$, common σ and the level of significance, *StatCalc* computes the power using (11.15).

Interval Estimation of $\mu_1 - \mu_2$

A $1 - \alpha$ confidence interval based on the test-statistic in (11.14) is given by

$$\bar{X}_1 - \bar{X}_2 \pm t_{n_1+n_2-2,1-\alpha/2}\sqrt{S_p^2\left(\frac{1}{n_1} + \frac{1}{n_2}\right)}, \qquad (11.16)$$

where $t_{n_1+n_2-2,1-\alpha/2}$ denotes the $1 - \frac{\alpha}{2}$ quantile of a t-distribution with $n_1 + n_2 - 2$ degrees of freedom. This confidence interval is valid only when $\sigma_1^2 = \sigma_2^2$.

The dialog box [StatCalc→Continuous → Normal → Two-Sample ...] computes the p-values for testing the difference between two normal means and confidence intervals for the difference between the means. The results are valid under the assumption that $\sigma_1^2 = \sigma_2^2$.

To compute a confidence interval for $\mu_1 - \mu_2$: Enter the values of the sample means, sample standard deviations, sample sizes, and the confidence level. Click [1-sided] to get one-sided lower and upper limits for $\mu_1 - \mu_2$. Click [2-sided] to get confidence interval for $\mu_1 - \mu_2$.

Example 11.20. *(Test about σ_1^2/σ_2^2)* A sample of 8 observations from a normal population with mean μ_1, and variance σ_1^2 produced a mean of 4 and standard deviation of 2.1. A sample of 11 observations from another normal population with mean μ_2, and variance σ_2^2 yielded a mean of 2 with standard deviation of 1.7. Since the inferential procedures given in this section are appropriate only when the population variances are equal, we first want to test that if the variances are indeed equal (see Section 11.5.1). The test for equality of variances yielded a p-value of $2 \times 0.262577 = 0.525154$, and hence the assumption of equality of population variances seems to be tenable.

To compute a 95% confidence interval for $\mu_1 - \mu_2$ using *StatCalc*, enter the sample means, standard deviations, sample sizes, and 0.95 for the confidence level. Click [1-sided] to get 0.484333 and 3.51567. This means that the interval (0.48433, ∞) would contain the difference $\mu_1 - \mu_2$ with 95% confidence. Furthermore, we can conclude that the interval $(-\infty, 3.51567)$ would contain the difference $\mu_1 - \mu_2$ with 95% confidence. Click [2-sided] to get 0.161782 and 3.83822. That is, the interval (0.161782, 3.83822) would contain the difference $\mu_1 - \mu_2$ with 95% confidence.

To compute p-values for testing $\mu_1 - \mu_2$: Select the dialog box [StatCalc$\rightarrow$ Continuous $\rightarrow$Normal $\rightarrow$ Two-Sample ...], enter the values of the sample sizes, sample means, and sample standard deviations, and click [p-values for] to get the p-values for a right-tailed test, left-tailed test, and two-tailed test.

Example 11.21. Suppose we want to test

$$H_0 : \mu_1 \leq \mu_2 \quad \text{vs.} \quad H_a : \mu_1 > \mu_2$$

at the level of 0.05 using the summary statistics given in Example 11.20. To compute the p-value, enter the summary statistics in the dialog box, and click on [p-values for] to get 0.0173486. Since the p-value is less than 0.05, we conclude that $\mu_1 > \mu_2$. *Power Calculation of Two-Sample t-test:* The dialog box [StatCalc$\rightarrow$Continuous $\rightarrow$ Normal $\rightarrow$ Two-Sample ...] computes the power of the two-sample t-test for

$$H_0 : \mu_1 \leq \mu_2 \quad \text{vs.} \quad Ha : \mu_1 > \mu_2$$

assuming that $\sigma_1^2 = \sigma_2^2$. To compute the power, enter the values of the level α of the test, the difference between the population means, the value of the common standard deviation σ and sample sizes n_1 and n_2. Click [Power]. Power of a two-tailed test can be computed by entering $\alpha/2$ for the level.

Example 11.22. *(Calculation of Power)* Suppose that the difference between two normal means is 1.5 and the common standard deviation is 2. It is desired to test

$$H_0 : \mu_1 \leq \mu_2 \quad \text{vs.} \quad H_0 : \mu_1 > \mu_2$$

at the level of significance 0.05. To compute the power when each sample size is 27, enter 0.05 for level, 1.5 for the mean difference, 2 for the common σ, 27 for n_1, and n_2; click [Power] to get 0.858742.

Sample Size Calculation: In practical applications, it is usually desired to compute the sample sizes required to attain a given power. This can be done by a trial-error method. Suppose in the above example we need to determine the sample sizes required to have a power of 0.90. By trying a few sample sizes more than 27, we can find the required sample size as 32 from each population. In this case, the power is 0.906942. Also, note that when $n_1 = 27$ and $n_2 = 37$, the power is 0.900729.

Example 11.23. A company, which employs thousands of computer programmers, wants to compare the mean difference between the salaries of the male and female programmers. A sample of 23 male programmers and a sample of 19 female programmers were selected, and programmers' salaries were recorded. Assume that the salary distributions for male and female programmers are normal. The summary statistics are given in the following table.

	Male	Female
sample size	23	19
mean	52.56	48.34 (in $1000)
SD	3.195	2.750 (in $1000)

a. Do these summary statistics indicate that the average salaries of the male programmers are higher than that of female programmers?

b. Construct a 95% confidence interval for the mean difference between the salaries of male and female programmers.

Solution: Since the salaries are from normal populations, a two-sample procedure for comparing normal means is appropriate for this problem. Furthermore, to choose between the two comparison methods (one assumes that the population variances are equal and the other is not), we need to test the equality of the population variances. Using the dialog box [StatCalc→Continuous → Normal → Two-Sample Case → Test and confidence interval for the Variance Ratio], we get the p-value for testing the equality of variances is $2 \times 0.261234 = 0.522468$, which is greater than any practical level of significance. Therefore, the assumption that the variances are equal is tenable, and we can use the two-sample t procedures for the present problem.

a. Let μ_1 denote the mean salaries of all male programmers, and μ_2 denote the mean salaries of all female programmers in the company. We want to test

$$H_0 : \mu_1 \leq \mu_2 \ \text{vs.} \ H_a : \mu_1 > \mu_2.$$

To compute the p-value for the above test, enter the sample sizes, means, and standard deviations, click [p-values for] to get 2.58684e-005. Since this p-value is much less than any practical levels, we reject the null hypothesis, and conclude that the mean salaries of male programmers is higher than that of female programmers.

b. To compute a 95% confidence interval for $\mu_1 - \mu_2$, enter 0.95 for the confidence level, and click [2-sided] to get 2.33847 and 6.10153. That is, the mean difference is somewhere between $2,338 and $6,101.

11.5.3 Inference for the Difference between Two Means when Variances Are Unknown and Arbitrary

The following method known as the Welch approximate degrees of freedom method can be used to make inferences about $\mu_1 - \mu_2$ when the population variances are unknown and arbitrary. This approximate method is based on the result that

$$\frac{\bar{X}_1 - \bar{X}_2}{\sqrt{\frac{S_1^2}{n_1} + \frac{S_2^2}{n_2}}} \sim t_f \ \text{approximately, with} \ f = \frac{\left(\frac{S_1^2}{n_1} + \frac{S_2^2}{n_2}\right)^2}{\left(\frac{S_1^4}{n_1^2(n_1-1)} + \frac{S_2^4}{n_2^2(n_2-1)}\right)}.$$

The hypothesis testing and interval estimation of $\mu_1 - \mu_2$ can be carried out as in Section 10.5.1 with the degrees of freedom f given above. For example, a $1 - \alpha$ confidence interval for $\mu_1 - \mu_2$ is given by

$$\bar{X}_1 - \bar{X}_2 \pm t_{f,1-\alpha/2} \sqrt{\frac{S_1^2}{n_1} + \frac{S_2^2}{n_2}}, \qquad (11.17)$$

where $t_{m,p}$ denotes the pth quantile of a t distribution with degrees of freedom f. This approximate method is commonly used, and the results based on this method are very accurate even for small samples.

The dialog box [StatCalc→Continuous → Normal → Two-Sample ...] uses the above approximate method for hypothesis testing and interval estimation of $\mu_1 - \mu_2$ without assuming equality of variances.

Example 11.24. A sample of 8 observations from a normal population with mean μ_1, and variance σ_1^2 produced $\bar{X}_1 = 4$ and standard deviation $S_1 = 2.1$. A sample of 11 observations from another normal population with mean μ_2 and variance σ_2^2 yielded $\bar{X}_2 = 2$ with standard deviation $S_2 = 5$.

To find the 95% Welch confidence interval for $\mu_1 - \mu_2$, select [StatCalc→Continuous → Normal →Two-Sample ...], enter the sample statistics and click [2-sided] to get $(-1.5985, 5.5985)$. To get one-sided limits, click [1-sided]. For this example, 95% one-sided lower limit is -0.956273, and 95% one-sided upper limit is 4.95627. Suppose we want to test $H_0 : \mu_1 - \mu_2 = 0$ vs. $H_0 : \mu_1 - \mu_2 \neq 0$. To compute the p-values using *StatCalc*, click [p-values for] to get $2 \times 0.126725 = 0.25345$. Thus, we cannot conclude that the means are significantly different.

Example 11.25. Let us consider the statistics in Example 11.20, and find the Welch confidence interval. Note that $n_1 = 8, \bar{X}_1 = 4$ and $s_1 = 2.1$; $n_2 = 11, \bar{X}_1 = 2$ and $s_1 = 1.7$. Using *StatCalc* as in the preceding example, we find the 95% Welch confidence interval as $(.0534, 3.947)$. The t-interval in Example 11.20 was calculated as $(.1618, 3.838)$, which is shorter than the Welch confidence interval. On the other hand, the data in Example 11.24 indicate that the population variances may not be equal, and 95% t-interval for Example 11.24 is $(-1.9848, 5.985)$, which is wider than the Welch confidence interval reported in the preceding example.

11.5.4   Comparison of Two Coefficients of Variation

Let $(\bar{X}_i, S_i^2)$ denote the mean and variance (unbiased estimate) of a random sample of size n_i from a $N(\mu_i, \sigma_i^2)$ distribution, $i = 1, 2$. Let $(\bar{x}_i, s_i^2)$ be an observed value of $(\bar{X}_i, S_i^2)$, $i = 1, 2$. Define $\tau_i = \sigma_i/\mu_i$, the coefficient of variation for the $N(\mu_i, \sigma_i^2)$ distribution. We shall describe a test for comparing τ_1 and τ_2, and the generalized variable approach to find confidence intervals for $\tau_1 - \tau_2$ or for the ratio τ_1/τ_2.

A Test for the Ratio of Coefficients of Variation

Consider testing

$$H_0 : \tau_1 \leq \tau_2 \quad \text{vs.} \quad H_a : \tau_1 > \tau_2. \qquad (11.18)$$

Forkman (2009) proposed a test based on a statistic that has an approximate F distribution. To outline the test, let $\hat{\tau}_i = S_i/\bar{X}_i$, where S_i^2 is the usual unbiased estimate of σ_i^2, $i = 1, 2$. The test statistic proposed by Forkman is given by

$$F = \frac{\hat{\tau}_1^2/[1 + \hat{\tau}_1^2(n_1 - 1)/n_1]}{\hat{\tau}_2^2/[1 + \hat{\tau}_2^2(n_2 - 1)/n_2]}. \tag{11.19}$$

Forkman has shown that the above statistic has an approximate F distribution with degrees of freedom $n_1 - 1$ and $n_2 - 1$. This test rejects H_0 in (11.18) when an observed value of F in (11.19) is greater than $F_{n_1-1,n_2-1;1-\alpha}$. Simulation studies by Krishnamoorthy and Lee (2013) indicated that this test is very accurate in controlling type I error rates, and has comparable powers with other tests.

Generalized Variable Approach

Let $(\bar{X}_i, S_i)$ denote the (mean, SD) based on a sample of size n_i from a $N(\mu_i, \sigma_i^2)$ distribution, $i = 1, 2$. For the two-sample case, let $(G_{\mu_i}, G_{\sigma_i})$ denote the generalized pivotal quantities (GPQs) for (μ_i, σ_i), $i = 1, 2$. Using the results of Section 2.7.2, a GPQ for $\tau_i = \sigma_i/\mu_i$ is given by

$$G_{\tau_i} = \frac{G_{\sigma_i}}{G_{\mu_i}} = \left(\frac{\bar{x}_i\sqrt{W_i}}{s_i} - \frac{Z_i}{\sqrt{n_i}} \right)^{-1}, \tag{11.20}$$

where $W_i = \chi_{n_i-1}^2/(n_i-1)$. Note that this GPQ for τ_i could be negative even though τ_i is assumed to positive. This is not a problem in finding a confidence interval (or a test) for the ratio, because a confidence interval for τ_1/τ_2 can be obtained from the one for τ_1^2/τ_2^2.

A GPQ for τ_1^2/τ_2^2 is given by

$$G_{\tau_1^2/\tau_2^2} = G_{\tau_1}^2/G_{\tau_2}^2 = \left(\frac{\bar{x}_2\sqrt{W_2}}{s_2} - \frac{Z_2}{\sqrt{n_2}} \right)^2 \Big/ \left(\frac{\bar{x}_1\sqrt{W_1}}{s_1} - \frac{Z_1}{\sqrt{n_1}} \right)^2. \tag{11.21}$$

The 100α percentile, and the $100(1-\alpha)$ percentile of $Q_{\tau_1^2/\tau_2^2}$, form a $1-2\alpha$ confidence interval for τ_1^2/τ_2^2, and the square root of this confidence interval is a $1-2\alpha$ confidence interval for τ_1/τ_2.

To find a confidence interval for the difference $\tau_1 - \tau_2$, we propose the GPQ

$$G_{\tau_1-\tau_2} = \sqrt{G_{\tau_1}^2} - \sqrt{G_{\tau_2}^2}. \tag{11.22}$$

For a given $(\bar{x}_1, s_1, \bar{x}_2, s_2)$, the distribution of $G_{\tau_1^2/\tau_2^2}$ (or that of $G_{\tau_1-\tau_2}$) does not depend on any unknown parameters, and so they can be estimated using Monte Carlo simulation as shown in Algorithm 11.1.

The generalized test for

$$H_0 : \tau_1 \leq \tau_2 \quad \text{vs.} \quad H_a : \tau_1 > \tau_2 \tag{11.23}$$

is described as follows. Let $G_{\tau_1^2/\tau_2^2;\alpha}$ denote the α quantile of $Q_{\tau_1^2/\tau_2^2}$. Note that $Q_{\tau_1^2/\tau_2^2;\alpha}$ is the $1 - \alpha$ lower confidence limit for τ_1^2/τ_2^2. The above null hypothesis is rejected if this lower confidence limit is greater than one, or equivalently,

$$P_1 = P\left(Q_{\tau_1^2/\tau_2^2} < 1 \right) \leq \alpha.$$

The above probability is referred to as the generalized p-value. The generalized p-value for a two-tailed test is given by $2\min\{P_1, 1 - P_1\}$. These generalized p-values and the generalized confidence intervals can be estimated using Monte Carlo simulation, as shown in the following algorithm.

Algorithm 11.1. Calculation of generalized confidence intervals

---

For given samples from $N(\mu_1, \sigma_1^2)$ and $N(\mu_2, \sigma_2^2)$, compute $(\bar{x}_1, s_1, \bar{x}_2, s_2)$.
For i = 1 to N
    Generate $Z_j \sim N(0, 1)$ and $W_j \sim \chi_{n_j-1}^2/(n_j - 1)$, $j = 1, 2$.
    Calculate $Q_i = G_{\tau_1}^2/G_{\tau_2}^2$ using (11.21).
    Set $I_i = 1$ if $Q_i < 1$; else set $I_i = 0$
(end do loop)

The proportion $P_1 = \frac{1}{N}\sum_{i=1}^{N} I_i$ is a Monte Carlo estimate of the generalized p-value for testing (11.23). The 100α percentile of these Q_is generated above is a $100\alpha\%$ lower confidence limit for τ_1^2/τ_2^2.

11.6   Tolerance Intervals

Let $X_1, \ldots, X_n$ be a sample from a normal population with mean μ and variance σ^2. Let $\bar{X}$ denote the sample mean and S denote the sample standard deviation.

11.6.1   Two-Sided Tolerance Intervals

For a given $0 < \beta < 1$, $0 < \gamma < 1$ and n, the tolerance factor k is to be determined so that the interval

$$\bar{X} \pm kS$$

would contain at least proportion β of the normal population with confidence γ. Mathematically, k should be determined so that

$$P_{\bar{X},S}\left\{P_X\left[X \in (\bar{X} - kS, \ \bar{X} + kS\,)|\bar{X}, S\right] \geq \beta\right\} = \gamma, \qquad (11.24)$$

where X also follows the $N(\mu, \sigma^2)$ distribution independently of the sample. An explicit expression for k is not available and has to be computed numerically. The k satisfies (11.24) and is called the *tolerance factor*, and $\bar{X} \pm kS$ is called a (β, γ) *tolerance interval* or a β content $-$ γ coverage tolerance interval. An approximate expression for k is given by

$$k \simeq \left(\frac{m\chi_{1,\beta}^2(1/n)}{\chi_{m,1-\gamma}^2}\right)^{1/2}, \qquad (11.25)$$

where $\chi_{1,p}^2(1/n)$ denotes the pth quantile of a noncentral chi-square distribution with df = 1, and noncentrality parameter $1/n$, $\chi_{m,1-\gamma}^2$ denotes the $(1 - \gamma)$th quantile of a central chi-square distribution with df = $m = n - 1$, the df associated with the sample variance. This approximation is extremely satisfactory even for small samples (as small as 3) if β and γ are greater than or equal to 0.95 [Wald and Wolfowitz, 1946].

The dialog box [StatCalc→Continuous → Normal→ Tolerance Intervals and Prediction Intervals] uses an exact method [see Kendall and Stuart, 1973, p. 134 or Krishnamoorthy and Mathew, 2009, Chapter 2] of computing the factor k.

11.6.2   One-Sided Tolerance Limits

The one-sided β content $-\gamma$ coverage upper tolerance limit is given by $\bar{X}+cS$, where the tolerance factor c is to be determined so that

$$P_{\bar{X},S}\{P_X[X \leq \bar{X} + cS|\bar{X}, S] \geq \beta\} = \gamma.$$

In this case, $c = \frac{1}{\sqrt{n}}t_{n-1,\gamma}(z_\beta\sqrt{n})$, where $t_{m,p}(\delta)$ denotes the $100p$th percentile of a noncentral t distribution with $df = m$ and noncentrality parameter δ, and z_p denotes the $100p$th percentile of the standard normal distribution. The quantity c is called tolerance factor. Thus, the (β, γ) upper tolerance limit is given by

$$\bar{X} + \frac{1}{\sqrt{n}}t_{n-1,\gamma}(z_\beta\sqrt{n})S.$$

That is, at least $100\beta\%$ of the data from the normal population are less than or equal to $\bar{X} + cS$ with confidence γ. The same tolerance factor c can be used to find the lower tolerance limits. That is,

$$\bar{X} - \frac{1}{\sqrt{n}}t_{n-1,\gamma}(z_\beta\sqrt{n})S,$$

is a (β, γ) lower tolerance limit. This means that at least $100\beta\%$ of the data are greater than or equal to $\bar{X} - cS$ with confidence γ.

It should be noted that $\bar{X} + cS$ is a $1 - \alpha$ upper confidence limit for the quantile $\mu + z_\beta\sigma$, and $\bar{X} - cS$ is a $1 - \alpha$ lower confidence limit for the quantile $\mu - z_\beta\sigma$, where z_β is the β quantile of the standard normal distribution.

Remark 11.2. The degrees of freedom associated with S^2 is $n - 1$. In some situations, the df associated with the S^2 could be different from $n - 1$. For example, in one-way analysis of variance, the df associated with the pooled sample variance S_p^2 is (total sample size $- g$), where g denotes the number of groups. If one is interested in computing (β, γ) tolerance interval of the form

$$\bar{X}_1 \pm k_1 S_p$$

for the first group, then for this case, the sample size is n_1 and the degrees of freedom associated with the pooled variance is

$$\sum_{i=1}^{g} n_i - g \,,$$

where n_i denotes the size of the sample from the ith group, $i = 1, \ldots, g$.

For a given n, df, $0 < \beta < 1$ and $0 < \gamma < 1$, the dialog box [StatCalc→ Continuous → Normal → Tolerance ...] computes one-sided as well as two-sided tolerance factors.

Example 11.26. When $n = 23$, df $= 22$, $\beta = 0.90$, and $\gamma = 0.95$, the one-sided tolerance factor is 1.86902, and the two-sided tolerance factor is 2.25125. To compute the factors, enter 23 for [Sample Size n], 22 for [DF], 0.90 for [Proportion p], and 0.95 for [Coverage Prob g]; click [1-sided] to get 1.86902, and click [2-sided] to get 2.25125.

Applications

The normal-based tolerance factors are applicable to a non-normal distribution if it has a one-to-one relation with a normal distribution. For example, if X follows a lognormal distribution, then $\ln(X)$ follows a normal distribution. Therefore, the factors given in the preceding sections can be used to construct tolerance intervals for a lognormal distribution. Specifically, if the sample $Y_1, \ldots, Y_n$ is from a lognormal distribution, then normal based methods for constructing tolerance intervals can be used after taking logarithmic transformation of the sample. If U is a (β, γ) upper tolerance limit based on the log-transformed data, then $\exp(U)$ is the (β, γ) upper tolerance limit for the lognormal distribution. Approximate tolerance intervals for a gamma distribution can be constructed using cube root transformation (see Section 16.6.3).

In many practical situations, one wants to assess the proportion of the data that fall in an interval or a region. For example, engineering products are usually required to satisfy certain tolerance specifications. The proportion of the products that are within the specifications can be assessed by constructing a suitable tolerance region based on a sample of products. As an example, consider the following *acceptance sampling plan*. A lot of items is submitted for inspection, and the lot will be accepted if at least 95% of the items are within the specification (L, U), where L is the lower specification limit and U is the upper specification limit. In order to save time and cost, typically a sample of items is inspected and a $(.95, .95)$ tolerance interval is constructed. If this tolerance interval falls in (L, U), then it can be concluded that at least 95% of the items in the lot are within the specification limits with 95% confidence, so the lot will be accepted.

In some situations, each item in the lot is required to satisfy only the lower specification. In this case, a $(.95, .95)$ lower tolerance limit is constructed and compared with the lower specification L. If the lower tolerance limit is greater than or equal to L, then the lot will be accepted.

Tolerance limits can be used to monitor exposure levels of employees to workplace contaminants. Specifically, if the upper tolerance limit based on exposure measurements from a sample of employees is less than a permissible exposure limit (PEL), then it indicates that a majority of the exposure measurements are within the PEL, and hence exposure monitoring might be reduced or terminated until a process change occurs. Such studies are feasible because the National Institute for Occupational Safety and Health provides PEL for many workplace chemicals (see Krishnamoorthy and Mathew, 2009, Chapter 2).

Example 11.27. Let us construct tolerance limits for the exposure data given in Example 11.1. We already verified that the data satisfy the normality assumption. Note that the sample size is 15 (df $= 15 - 1 = 14$), the sample mean is 78.5, and the standard deviation is 15.9. The tolerance factor for a $(.95, .95)$ tolerance interval is 2.96494. Using these numbers, we compute the tolerance interval as

$$78.5 \pm 2.96494 \times 15.9 = (31.4, \ 125.6).$$

That is, at least 95% of the exposure measurements fall between 31.4 and 125.6 with 95% confidence. The tolerance factor for a (.95, .95) one-sided limit is 2.566. The one-sided upper limit is $78.5 + 2.566 \times 15.9 = 119.3$. That is, at least 95% of the exposure measurements are below 119.3 with 95% confidence. The one-sided lower tolerance limit is $78.5 - 2.566 \times 15.9 = 37.7$. That is, at least 95% of the exposure measurements are above 37.7.

11.6.3 Equal-Tailed Tolerance Intervals

Let $X_1, \ldots, X_n$ be a sample from a normal population with mean μ and variance σ^2. Let $\bar{X}$ denote the sample mean and S denote the sample standard deviation. The β content $- \gamma$ coverage equal-tailed tolerance interval (L, U) is constructed so that it would contain at least $100\beta\%$ of the "center data" of the normal population. That is, (L, U) is constructed such that no more than $100\frac{(1-\beta)}{2}\%$ of the data are less than L, and no more that $100\frac{(1-\beta)}{2}\%$ of the data are greater than U with confidence γ. This amounts to constructing (L, U) so that it would contain

$$\left(\mu - z_{\frac{1+\beta}{2}}\sigma, \ \mu + z_{\frac{1+\beta}{2}}\sigma \right)$$

with confidence γ. Toward this, we consider the intervals of the form $(\bar{X} - kS, \ \bar{X} + kS)$, where k is to be determined so that

$$P\left(\bar{X} - kS < \mu - z_{\frac{1+\beta}{2}}\sigma \ \text{and} \ \mu + z_{\frac{1+\beta}{2}}\sigma < \bar{X} + kS \right) = \gamma.$$

The dialog box [StatCalc→Continuous→Normal→Tolerance Intervals and Prediction Intervals] uses an exact method due to Owen (1964) for computing the tolerance factor k satisfying the above probability requirement.

Example 11.28. *(Equal-tailed tolerance intervals)* In order to understand the difference between the tolerance interval and the equal-tailed tolerance interval, let us consider Example 11.27, where we constructed the (.95, .95) tolerance interval as (31.4, 125.6). Note that this interval would contain at least 95% of the data (not necessarily center data) with 95% confidence. Also, for this example, the sample size is 15, the sample mean is 78.5 and the standard deviation is 15.9. To compute the (.95, .95) equal-tailed tolerance factor, enter 15 for [Sample Size n], 0.95 for [Proportion p], and 0.95 for [Coverage Prob g]; click [Factors for equal-tailed TI]] to get 3.216. The (.95, .95) equal-tailed tolerance interval is $78.5 \pm 3.216 \times 15.9 = (27.37, 129.6)$. We also observe that this equal-tailed tolerance interval is wider than the tolerance interval (31.4, 125.6).

11.6.4 Simultaneous Hypothesis Testing for Quantiles

Let $X_1, \ldots, X_n$ be a sample from a normal population with mean μ and variance σ^2. Let $\bar{X}$ denote the sample mean and S denote the sample standard deviation. Owen (1964) pointed out an acceptance sampling plan, where a lot of items will be accepted if the sample provides evidence in favor of the alternative hypothesis given below:

$$H_0 : H_a^c \quad \text{vs.} \quad H_a : L < \mu - z_{\frac{1+\beta}{2}}\sigma \ \text{and} \ \mu + z_{\frac{1+\beta}{2}}\sigma < U,$$

where L and U are specified numbers, and β is a number in $(0, 1)$, usually close to 1. Note that the lot is not acceptable if either

$$\mu - z_{\frac{1+\beta}{2}}\sigma \leq L, \ U \leq \mu - z_{\frac{1+\beta}{2}}\sigma \ \text{ or } \ \mu - z_{\frac{1+\beta}{2}}\sigma \leq L \ \text{ and } \ U \leq \mu - z_{\frac{1+\beta}{2}}\sigma.$$

The null hypothesis will be rejected at level α, if

$$L < \bar{X} - kS \ \text{ and } \ \bar{X} + kS < U,$$

where k is to be determined such that

$$P(L < \bar{X} - kS \text{ and } \bar{X} + kS < U | H_0) = \alpha.$$

Notice that the factor k is determined in such a way that the probability of accepting an unacceptable lot (rejecting H_0 when it is true) is no more than α.

The dialog box [StatCalc→Continuous → Normal → Tolerance Intervals and Prediction Intervals] uses an exact method due to Owen (1964) for computing the factor k satisfying the above probability requirement.

Example 11.29. *(Test for quantiles)* Let us use the summary statistics in Example 11.27 for illustrating above quantile test. Note that $n = 15$, $\bar{X} = 78.5$ and $S = 15.9$. We would like to test if the lower 2.5th percentile is greater than 30 and the upper 2.5th percentile is less than 128 at the level of 0.05. That is, our

$$H_0 : H_a^c \ \text{ vs. } \ H_a : 30 < \mu - z_{.975}\sigma \text{ and } \mu + z_{.975}\sigma < 128.$$

Note that $(1 + \beta)/2 = 0.975$ implies that $\beta = 0.95$. To find the factor k, enter 15 for the sample size, 0.95 for the proportion p, 0.05 for the level, and click [2-sided] to get $k = 2.61584$. Thus, $\bar{X} - kS = 36.9082$ and $\bar{X} + kS = 120.092$. Thus, we have enough evidence to conclude that the lower 2.5th percentile of the normal distribution is greater than 30 and the upper 2.5th percentile of the normal distribution is less than 128.

11.7   Inference Based on Censored Samples

Consider a sample of size n with m censored observations from a $N(\mu, \sigma^2)$ distribution. The measurements on these m censored observations are not recorded, because the values (of the variable of interest) of these m units could be less than the trace level (detection limit of a device or a laboratory method). This type of censoring is referred to as the type I left-censoring. If an experiment is terminated at a predetermined mission time, then the censoring is referred to as the time-censoring or type I right-censoring. Note that the value of m is unknown prior to the experiment and is a random variable. In some situations, an experiment may be carried out until a fixed number of items fail, and this type of censoring is referred to as the type II right-censoring (failure-censoring). In type II censoring, the values of the items that were not failed are only known to be larger than the value of the last item failed.

In the sequel, we shall describe some inferential methods for the mean, variance, and percentiles of a normal distribution based on a type II left-censored sample.

Inference based on a type II right-censored sample can be obtained by converting the right-censored sample to left-censored sample. This conversion can be done by multiplying the right-censored sample by -1.

Consider a left-censored sample of size n with m censored values. Denote the uncensored sample by

$$X_1, ..., X_{n-m},$$

X_1 being the smallest, X_2 the second smallest, and so on. Define

$$\bar{X}_d = \frac{1}{n-m} \sum_{i=1}^{n-m} X_i \text{ and } S_d^2 = \frac{1}{n-m} \sum_{i=1}^{n} (X_i - \bar{X}_d)^2. \tag{11.26}$$

The log-likelihood function, after omitting a constant term, can be written as

$$l(\mu, \sigma) = m \ln \Phi(z^*) - (n - m) \ln \sigma - \frac{(n-m)(s_d^2 + (\bar{x}_d - \mu)^2)}{2\sigma^2}, \tag{11.27}$$

where $z^* = \frac{X_1 - \mu}{\sigma}$ if the sample is type II left-censored and is $(DL - \mu)/\sigma$, if the sample includes a single detection limit; that is, the sample is type I left-censored. Now let $\hat{\mu}$ and $\hat{\sigma}$ denote the maximum likelihood estimates of μ and σ, respectively, obtained by maximizing the above log-likelihood function. These can be numerically obtained as follows.

Computation of the MLEs

The method by Cohen (1961) is commonly used to find the maximum likelihood estimates (MLEs) based on a censored normal sample. An alternative approach by Krishnamoorthy and Xie (2011), which appears to be slightly faster than Cohen's method, determines the MLEs as the roots of the equations

$$\hat{\sigma}^2(\mu) = S_d^2 + (\bar{X}_d - \mu)^2 - (X_1 - \mu)(\bar{X}_d - \mu) \tag{11.28}$$

and

$$\frac{(n-m)(\bar{X}_d - \mu)}{\hat{\sigma}(\mu)} - m \frac{\phi(\hat{z}^*)}{\Phi(\hat{z}^*)} = 0, \tag{11.29}$$

where $\bar{X}_d$ and S_d^2 are as defined in (11.26), $\hat{z}^* = (X_1 - \mu)/\hat{\sigma}(\mu)$. Note that, for a given sample, the above equation is a function of μ only. The value of μ that satisfies (11.29) is the MLE $\hat{\mu}$ of μ, and the corresponding $\hat{\sigma}(\hat{\mu})$ is the MLE of σ. The root of (11.29) can be found using a root bracketing (bisection) method with the bracketing interval, for example, $(\bar{X}_d - 3S_d, \bar{X}_d)$.

Remark 11.3. The MLEs based on a sample with m nondetects below a single detection limit DL can be obtained by solving equations (11.28) and (11.29) with X_1 replaced by DL.

The R function 11.1 calculates the MLEs for a given vector x of detected observations, the number of nondetects m, and the value of the detection limit dl.

Pivotal Quantities

Consider a type II singly left-censored sample of size n with m censored observations. As the family of normal distributions is a location-scale family, and the MLEs are equivariant estimators, Result 2.7.1 implies that

$$\frac{\hat{\mu} - \mu}{\sigma} \sim \hat{\mu}^* \text{ and } \frac{\hat{\sigma}}{\sigma} \sim \hat{\sigma}^*, \tag{11.30}$$

where the notation $\sim$ means "distributed as," and $\widehat{\mu}^*$ and $\widehat{\sigma}^*$ are the MLEs based on a type II censored sample of size n (with m censored values) from a $N(0,1)$ distribution. That is, the MLEs $\widehat{\mu}^*$ and $\widehat{\sigma}^*$ are the values of μ and σ, respectively, that maximize the log-likelihood function based on a random sample of size n with m censored values from a $N(0,1)$ distribution. As a result, the distributions of the above quantities can be evaluated empirically by generating censored samples of size n from a $N(0,1)$ distribution. The empirical distributions can be used to find confidence intervals for various parameters, such as the mean, variance, and percentiles as shown in the sequel.

11.7.1   Confidence Intervals for the Mean and Variance

It follows from (11.30) that $\frac{\widehat{\mu}-\mu}{\widehat{\sigma}} \sim \frac{\widehat{\mu}^*}{\widehat{\sigma}^*}$. Let c_l and c_u be determined so that

$$P\left(c_l \le \frac{\widehat{\mu}^*}{\widehat{\sigma}^*} \le c_u\right) = 1 - \alpha, \qquad (11.31)$$

where $1 - \alpha$ is the confidence level of the desired confidence interval for μ. Then

$$(\widehat{\mu} - c_u \widehat{\sigma}, \widehat{\mu} - c_l \widehat{\sigma}) \qquad (11.32)$$

is an approximate $1 - \alpha$ confidence interval for μ. A confidence interval for variance σ^2 can be obtained similarly. Let v_l and v_u be determined so that

$$P(v_l \le \widehat{\sigma}^{*2} \le v_u) = 1 - \alpha, \qquad (11.33)$$

where $\widehat{\sigma}^*$ is as defined in (11.30). Then it follows from (11.30) that

$$\left(\frac{\widehat{\sigma}^2}{v_u}, \frac{\widehat{\sigma}^2}{v_l}\right),$$

is an approximate $1 - \alpha$ confidence interval for σ^2.

Algorithm 11.2. Estimation of (c_l, c_u) satisfying (11.31), and (v_l, v_u) satisfying (11.33)

1. Generate a sample of size n from $N(0,1)$ distribution.
2. Sort the random numbers generated in the previous step, and discard the smallest m numbers.
3. Based on the above censored sample, compute the MLEs $\widehat{\mu}^*$ and $\widehat{\sigma}^*$ using the log-likelihood function (11.27).
4. Set $Q_\mu^* = \widehat{\mu}^*/\widehat{\sigma}^*$ and $Q_{\sigma^2}^* = \widehat{\sigma}^{*2}$.
5. Repeat steps 1–4 for a large number of times, say, 10000.

Let $Q_{\mu;1-\alpha}^*$ denote the $100(1 - \alpha)$ percentile of 10,000 Q_μ^*s generated above. Then, $c_l = Q_{\mu;\alpha/2}^*$ and $c_u = Q_{\mu;1-\alpha/2}^*$. Similarly, $v_l = Q_{\sigma^2;\alpha/2}^*$ and $v_u = Q_{\sigma^2;1-\alpha/2}^*$.

The dialog box [StatCalc→Continuous → Normal → Censored Samples] calculates the values (c_l, c_u) and (v_l, v_u) using the above algorithm with 100,000 simulation runs. Furthermore, for a given value of X_1 (type II censored), detection limit

DL, or mission time (when type I censored), the mean and standard deviation of the uncensored data, sample size, and the number of censored data, the aforementioned dialog box computes confidence intervals for the mean and for the variance.

11.7.2 Tolerance Intervals

One-sided tolerance limits are one-sided confidence limits for appropriate percentile of the population interest. For a $N(\mu, \sigma^2)$ distribution, the 100β percentile is

$$\xi_\beta = \mu + z_\beta \sigma, \tag{11.34}$$

where z_β is the β quantile of the standard normal distribution. Using the distributional results in (11.30), it can be readily verified that

$$Q_\beta = \frac{\xi_\beta - \widehat{\mu}}{\widehat{\sigma}} = \frac{\mu - \widehat{\mu}}{\widehat{\sigma}} + z_\beta \frac{\sigma}{\widehat{\sigma}} \sim \frac{z_\beta - \widehat{\mu}^*}{\widehat{\sigma}^*}, \tag{11.35}$$

where $\widehat{\mu}^*$ and $\widehat{\sigma}^*$ are the MLEs based on a type II censored sample of size n with m censored values from a $N(0, 1)$ distribution.

For $0 < \alpha < .5$ and $0 < \beta < .5$, let $Q_{\beta;1-\alpha}$ denote the $100(1 - \alpha)$ percentile of Q_β in (11.35). Then

$$\widehat{\mu} + Q_{\beta;1-\alpha}\widehat{\sigma} \tag{11.36}$$

is a $1 - \alpha$ upper confidence limit for ξ_β, or equivalently, $(\beta, 1 - \alpha)$ upper tolerance limit for the sampled normal population. The $(\beta, 1 - \alpha)$ lower tolerance limit is constructed similarly as

$$\widehat{\mu} + Q_{1-\beta;\alpha}\widehat{\sigma}. \tag{11.37}$$

To find a $(\beta, 1 - \alpha)$ tolerance interval of the form $\widehat{\mu} \pm k\widehat{\sigma}$, the factor k is to be determined so that

$$P_{\widehat{\mu}^*,\widehat{\sigma}^*} \left\{ P_Z \left(\widehat{\mu}^* - k\widehat{\sigma}^* \leq Z \leq \widehat{\mu}^* + k\widehat{\sigma}^* \middle| \widehat{\mu}^*, \widehat{\sigma}^* \right) \geq \beta \right\} = 1 - \alpha, \tag{11.38}$$

where $Z \sim N(0, 1)$ and $(\widehat{\mu}^*, \widehat{\sigma}^*)$ is as defined in (11.35). This condition is further simplified (see Krishnamoorthy and Xie, 2011), and the k should be determined so that

$$P_{\widehat{\mu}^*,\widehat{\sigma}^*} \left(v/\widehat{\sigma}^* \leq k \right) = 1 - \alpha, \tag{11.39}$$

R function 11.1. Computation of the MLEs based on a censored sample[a]

```
mles.norm.cens = function(x, m, dl){
nm = length(x)
n = m + nm
l = 1
xbd = mean(x)
sqd = var(x)
xl = xbd-3*sqrt(sqd)
xu = xbd
eqnu = function(uh, xbd, sqd, n, m, dl){
sigsqh = sqd +(xbd-dl)*(xbd-uh)
sh = sqrt(sigsqh); zs = (dl-uh)/sh
ans = (n-m)*(xbd-uh)/sh-m*dnorm(zs)/pnorm(zs)
return(ans)}
repeat{
fxl = eqnu(xl, xbd, sqd, n, m, dl)
fxu = eqnu(xu, xbd, sqd, n, m, dl)
if(fxl*fxu < 0.0){break}
xl = xl - 0.1}
c0 = (xu*fxl-xl*fxu)/(fxl-fxu)
repeat{
c1 = (xu*fxl-xl*fxu)/(fxl-fxu)
if(eqnu(c1, xbd, sqd, n, m, dl)*fxu < 0.0){
xl = c1
fxl = eqnu(xl, xbd, sqd, n, m, dl)
if(eqnu(c0, xbd, sqd, n, m, dl)*eqnu(c1, xbd, sqd, n, m, dl) > 0.0)+
{fxu = fxu/2.0}}
else{xu = c1; fxu = eqnu(xu, xbd, sqd, n, m, dl)
if(eqnu(c0, xbd, sqd, n, m, dl)*eqnu(c1, xbd, sqd, n, m, dl) > 0.0)+
{fxl = fxl/2.0}}
c0 = c1; ans = abs(eqnu(c1, xbd, sqd, n, m, dl))
if(l >= 30 | ans <= 1.0e-7){break}
l = l + 1}
mlemu = c1; mlesqh = sqd + (xbd-mlemu)**2-(dl-mlemu)*(xbd-mlemu)
return(c(mlemu,mlesqh))}
```

[a]Electronic version of this R function can be found in HBSDA.r located in *StatCalc* directory.

where v satisfies

$$\Phi(\widehat{\mu}^* + v) - \Phi(\widehat{\mu}^* - v) = p, \tag{11.40}$$

and Φ denotes the standard normal distribution function.

Krishnamoorthy and Xie (2011) have provided an algorithm to compute the value of k satisfying (11.39). The dialog box [StatCalc→Continuous → Normal → Censored Samples] calculates the values of $Q_{\beta;1-\alpha}$ for computing one-sided tolerance limits and the factor k, satisfying (11.39), for computing tolerance intervals.

11.7.3   Inference Based on Type I Censored Samples

All inferential methods described in earlier sections are exact, except for simulation errors, when the samples are type II censored. If the samples are type I censored, then the aforementioned factors for one-sided tolerance limits or tolerance intervals can be used to find approximate tolerance intervals, and confidence intervals for the mean and variance (see Schmee, Gladstein and Nelson, 1985, and Krishnamoorthy, Mallick and Mathew, 2011). For more accurate methods, see Krishnamoorthy and Xu (2011).

In most applications, the samples are type I right-censored or left-censored. For instance, in analysis of exposure or pollution data, the samples often include detection limits. If a sample includes a single detection limit (DL), the contaminant concentrations below DL are not detected, as a result, the resulting sample is type I left-censored. On the other hand, in survival analysis the samples are often type I right-censored, because the measurements on test units are not observed after the mission time. The method for finding the MLEs for left-censored samples can be used to find the MLEs for right censored samples; multiplying the right-censored sample by -1, the sample can be converted to left-censored. This conversion results into $-\widehat{\mu}$, and by changing the sign, we obtain the MLE of μ based on the right-censored sample. Note that the conversion does not affect the MLE of σ.

The dialog box [StatCalc→Continuous → Normal → Censored Samples] calculates confidence intervals for the mean and variance, one-sided tolerance limits, and tolerance intervals for type I right-censored samples. This is done by entering 1 for [left-censored] and 2 for [right-censored] in the dialog box cited above.

Example 11.30. A batch of 25 AA size batteries was put on test to estimate the average life hours. After 5 hours, the test was terminated, and by that time 15 batteries were dead with lifetimes (in minutes)

> 281 282 286 287 289 291 292 292 292 293 293 294 294 296 297

Normal Q–Q plot for uncensored data indicates that the normality assumption is tenable. The mean and standard deviation of the uncensored data are

$$\bar{X} = 290.6 \ \text{ and } \ S = \frac{1}{14}\sqrt{\sum_{i=1}^{15}(X_i - \bar{X})^2} = 4.733.$$

Note that the sample is type I right-censored with $n = 25$ and the number of censored values $m = 10$.

The R function 11.1 calculates the MLEs based on a left-censored sample. To use this function to find the MLEs based on a right-censored sample, we can call the function as follows.

```
x = c(281,282,286,287,289,291,292,292,292,293,293,294,294,296,297)
dl = 300
m = 10
mles.norm.cens(-x, 10, -300)
[1] -296.8114    80.7872
```

Note that $-x$ with $-$dl is a left-censored sample. Thus, the MLEs are $\widehat{\mu} = 296.81$ and $\widehat{\sigma}^2 = 80.79$.

To find a 95% confidence interval for the mean life hour, select the dialog box [StatCalc→Continuous → Normal → Censored Samples], enter 2 to indicate right-censored, 25 for [Sample Size], 10 for [No. censored], 300 for [X1 or DL], 290.6 for [Mean(uncens)], 4.733 for [SD(uncens)], .95 for confidence level, and click on [CIs] to get (290.7, 300.9). That is, on average the battery will last between 291 and 301 minutes. Note also that the MLE for the mean is 296.775 and the MLE of σ is 8.8855. Further, a 95% confidence interval for σ^2 is (39.2, 210.7).

To find the (.90, .95) lower tolerance limits for the lifetime distribution of the batteries, enter .90 for [Content Level p] and .95 for [Coverage Level] while keeping other entries (in the preceding paragraph); click on [Low Tol Lim] to get 275.9. This means that at least 90% of such batteries will last about 276 minutes with confidence 95%. To find (.90, .95) two-sided tolerance intervals, click on [Tol Interval] to get (272.947, 320.603). That is, at least 90% of batteries will last 273 to 321 minutes with confidence 95%.

Example 11.31. To illustrate the methods for type I left-censored data, let us create a censored sample from the exposure data in Example 11.1. The complete sorted data is

54 59 63 66 67 69 75 77 79 82 89 93 98 102 104

The data include 15 uncensored values with mean $\bar{X} = 78.4667$ and the standard deviation $S = 15.87874$. To compute the (.95, .95) one-sided upper tolerance limit, we find the required factor using the dialog box [StatCalc→Continuous → Normal → Tolerance Intervals and Prediction Intervals] as 2.566. Thus, the (.95, .95) one-sided upper tolerance limit for the exposure distribution is

$$\bar{X} + 2.566 \times S = 78.4667 + 2.566 \times 15.8784 = 119.21.$$

The (.95, .95) tolerance interval was similarly computed as (27.40, 129.53).

Let us now assume a detection limit of 68, and pretend that the data below 68 are nondetects. The resulting sample is type I left-censored with the censoring value 68, number of nondetects is 5. The detected values are $69, 75, 77, 79, 82, 89, 93, 98, 102, 104$ with the mean $\bar{X} = 86.8$ and the standard deviation $S = 12.1637$. The MLEs of μ and σ^2 can be computed using R function 11.1 as follows.

```
x = c(69,  75,  77,  79,  82,  89,  93,  98, 102, 104)
m = 5
dl = 68
mles.norm.cens(x,m,dl)
[1]  76.52146 341.19208
```

To compute a (.95, .95) upper tolerance limit based on this censored sample, select the dialog box [StatCalc→Continuous → Normal → Censored Samples], enter 1 in the first edit box to indicate that the sample is left-censored, 15 for [Sample Size], 5 for [No. Censored], 68 for [X1 or DL], 86.8 for [Mean (uncens)], 12.1637 for [SD(uncens)], .95 for [Cont Level], .95 for [Coverage level], and click on [Upp Tol Lim] to get 129.4. Note that the upper tolerance limit based on all 15 measurements is 119.21. The increase in the upper tolerance limit is due to the loss of 5 data points which we assumed to be nondetects. To find the (.95, .95) two-sided tolerance interval, click on [Tol Interval] to get (9.79,143.53) which is wider than the one based on all 15 measurements.

For this example, the 95% confidence interval for the mean of the exposure distribution is (60.37, 87.93) based on the censored sample, and is (76.20, 80.74) based on all 15 measurements. As the calculations of the factors and critical values are based on simulation, a reader may get slightly different results from those reported here.

11.8 Properties and Results

1. Let $X_1, \ldots, X_n$ be independent normal random variables with $X_i \sim N(\mu_i, \sigma_i^2)$, $i = 1, 2, 3, \ldots, n$. Then

$$\sum_{i=1}^{n} a_i X_i \sim N\left(\sum_{i=1}^{n} a_i \mu_i, \sum_{i=1}^{n} a_i^2 \sigma_i^2\right),$$

where $a_1, \ldots, a_n$ are constants.

2. Let U_1 and U_2 be independent uniform(0,1) random variables. Then

$$X_1 = \cos(2\pi U_1)\sqrt{-2\ln(U_2)},$$
$$X_2 = \sin(2\pi U_1)\sqrt{-2\ln(U_2)}$$

are independent standard normal random variables [Box–Muller transformation].

3. Let Z be a standard normal random variable. Then Z^2 is distributed as a chi-square random variable with df = 1.

4. Let X and Y be independent normal random variables with common variance but possibly different means. Define $U = X + Y$ and $V = X - Y$. Then U and V are independent normal random variables.

5. The sample mean
$$\bar{X} \text{ and } \{(X_1 - \bar{X}), \ldots, (X_n - \bar{X})\}$$
are statistically independent.

6. Let $X_1, \ldots, X_n$ be independent $N(\mu, \sigma^2)$ random variables. Then

$$V^2 = \sum_{i=1}^{n} (X_i - \bar{X})^2 \text{ and } \left\{\frac{(X_1 - \bar{X})}{V}, \ldots, \frac{(X_n - \bar{X})}{V}\right\}$$

are statistically independent.

7. Stein's (1981) Lemma: If X follows a normal distribution with mean μ and standard deviation σ, then

$$E(X - \mu)h(X) = \sigma E\left[\frac{\partial h(X)}{\partial X}\right],$$

provided the indicated expectations exist.

8. Let $X_1, \ldots, X_n$ be independent identically distributed normal random variables. Then

$$\frac{\sum_{i=1}^{n}(X_i - \bar{X})^2}{\sigma^2} \sim \chi^2_{n-1} \quad \text{and} \quad \frac{\sum_{i=1}^{n}(X_i - \mu)^2}{\sigma^2} \sim \chi^2_n.$$

11.9   Relation to Other Distributions

1. Cauchy: If X and Y are independent standard normal random variables, then $U = X/Y$ follows the Cauchy distribution (Chapter 26) with probability density function

$$f(u) = \frac{1}{\pi(1 + u^2)}, \quad -\infty < u < \infty.$$

2. F Distribution: If X and Y are independent standard normal random variables, then X^2 and Y^2 are independent and distributed as a χ^2 random variable with df $= 1$. Also $F = (X/Y)^2$ follows an $F_{1,1}$ distribution.

3. Gamma: Let Z be a standard normal random variable. Then

$$P(0 < Z \leq z) = P(Y < z^2)/2,$$

and

$$P(Y \leq y) = 2P(Z \leq \sqrt{y}) - 1, \quad y > 0,$$

where Y is a gamma random variable with shape parameter 0.5 and scale parameter 2.

4. Lognormal: A random variable Y is said to have a lognormal distribution with parameters μ and σ if $\ln(Y)$ follows a normal distribution. Therefore,

$$P(Y \leq x) = P(\ln(Y) \leq \ln(x)) = P(X \leq \ln(x)),$$

where X is the normal random variable with mean μ and standard deviation σ.

For more results and properties, see Patel and Read (1981).

11.10   Random Number Generation

Algorithm 11.3. Normal variate generator

Generate U_1 and U_2 from uniform(0,1) distribution. Set

$$X_1 = \cos(2\pi U_1)\sqrt{-2\ln(U_2)}$$
$$X_2 = \sin(2\pi U_1)\sqrt{-2\ln(U_2)}.$$

Then X_1 and X_2 are independent N(0, 1) random variables. There are several other methods available for generating normal random numbers (see Kennedy and Gentle 1980, Section 6.5). The above Box–Muller transformation is simple to implement and is satisfactory if it is used with a good uniform random number generator.

The following algorithm due to Kinderman and Ramage (1976; correction Vol. 85, p. 272) is faster than the Box–Muller transformations. For better accuracy, double precision may be required.

Algorithm 11.4. Normal variate generator[1]

In the following, u, v, and w are independent uniform(0, 1) random numbers.

The output x is a N(0, 1) random number.

Set g = 2.21603 58671 66471

$$f(t) = \frac{1}{\sqrt{2\pi}} \exp(-t^2/2) - 0.180025191068563(g - |t|), \quad |t| < g$$

```
        Generate u
        If u < 0.88407 04022 98758, generate v
        return x = g*(1.3113 16354 44180*u + v + 1)

        If u < 0.97331 09541 73898 go to 4

3       Generate v and w
        Set t = g**2/2 - ln(w)
        If v**2*t > g**2/2, go to 3

        If u < 0.98665 54770 86949, return x = sqrt(2*t)
        else return x = - sqrt(2*t)

4       If u < 0.95872 08247 90463 goto 6

5       Generate v and w
        Set z = v - w
            t = g - 0.63083 48019 21960*min(v, w)
        If max(v, w) <= 0.75559 15316 67601 goto 9
        If 0.03424 05037 50111*abs(z) <= f(t), goto 9
        goto 5

6       If u < 0.91131 27802 88703 goto 8

7       Generate v and w
        Set z = v - w
            t = 0.47972 74042 22441 + 1.10547 36610 22070*min(v, w)
        If max(v, w) <= 0.87283 49766 71790, goto 9
        If 0.04926 44963 73128*abs(z) <= f(t), goto 9
        goto 7

8       Generate v and w
```

---

[1]Reproduced with permission from the American Statistical Association.

```
     Set z = v - w
          t = 0.47972 74042 22441 - 0.59950 71380 15940*min(v, w)
     If max(v, w) <= 0.80557 79244 23817 goto 9
     If t >= 0 and 0.05337 75495 06886*abs(z) <= f(t), goto 9
     goto 8

9    If z < 0, return x = t
     else return x = -t
```

11.11   Computation of the Distribution Function

Method 1

For $z > 0$, the following simple approximation given in Abramowitz and Stegun (1964) can be used.

$$\Phi(x) = 1 - \phi(x)(b_1 t + b_2 t^2 + b_3 t^3 + b_4 t^4 + b_5 t^5) + \epsilon(x), \quad t = (1 + b_0 x)^{-1},$$

where $b_0 = 0.2316419$, $b_1 = 0.319381530$, $b_2 = -0.356563782$, $b_3 = 1.781477937$, $b_4 = -1.821255978$, $b_5 = 1.330274429$, and $|\epsilon(x)| \leq 7.5 \times 10^{-8}$.

Method 2

For $0 < z < 7$, the following polynomial approximation can be used to compute the cumulative distribution function.

$$\Phi(z) = e^{\frac{-z^2}{2}} \frac{P_7 z^7 + P_6 z^6 + P_5 z^5 + P_4 z^4 + P_3 z^3 + P_2 z^2 + P_1 z + P_0}{Q_8 z^8 + Q_7 z^7 + Q_6 z^6 + Q_5 z^5 + Q_4 z^4 + Q_3 z^3 + Q_2 z^2 + Q_1 z + Q_0},$$

where

$P_0 = 913.167442114755700$, $P_1 = 1024.60809538333800$, $P_2 = 580.109897562908800$,
$P_3 = 202.102090717023000$, $P_4 = 46.0649519338751400$, $P_5 = 6.81311678753268400$,
$P_6 = 6.047379926867041E - 1$, $P_7 = 2.493381293151434E - 2$,
and
$Q_0 = 1826.33488422951125$, $Q_1 = 3506.420597749092$, $Q_2 = 3044.77121163622200$,
$Q_3 = 1566.104625828454$, $Q_4 = 523.596091947383490$, $Q_5 = 116.9795245776655$,
$Q_6 = 17.1406995062577800$, $Q_7 = 1.515843318555982$,
$Q_8 = 6.25E - 2$.

R function 11.2.   Calculation of the standard normal cdf[a]

```
normcdf <- function(x)
{
p <- c(913.167442114755700, 1024.60809538333800,
         580.109897562908800, 202.102090717023000,
         46.0649519338751400, 6.81311678753268400,
         6.047379926867041E-1,2.493381293151434E-2)
q <- c(1826.33488422951125, 3506.420597749092,
         3044.77121163622200, 1566.104625828454,
         523.596091947383490, 116.9795245776655,
         17.1406995062577800, 1.515843318555982,
         6.25E-2)
sqr2pi <- 2.506628274631001;
if(x > 0.0){z <- x; check <- 1.0}
else{z <- -x; check <- 0.0;}
if (z > 32.0){
if (x > 0.0){prb <- 1.0}
else{prb <- 0.0}}
first <- exp(-0.5*z*z)
phi <- first/sqr2pi
if (z < 7.0){
prb <- (first*((((((p[8]*z + p[7])*z + p[6])*z + p[5])*z + p[4])*z
     + p[3])*z + p[2])*z + p[1])/(((((((q[9]*z + q[8])*z + q[7])*z
     + q[6])*z + q[5])*z + q[4])*z + q[3])*z + q[2])*z + q[1]))
}
else{prb <- (phi/(z + 1.0/(z + 2.0/(z + 3.0/(z + 4.0/
     (z + 5.0/(z + 6.0/(z + 7.0)))))))))}
if (check == 1.0){prb <- 1.0 - prb}
return(prb)
}
```

---

[a]Electronic version of this R function can be found in HBSDA.r, located in *StatCalc* directory.

For $z \geq 7$, the following continued fraction can be used to compute the probabilities.

$$\Phi(z) = 1 - \varphi(z) \left[\frac{1}{z+} \frac{1}{z+} \frac{2}{z+} \frac{3}{z+} \frac{4}{z+} \frac{5}{z+} \cdots \right],$$

where $\varphi(z)$ denotes the standard normal density function. The above method is supposed to give 14 decimal accurate probabilities. [Hart et al., 1968, p. 137].
R function 11.2, based on the above computational method, evaluates the standard normal cdf.

12

Chi-Square Distribution

12.1 Description

Let $X_1, \ldots, X_n$ be independent standard normal random variables. The distribution of $X = \sum_{i=1}^{n} X_i^2$ is called the chi-square distribution with degrees of freedom (df) n, and its probability density function is given by

$$f(x|n) = \frac{1}{2^{n/2}\Gamma(n/2)} e^{-x/2} x^{n/2-1}, \quad x > 0, \ n > 0. \qquad (12.1)$$

The chi-square random variable with df $= n$ is denoted by χ_n^2. Since the probability density function is valid for any $n > 0$, alternatively, we can define the chi-square distribution as the one with the probability density function (12.1). This latter definition holds for any $n > 0$.

The cdf is given by

$$F(x|n) = \frac{1}{2^{n/2}\Gamma(n/2)} \int_0^x e^{-t/2} t^{n/2-1} dt, \quad n > 0.$$

An infinite series expression for the cdf is given in Section 11.5.1. Alternatively, the cdf may be calculated using the relation that $P(X \leq x|n) = P\left(Y \leq \frac{x}{2}\right)$, where Y is a gamma random variable with the shape parameter $a = n/2$ and the scale parameter $b = 1$. Plots in Figure 12.1 indicate that, for large degrees of freedom n, the chi-square distribution is symmetric about its mean. Furthermore, χ_a^2 is stochastically larger than χ_b^2 for $a > b$.

12.2 Moments

Mean:	n
Variance:	$2n$
Mode:	$n - 2, \quad n > 2$
Coefficient of Variation:	$\sqrt{\frac{2}{n}}$

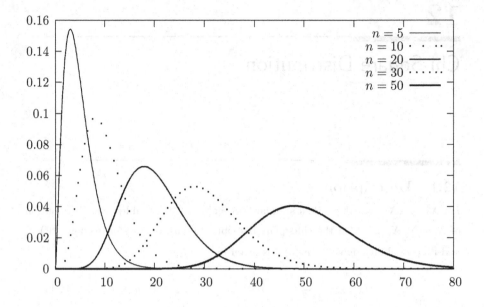

FIGURE 12.1: Chi-square probability density functions

Coefficient of Skewness:	$2\sqrt{\frac{2}{n}}$
Coefficient of Kurtosis:	$3 + \frac{12}{n}$
Mean Deviation:	$\frac{n^{n/2}e^{-n/2}}{2^{n/2-1}\Gamma(n/2)}$
Moment Generating Function:	$(1 - 2t)^{-n/2}$
Moments about the Origin:	$E[(\chi_n^2)^k] = 2^k \prod_{i=0}^{k-1}(n/2 + i), \quad k = 1, 2, \cdots$
$E\left(\ln(\chi_n^2)\right)$:	$\psi(n/2) + \ln(2)$

12.3   Probabilities, Percentiles, and Moments

The dialog box [StatCalc→Continuous → Chi-sqr] computes the probabilities and percentiles of a chi-square distribution. For the df greater than 100,000, a normal approximation to the chi-square distribution is used to compute the distribution function as well as the percentiles.

To compute probabilities: Enter the value of the df, and the value of x at which the cdf is to be computed; click P(X <= x). For example, when df = 13.0 and x = 12.3,

$$P(X \leq 12.3) = 0.496789 \text{ and } P(X > 12.3) = 0.503211.$$

To compute percentiles: Enter the values of the df and the cumulative probability, and click [x]. As an example, when df = 13.0 and the cumulative probability = 0.95, the 95th percentile is 22.362. That is, $P(X \leq 22.362) = 0.95$.

To compute the df: Enter the values of the cumulative probability and x, and click [DF]. For x = 6.0 and the cumulative probability = 0.8, the value of DF is 4.00862.

To compute moments: Enter the value of the df and click [M].

12.4   Applications

The chi-square distribution is also called the variance distribution by some authors, because the variance of a random sample from a normal distribution follows a chi-square distribution. Specifically, if $X_1, \ldots, X_n$ is a random sample from a normal distribution with mean μ and variance σ^2, then

$$\frac{\sum_{i=1}^{n} (X_i - \bar{X})^2}{\sigma^2} = \frac{(n-1)S^2}{\sigma^2} \sim \chi^2_{n-1}.$$

This distributional result is useful to make inferences about σ^2 (see Section 11.4.5).

In categorical data analysis consists of an $r \times c$ table, the usual test statistic,

$$T = \sum_{i=1}^{r} \sum_{j=1}^{c} \frac{(O_{ij} - E_{ij})^2}{E_{ij}} \sim \chi^2_{(r-1) \times (c-1)} \quad \text{approximately,}$$

where O_{ij} and E_{ij} denote, respectively, the observed and expected cell frequencies. The null hypothesis of independent attributes will be rejected at a level of significance α, if an observed value of T is greater than $(1 - \alpha)$th quantile of a chi-square distribution with df = $(r-1) \times (c-1)$.

The chi-square statistic

$$\sum_{i=1}^{k} \frac{(O_i - E_i)^2}{E_i}$$

can be used to test whether a frequency distribution fits a specific model. See Section 2.4.2 for more details.

12.5   Properties and Results

12.5.1   Properties

1. If $X_1, \ldots, X_k$ are independent chi-square random variables with degrees of freedom $n_1, \ldots, n_k$, respectively, then

$$\sum_{i=1}^{k} X_i \sim \chi_m^2 \quad \text{with } m = \sum_{i=1}^{k} n_i.$$

2. Let Z be a standard normal random variable. Then $Z^2 \sim \chi_1^2$.

3. Let $F(x|n)$ denote the cdf of χ_n^2. Then

   a. $F(x|n) = \frac{1}{\Gamma(n/2)} \sum_{i=0}^{\infty} \frac{(-1)^i (x/2)^{n/2+i}}{i! \Gamma(n/2+i)}$,

   b. $F(x|n+2) = F(x|n) - \frac{(x/2)^{n/2} e^{-x/2}}{\Gamma(n/2+1)}$,

   c. $F(x|2n) = 1 - 2 \sum_{k=1}^{n} f(x|2k)$,

   d. $F(x|2n+1) = 2\Phi(\sqrt{x}) - 1 - 2 \sum_{k=1}^{n} f(x|2k+1)$,

   where $f(x|n)$ is the probability density function of χ_n^2, and Φ denotes the cdf of the standard normal random variable [(a) Abramowitz and Stegun 1965, p. 941; (b) and (c) Peizer and Pratt 1968; (d) Puri 1973].

4. Let $\mathbf{Z}' = (Z_1, \ldots, Z_m)'$ be a random vector whose elements are independent standard normal random variables, and A be an $m \times m$ symmetric matrix with rank $= k$. Then

$$Q = Z'AZ = \sum_{i=1}^{m} \sum_{j=1}^{m} a_{ij} Z_i Z_j \sim \chi_k^2$$

   if and only if A is an idempotent matrix, that is, $A^2 = A$.

5. Cochran's Theorem: Let $\mathbf{Z}$ be as defined in (4) and A_i be an $m \times m$ symmetric matrix with rank$(A_i) = k_i$, $i = 1, 2, \ldots, r$. Let

$$Q_i = \mathbf{Z}' A_i \mathbf{Z}, \quad i = 1, 2, \ldots, r$$

   and

$$\sum_{i=1}^{m} Z_i^2 = \sum_{i=1}^{r} Q_i.$$

   Then $Q_1, \ldots, Q_r$ are independent with $Q_i \sim \chi_{k_i}^2$, $i = 1, 2, \ldots, r$, if and only if

$$\sum_{i=1}^{r} k_i = m.$$

6. For any real valued function f,

$$E[(\chi_n^2)^k f(\chi_n^2)] = \frac{2^k \Gamma(n/2+k)}{\Gamma(n/2)} E[f(\chi_{n+2k}^2)],$$

   provided the indicated expectations exist.

7. Haff's (1979) Identity: Let f and h be real valued functions, and X be a chi-square random variable with df $= n$. Then

$$E[f(X)h(X)] = 2E\left[f(X)\frac{\partial h(X)}{\partial X}\right] + 2E\left[\frac{\partial f(X)}{\partial X}h(X)\right] + (n-2)E\left[\frac{f(X)h(X)}{X}\right],$$

provided the indicated expectations exist.

12.5.2 Relation to Other Distributions

1. F and Beta: Let X and Y be independent chi-square random variables with degrees of freedoms m and n, respectively. Then

$$\frac{(X/m)}{(Y/n)} \sim F_{m,n}.$$

Furthermore, $\frac{X}{X+Y} \sim$ beta$(m/2, n/2)$ distribution.

2. Beta: If $X_1, \ldots, X_k$ are independent chi-square random variables with degrees of freedoms $n_1, \ldots, n_k$, respectively. Define

$$W_i = \frac{X_1 + \ldots + X_i}{X_1 + \ldots + X_{i+1}}, \quad i = 1, 2, \ldots, k-1.$$

The random variables $W_1, \ldots, W_{k-1}$ are independent with

$$W_i \sim \text{beta}\left(\frac{m_1 + \ldots + m_i}{2}, \frac{m_{i+1}}{2}\right), \quad i = 1, 2, \ldots, k-1.$$

3. Gamma: The gamma distribution with shape parameter a and scale parameter b specializes to the chi-square distribution with df $= n$ when $a = n/2$ and $b = 2$. That is, gamma$(n/2, 2) \sim \chi_n^2$.

4. Poisson: Let χ_n^2 be a chi-square random variable with even degrees of freedom n. Then

$$P(\chi_n^2 > x) = \sum_{k=0}^{n/2-1} \frac{e^{-x/2}(x/2)^k}{k!}$$

[see Section 16.1].

5. t distribution: See Section 14.5.1.

6. Laplace: See Section 21.8.

7. Uniform: See Section 10.4.

12.5.3 Approximations

1. Let Z denote the standard normal random variable.

   a. $P(\chi_n^2 \leq x) \simeq P(Z \leq \sqrt{2x} - \sqrt{2n-1})$, $n > 30$.

   b. $P(\chi_n^2 \leq x) \simeq P\left(Z \leq \sqrt{\frac{9n}{2}}\left[\left(\frac{x}{n}\right)^{1/3} - 1 + \frac{2}{9n}\right]\right)$.

c. Let X denote the chi-square random variable with df $= n$. Then

$$\frac{X - n + 2/3 - 0.08/n}{|X - n + 1|} \left((n - 1) \ln \left(\frac{n - 1}{X} \right) + X - n + 1 \right)^{1/2}$$

is approximately distributed as a standard normal random variable [Peizer and Pratt 1968].

2. Wilson–Hilferty (1931) approximation:

$$\left(\frac{\chi_n^2}{n} \right)^{\frac{1}{3}} \sim N \left(1 - \frac{2}{9n}, \frac{2}{9n} \right), \text{ approximately.}$$

3. Let $\chi_{n,p}^2$ denote the pth quantile of a χ_n^2 distribution, and z_p denote the pth quantile of the standard normal distribution. Then

a. $\chi_{n,p}^2 \simeq \frac{1}{2} \left(z_p + \sqrt{2n - 1} \right)^2, \quad n > 30.$

b. $\chi_{n,p}^2 \simeq n \left(1 - \frac{2}{9n} + z_p \sqrt{\frac{2}{9n}} \right)^3.$

The approximation (b) is satisfactory even for small n [Wilson and Hilferty, 1931].

12.6 Random Number Generation

For smaller degrees of freedom, the following algorithm is reasonably efficient.

Algorithm 12.1. Chi-square variate generator

Generate $U_1, \ldots, U_n$ from uniform(0, 1) distribution.
Set $X = -2(\ln U_1 + \ldots + \ln U_n)$.
Then X is a chi-square random number with df $= 2n$. To generate chi-square random numbers with odd df, add one Z^2 to X, where $Z \sim N(0, 1)$.

Since the chi-square distribution is a special case of the gamma distribution with the shape parameter $a = n/2$, and the scale parameter $b = 2$, the algorithms for generating gamma variates can be used to generate the chi-square variates (see Section 16.9).

12.7 Computation of the Distribution Function

The distribution function and the percentiles of the chi-square random variable can be evaluated as a special case of the gamma$(n/2, 2)$ distribution (see Section 15.8). Specifically,

$$P(\chi_n^2 \leq x|n) = P(Y \leq x|n/2, 2),$$

where Y is a gamma$(n/2, 2)$ random variable.

13

F Distribution

13.1 Description

Let X and Y be independent chi-square random variables with degrees of freedoms (dfs) m and n, respectively. The distribution of the ratio

$$F_{m,n} = \frac{\left(\frac{X}{m}\right)}{\left(\frac{Y}{n}\right)}$$

is called the F distribution with the numerator df $= m$ and the denominator df $= n$. The probability density function of an $F_{m,n}$ distribution is given by

$$f(x|m,n) = \frac{\Gamma\left(\frac{m+n}{2}\right)}{\Gamma\left(\frac{m}{2}\right)\Gamma\left(\frac{n}{2}\right)} \frac{\left(\frac{m}{2}\right)^{m/2} x^{m/2-1}}{\left(\frac{n}{2}\right)^{m/2}\left[1+\frac{mx}{n}\right]^{m/2+n/2}}, \quad m>0, n>0, x>0.$$

Let S_i^2 denote the variance of a random sample of size n_i from a $N(\mu_i, \sigma^2)$ distribution, $i = 1, 2$. Then the variance ratio

$$\frac{S_1^2}{S_2^2} \sim F_{n_1-1, n_2-1}$$

distribution. For this reason, the F distribution is also known as the variance ratio distribution.

We observe from the plots of pdfs in Figure 13.1 that the F distribution is always skewed to right; also, for equally large values of m and n, the F distribution is approximately symmetric about unity.

13.2 Moments

Mean:	$\frac{n}{n-2}$
Variance:	$\frac{2n^2(m+n-2)}{m(n-2)^2(n-4)}, \quad n > 4.$
Mode:	$\frac{n(m-2)}{m(n+2)}, \quad m > 2.$
Moment Generating Function:	does not exist.

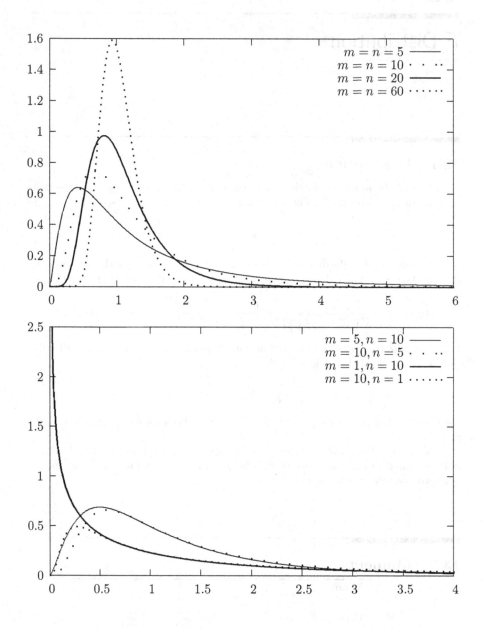

FIGURE 13.1: The probability density functions of $F_{m,n}$

Coefficient of Variation:	$\dfrac{\sqrt{2(m+n-2)}}{\sqrt{m(n-4)}}$,	$n > 4$.
Coefficient of Skewness:	$\dfrac{(2m+n-2)\sqrt{8(n-4)}}{(n-6)\sqrt{m(m+n-2)}}$,	$n > 6$.
Coefficient of Kurtosis:	$3 + \dfrac{12[(n-2)^2(n-4)+m(m+n-2)(5n-22)]}{m(n-6)(n-8)(m+n-2)}$,	$n > 8$.
Moments about the Origin:	$\dfrac{\Gamma(m/2+k)\Gamma(n/2-k)}{\Gamma(m/2)\Gamma(n/2)}(n/m)^k$, $n > 2k, \quad k = 1, 2, \ldots$	

13.3 Probabilities, Percentiles, and Moments

The dialog box [StatCalc→Continuous→F] computes probabilities, percentiles, moments, and also the degrees of freedoms when other parameters are given.

To compute probabilities: Enter the numerator df, denominator df, and the value x at which the cdf is to be evaluated; click [P(X <= x)]. For example, when the numerator df = 3.3, denominator df = 44.5, and the observed value $x = 2.3$, $P(X \le 2.3) = 0.915262$ and $P(X > 2.3) = 0.084738$.

To compute percentiles: Enter the values of the degrees of freedoms and the cumulative probability; click [x]. For example, when the numerator df = 3.3, denominator df = 44.5, and the cumulative probability = 0.95, the 95th percentile is 2.73281. That is, $P(X \le 2.73281) = 0.95$.

To compute other parameters: *StatCalc* also computes the df when other values are given. For example, when the numerator df = 3.3, cumulative probability = 0.90, $x = 2.3$, and the value of the denominator df = 22.4465. To find this value, enter other known values in appropriate edit boxes, and click on [Den DF].

To compute moments: Enter the values of the numerator df, denominator df, and click [M].

13.4 Properties and Results

13.4.1 Identities

1. For $x > 0$, $P(F_{m,n} \le x) = P(F_{n,m} \ge 1/x)$.

2. If $F_{m,n,p}$ is the pth quantile of an $F_{m,n}$ distribution, then

$$F_{n,m,1-p} = \frac{1}{F_{m,n,p}}.$$

13.4.2   Relation to Other Distributions

1. Binomial: Let X be a binomial(n, p) random variable. For a given k

$$P(X \geq k | n, p) = P\left(F_{2k, 2(n-k+1)} \leq \frac{(n-k+1)p}{k(1-p)} \right).$$

2. Beta: Let $X = F_{m,n}$. Then

$$\frac{mX}{n + mX}$$

follows a beta$(m/2, n/2)$ distribution.

3. Student's t: Consider a Student's t random variable with df $= n$, say, t_n. Then t_n^2 is distributed as $F_{1,n}$.

4. Laplace: See Section 21.8.

13.4.3   Series Expansions

For $y > 0$, let $x = \frac{n}{n+my}$.

1. For even m and any positive integer n,

$$
\begin{aligned}
P(F_{m,n} \leq y) \;=\; & 1 - x^{(m+n-2)/2}\left\{ 1 + \frac{m+n-2}{2}\left(\frac{1-x}{x} \right) \right.\\
& + \frac{(m+n-2)(m+n-4)}{2 \cdot 4}\left(\frac{1-x}{x} \right)^2 \\
& \left. + \frac{(m+n-2)\cdots(n+2)}{2 \cdot 4 \cdots (m-2)}\left(\frac{1-x}{x} \right)^{(m-2)/2} \right\}.
\end{aligned}
$$

2. For even n and any positive integer m,

$$
\begin{aligned}
P(F_{m,n} \leq y) \;=\; & (1-x)^{(m+n-2)/2}\left\{ 1 + \frac{m+n-2}{2}\left(\frac{x}{1-x} \right) \right.\\
& + \frac{(m+n-2)(m+n-4)}{2 \cdot 4}\left(\frac{x}{1-x} \right)^2 + \cdots \\
& \left. + \frac{(m+n-2)\cdots(m+2)}{2 \cdot 4 \cdots (n-2)}\left(\frac{x}{1-x} \right)^{(n-2)/2} \right\}.
\end{aligned}
$$

3. Let $\theta = \arctan\left(\sqrt{\frac{my}{n}} \right)$. For odd n,

     (a) $P(F_{1,1} \leq y) = \frac{2\theta}{\pi}$.

     (b) $P(F_{1,n} \leq y) = \frac{2}{\pi}\left\{ \theta + \sin(\theta)\left[\cos(\theta) + \frac{2}{3}\cos^3(\theta) + \ldots + \frac{2 \cdot 4 \cdots (n-3)}{3 \cdot 5 \cdots (n-2)}\cos^{n-2}(\theta) \right] \right\}$.

(c) For odd m and any positive integer n,

$$
\begin{aligned}
P(F_{m,n} \le y) \;=\; & \frac{2}{\pi}\left\{\theta + \sin(\theta)\left[\cos(\theta) + \frac{2\cos^3(\theta)}{3} + \cdots\right.\right.\\
& + \left.\left. \frac{2\cdot 4\cdots(n-3)}{3\cdot 5\cdots(n-2)}\cos^{n-2}(\theta)\right]\right\}\\
& - \frac{2[(n-1)/2]!}{\sqrt{\pi}\,\Gamma(n/2)}\sin(\theta)\cos^n(\theta)\times\left\{1 + \frac{n+1}{3}\sin^2(\theta) + \cdots\right.\\
& + \left. \frac{(n+1)(n+3)\cdots(m+n-4)}{3\cdot 5\cdots(m-2)}\sin^{m-3}(\theta)\right\}.
\end{aligned}
$$

[Abramowitz and Stegun. 1965, p. 946]

13.4.4 Approximations

1. For large m, $\frac{n}{F_{m,n}}$ is distributed as χ_n^2. For large n, $mF_{m,n}$ is distributed as χ_m^2.

2. Let $M = n/(n-2)$. For large m and n, $\dfrac{F_{n,m}-M}{M\sqrt{\frac{2(m+n-2)}{m(n-4)}}}$ is distributed as the standard normal random variable. This approximation is satisfactory only when both degrees of freedoms are greater than or equal to 100.

3. The distribution of

$$
Z = \frac{\sqrt{(2n-1)mF_{m,n}/n} - \sqrt{2m-1}}{\sqrt{1 + mF_{m,n}/n}}
$$

is approximately standard normal. This approximation is satisfactory even for small degrees of freedoms.

4. $\dfrac{F^{1/3}\left(1-\frac{2}{9n}\right)-\left(1-\frac{2}{9m}\right)}{\sqrt{\frac{2}{9m}+F^{2/3}\frac{2}{9n}}} \sim N(0,1)$ approximately.

[Abramowitz and Stegun, 1965, p. 947]

13.5 Random Number Generation

Algorithm 13.1. *F* variate generator

For a given m and n:
Generate X from gamma$(m/2, 2)$ (see Section 16.9)
Generate Y from gamma$(n/2, 2)$
Set $F = nX/(mY)$.

F is the desired random number from the F distribution with numerator df $= m$, and the denominator df $= n$.

Algorithm 13.2. *F* variate generator

Generate Y from a beta$(m/2, n/2)$ distribution (see Section 16.7), and set

$$F = \frac{nY}{m(1-Y)}.$$

F is the desired random number from the F distribution with numerator df $= m$, and the denominator df $= n$.

13.6   A Computational Method for Probabilities

For smaller degrees of freedoms, the distribution function of $F_{m,n}$ random variable can be evaluated using the series expansions given in Section 12.4. For other degrees of freedoms, the algorithm for evaluating the beta distribution can be used. Probabilities can be computed using the relation that

$$P(F_{m,n} \leq x) = P\left(Y \leq \frac{mx}{n+mx}\right),$$

where Y is the beta$(m/2, n/2)$ random variable. The pth quantile of an $F_{m,n}$ distribution can be computed using the relation that

$$F_{m,n,p} = \frac{n\, B_{m/2,n/2;p}}{m(1 - B_{m/2,n/2;p})},$$

where $B_{a,b;p}$ denotes the pth quantile of a beta(a, b) distribution.

14

Student's t Distribution

14.1 Description

Let Z and S be independent random variables such that

$$Z \sim N(0,1) \quad \text{and} \quad nS^2 \sim \chi_n^2.$$

The distribution of $t = Z/S$ is called Student's t distribution with df $= n$. The Student's t random variable with df $= n$ is commonly denoted by t_n, and its probability density function is

$$f(x|n) = \frac{\Gamma[(n+1)/2]}{\Gamma(n/2)\sqrt{n\pi}} \frac{1}{(1+x^2/n)^{(n+1)/2}}, \quad -\infty < x < \infty, \ n \geq 1.$$

Probability density plots of t_n are given in Figure 14.1 for various degrees of freedoms. We observe from the plots that for large n, the curve of t_n is approaching the standard normal curve.

Series expansions for computing the cumulative distribution function (cdf) of t_n are given in Section 14.5.3. The cdf can also be computed using the calculation for the cdf of F or of the beta random variable (see Section 14.5.2).

This t-distribution arises when estimating the mean of a normal distribution based on a small sample. Specifically, if $(\bar{X}, S)$ denote the (mean, SD) of a random sample of size n from a normal population with mean μ, then

$$\frac{\bar{X} - \mu}{S/\sqrt{n}} \sim t_{n-1}.$$

This distribution also plays an important role in many commonly used statistical analyses, including linear regression analysis.

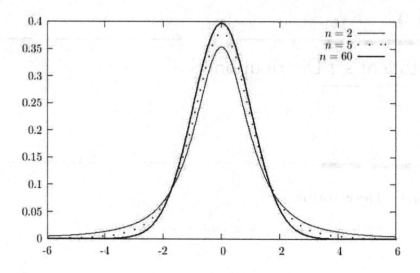

FIGURE 14.1: The probability density functions of t_n

14.2 Moments

Mean:	0 for $n > 1$; undefined for $n = 1$
Variance:	$n/(n-2), \quad n > 2$
Median:	0
Mode:	0
Mean Deviation:	$\frac{\sqrt{n}\ \Gamma((n-1)/2)}{\sqrt{\pi}\ \Gamma(n/2)}$
Coefficient of Skewness:	0
Coefficient of Kurtosis:	$\frac{3(n-2)}{(n-4)}, \quad n > 4$
Moment Generating Function:	does not exist
Moments about the Origin:	$E(t_n^k) = \begin{cases} 0 \text{ for odd } k < n, \\ \frac{1 \cdot 3 \cdot 5 \cdots (k-1)}{(n-2)(n-4)\cdots(n-k)} n^{k/2} \\ \text{for even } k < n \end{cases}$

14.3   Probabilities, Percentiles, and Moments

The dialog box [StatCalc→Continuous→ Student t] computes probabilities, percentiles, moments and also the degrees of freedom (df) for given other values.
To compute probabilities: Enter the value of the df, and the observed value x; click [x]. For example, when df $= 12.0$ and the observed value $x = 1.3$,

$$P(X \leq 1.3) = 0.890991 \text{ and } P(X > 1.3) = 0.109009.$$

To compute percentiles: Enter the value of the degrees of freedom, and the cumulative probability; click [x]. For example, when df $= 12.0$, and the cumulative probability $= 0.95$, the 95th percentile is 1.78229. That is, $P(X \leq 1.78229) = 0.95$.

To compute the DF: Enter the value of x, and the cumulative probability; click [DF]. For example, when $x = 1.3$, and the cumulative probability $= 0.9$, the value of DF $= 46.5601$.
To compute moments: Enter the value of the df and click [M].

14.4   Distribution of the Maximum of Several $|t|$ Variables

Let $X_1, \ldots, X_k$ be independent normal random variables with mean μ and common standard deviation σ. Let mS^2/σ^2 follow a chi-square distribution with df $= m$. The dialog box [StatCalc→Continuous→Student's t→ Max $|t|$] computes the distribution function of

$$X = \max_{1 \leq i \leq k} \left\{ \frac{|X_i|}{S} \right\} = \max_{1 \leq i \leq k} \{|t_i|\}, \tag{14.1}$$

where $t_1, \ldots, t_k$ are Student's t variables with df $= m$. The percentiles of X are useful for constructing simultaneous confidence intervals for the treatment effects and orthogonal estimates in the analysis of variance, and to test extreme values.

14.4.1   Applications

One-Way Analysis of Variance

Suppose we want to compare the effects of k treatments in a one-way analysis of variance setup based on the following summary statistics:

treatments	1	...	k
sample sizes	n_1	...	n_k
sample means	$\bar{X}_1$	...	$\bar{X}_k$
sample variances	S_1^2	...	S_k^2

Let $n = \sum\limits_{i=1}^{k} n_i$, and $S_p^2 = \sum\limits_{i=1}^{k} \frac{(n_i-1)S_i^2}{n-k}$ be the pooled sample variance, and

$$\bar{\bar{X}} = \frac{\sum\limits_{i=1}^{k} n_i \bar{X}_i}{n}$$

be the pooled sample mean.

For testing $H_0 : \mu_1 = \dots = \mu_k$ vs. $H_a : \mu_i \neq \mu_j$ for some $i \neq j$, the F statistic is given by

$$\frac{\sum\limits_{i=1}^{k} n_i (\bar{X}_i - \bar{\bar{X}})^2 / (k-1)}{S_p^2},$$

which follows an F distribution with numerator df $= k - 1$ and the denominator df $= n - k$. For an observed value F_0 of the F statistic, the null hypothesis will be rejected if $F_0 > F_{k-1,n-k,1-\alpha}$, where $F_{k-1,n-k,1-\alpha}$ denotes the $(1-\alpha)$th quantile of an F distribution with the numerator df $= k - 1$, and the denominator df $= n - k$. Once the null hypothesis is rejected, it may be desired to estimate all the treatment effects simultaneously.

Simultaneous Confidence Intervals for the Treatment Means

It can be shown that

$$\sqrt{n_1}(\bar{X}_1 - \mu_1)/\sigma, \dots, \sqrt{n_k}(\bar{X}_k - \mu_k)/\sigma$$

are independent standard normal random variables, and they are independent of

$$\frac{(n-k)S_p^2}{\sigma^2} \sim \chi^2_{n-k}.$$

Define

$$Y = \max_{1 \leq i \leq k} \left\{ \frac{\sqrt{n_i}|(\bar{X}_i - \mu_i)|}{S_p} \right\}.$$

Then, Y is distributed as X in (14.1). Thus, if c denotes the $(1-\alpha)$th quantile of Y, then

$$\bar{X}_1 \pm c\frac{S_p}{\sqrt{n_1}}, \dots, \bar{X}_k \pm c\frac{S_p}{\sqrt{n_k}} \tag{14.2}$$

are exact simultaneous confidence intervals for $\mu_1, \dots, \mu_k$.

14.4.2   Percentiles of Max$\{|t_1|, \dots, |t_k|\}$

The dialog box [StatCalc→Continuous→Student's t→Distribution of max$\{|t_1|, \dots, |t_k|\}$] computes the cumulative probabilities, and the percentiles of X defined in (14.1).

To compute probabilities: Enter the values of the number of groups k, df, and the observed value x of X defined in (14.1); click [P(X <= x)]. For example, when $k = 4$, df $= 45$ and $x = 2.3$, $P(X \leq 2.3) = 0.900976$ and $P(X > 2.3) = 0.099024$.

To compute percentiles: Enter the values of k, df, and the cumulative probability; click [x]. For example, when $k = 4$, df $= 45$, and the cumulative probability is 0.95, the 95th percentile is 2.5897. That is, $P(X \leq 2.5897) = 0.95$.

Example 14.1. Consider the one-way ANOVA model with the following summary statistics:

treatments	1	2	3
sample sizes	11	9	14
sample means	5	3	7
sample variances	4	3	6

The pooled variance S_p^2 is computed as 4.58. Let us compute 95% simultaneous confidence intervals for the mean treatment effects. To get the critical point using *StatCalc*, select the dialog box [StatCalc→Continuous→Student's t→Distribution of $\max\{|t_1|, ..., |t_k|\}$], enter 3 for k, $11 + 9 + 14 - 3 = 31$ for df, 0.95 for [P(X <= x)], and click [x]. The required critical point is 2.5178, and the 95% simultaneous confidence intervals for the mean treatment effects based on (14.2) are

$$5 \pm 2.5178\sqrt{\frac{4.58}{11}}, \quad 3 \pm 2.5178\sqrt{\frac{4.58}{9}}, \quad 7 \pm 2.5178\sqrt{\frac{4.58}{14}}.$$

14.5 Properties and Results

14.5.1 Properties

1. The t distribution is symmetric about 0. That is,

$$P(-x \le t < 0) = P(0 < t \le x).$$

2. Let X and Y be independent chi-square random variables with dfs 1 and n, respectively. Let I be a random variable independent of X and Y such that $P(I = 1) = P(I = -1) = 1/2$. Then

$$I\sqrt{\frac{X}{Y/n}} \sim t_n.$$

3. If X and Y are independent chi-square random variables with df $= n$, then

$$\frac{0.5\sqrt{n}(X - Y)}{\sqrt{XY}} \sim t_n.$$

14.5.2 Relation to Other Distributions

1. Let $F_{1,n}$ denote the F random variable with the numerator df $= 1$, and the denominator df $= n$. Then, for any $x > 0$,

   a. $P(t_n^2 \le x) = P(F_{1,n} \le x)$

   b. $P(F_{1,n} \le x) = 2P(t_n \le \sqrt{x}) - 1$

   c. $P(t_n \le x) = \frac{1}{2}\left[P(F_{1,n} \le x^2) + 1\right]$

2. Let $t_{n,\alpha}$ denote the αth quantile of Student's t distribution with df $= n$. Then

   a. $F_{n,n,\alpha} = 1 + \frac{2(t_{n,\alpha})^2}{n} + \frac{2t_{n,\alpha}}{\sqrt{n}}\sqrt{1 + \frac{(t_{n,\alpha})^2}{n}}$

   b. $t_{n,\alpha} = \frac{\sqrt{n}}{2}\left(\frac{F_{n,n,\alpha}-1}{\sqrt{F_{n,n,\alpha}}}\right).$ [Cacoullos, 1965]

3. Relation to beta distribution: (see Section 16.6.2)

14.5.3   Series Expansions for Cumulative Probability

1. For odd n,

$$P(t_n \leq x) = 0.5 + \frac{\arctan(c)}{\pi} + \frac{cd}{\pi}\sum_{k=0}^{(n-3)/2} a_k d^k,$$

and for even n,

$$P(t_n \leq x) = 0.5 + \frac{0.5c\sqrt{d}}{\pi}\sum_{k=0}^{(n-2)/2} b_k d^k,$$

where

$$a_0 = 1, \quad b_0 = 1,$$
$$a_k = \frac{2ka_{k-1}}{2k+1}, \quad b_k = \frac{(2k-1)b_{k-1}}{2k},$$
$$c = x/\sqrt{n}, \text{ and } d = \frac{n}{n+x^2}.$$

                                                [Owen, 1968]

2. Let $x = \arctan(t/\sqrt{n})$. Then, for $n > 1$ and odd,

$$
\begin{aligned}
P(|t_n| \leq t) = {} & \frac{2}{\pi}\left[x + \sin(x)\left(\cos(x) + \frac{2}{3}\cos^3(x) + \ldots\right.\right. \\
& + \left.\left. \frac{2 \cdot 4 \cdot \ldots \cdot (n-3)}{1 \cdot 3 \cdot \ldots \cdot (n-2)}\cos^{n-2}(x)\right)\right].
\end{aligned}
$$

For even n,

$$
\begin{aligned}
P(|t_n| \leq t) = {} & \sin(x)\left[1 + \frac{1}{2}\cos^2(x) + \frac{1 \cdot 3}{2 \cdot 4}\cos^4(x) + \ldots \right. \\
& + \left. \frac{1 \cdot 3 \cdot 5 \ldots (n-3)}{2 \cdot 4 \cdot 6 \ldots (n-2)}\cos^{n-2}(x)\right],
\end{aligned}
$$

and $P(|t_1| \leq t) = \frac{2x}{\pi}$.   [Abramowitz and Stegun 1965, p. 948]

An Approximation

$$P(t_n \leq t) \simeq P\left(Z \leq \frac{t\left(1 - \frac{1}{4n}\right)}{\sqrt{1 + \frac{t^2}{2n}}}\right),$$

where Z is the standard normal random variable.

14.6 Random Number Generation

Algorithm 14.1. t-random variate generator

Generate Z from $N(0,1)$
Generate S from gamma$(n/2, 2)$
Set $x = \dfrac{Z}{\sqrt{S/n}}$.

Then, x is a Student's t random variate with df $= n$.

14.7 Computation of the Distribution Function

For small integer dfs, the series expansions in Section 14.5.3 can be used to compute the cumulative probabilities. For other dfs, use the relation that, for $x > 0$,

$$P(t_n \leq x) = \frac{1}{2}\left[P\left(Y \leq \frac{x^2}{n+x^2}\right) + 1\right],$$

where Y is a beta$(1/2, n/2)$ random variable. If x is negative, then $P(t_n \leq x) = 1 - P(t_n \leq |x|)$.

15

Exponential Distribution

15.1 Description

A classical situation in which an exponential distribution arises is as follows: Consider a Poisson process with mean λ where we count the events occurring in a given interval of time or space. Let X denote the waiting time until the first event to occur. Then, for a given $x > 0$,

$$
\begin{aligned}
P(X > x) &= P(\text{no event in } (0, x)) \\
&= \exp(-x\lambda),
\end{aligned}
$$

and hence

$$P(X \leq x) = 1 - \exp(-x\lambda). \tag{15.1}$$

The above distribution is the exponential distribution with mean waiting time $b = 1/\lambda$. The probability density function (pdf) is given by

$$f(x|b) = \frac{1}{b} \exp\left(-\frac{x}{b}\right), \quad x > 0,\ b > 0. \tag{15.2}$$

Suppose that the waiting time is known to exceed a threshold value a, then the pdf is given by

$$f(x|a, b) = \frac{1}{b} \exp\left(-\frac{x - a}{b}\right), \quad x > a,\ b > 0. \tag{15.3}$$

The distribution with the above pdf is called the two-parameter exponential distribution, and we referred to it as exponential(a, b). The cumulative distribution function is given by

$$F(x|a, b) = 1 - \exp(-(x - a)/b), \quad x > a,\ b > 0. \tag{15.4}$$

15.2 Moments

Mean:	$b + a$	Median	$a - \ln(.5)b$
Variance:	b^2	Mode:	a

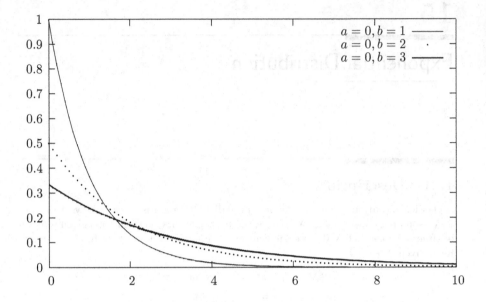

FIGURE 15.1: Exponential probability density functions

Coefficient of Variation:	$\frac{b}{b+a}$
Coefficient of Skewness:	2
Coefficient of Kurtosis:	9
Moment Generating Function:	$(1 - bt)^{-1}, \ t < \frac{1}{b}$ when $a = 0$
Moments about the Origin:	$E(X^k) = b^k \Gamma(k+1) = b^k k!, \ k = 1, 2, \ldots; \ a = 0$

15.3   Probabilities, Percentiles, and Moments

The dialog box [StatCalc→Continuous→Exponential→Probabilities...] computes the tail probabilities, percentiles, moments, and other parameters of an exponential distribution.

To compute probabilities: Enter the values of the shape parameter a, scale parameter b, and the observed value x; click on [P(X <= x)]. When $a = 1.1$, $b = 1.6$, and $x = 2$, $P(X \le 2) = 0.430217$ and $P(X > 2) = 0.569783$.

To compute percentiles: Enter the values of a, b, and the cumulative probability; click [x]. For example, when $a = 2$, $b = 3$, and the cumulative probability $= 0.05$, the 5th percentile is 2.15388. That is, $P(X \leq 2.15388) = 0.05$.

To compute other parameters: Enter the values of the cumulative probability, one of the parameters, and a positive value for x; click on the parameter that is missing. For example, when $b = 3$, $x = 7$, and $P(X \leq x) = 0.9$, the value of the location parameter $a = 0.0922454$.

To compute moments: Enter the values of a and b; click [M].

15.4   Estimation

Let $X_1, \ldots, X_n$ be a sample of observations from an exponential distribution with pdf in (15.3).

MLEs and Their Distributions

The maximum likelihood estimators (MLEs) of a and b are given by

$$\widehat{a} = X_{(1)} \quad \text{and} \quad \widehat{b} = \frac{1}{n}\sum_{i=1}^{n}(X_i - X_{(1)}) = \bar{X} - X_{(1)}, \tag{15.5}$$

where $X_{(1)}$ is the smallest of the X_is. The MLEs $\widehat{a}$ and $\widehat{b}$ are independent with

$$\frac{2n(\widehat{a} - a)}{b} \sim \chi_2^2 \quad \text{and} \quad \frac{2n\widehat{b}}{b} \sim \chi_{2n-2}^2. \tag{15.6}$$

15.5   Confidence Intervals

Location and Scale Parameter

The pivotal quantity $2n\widehat{b}/b$ in (15.6) can be used to make inference on b. In particular, a $1 - \alpha$ confidence interval for b is given by

$$\left(\frac{2n\widehat{b}}{\chi_{2n-2,1-\alpha/2}^2}, \ \frac{2n\widehat{b}}{\chi_{2n-2,\alpha/2}^2} \right).$$

It follows from (15.6) that

$$\frac{\widehat{a} - a}{\widehat{b}} \sim \frac{1}{n-1} F_{2,2n-2}.$$

A $1 - \alpha$ confidence interval for a (based on the above distributional result) is given by

$$\left(\widehat{a} - \frac{\widehat{b}}{n-1} F_{2,2n-2,1-\alpha/2}, \ \widehat{a} - \frac{\widehat{b}}{n-1} F_{2,2n-2,\alpha/2} \right),$$

where $F_{a,b;p}$ denotes the pth quantile of an F distribution with the numerator df $= a$, and the denominator df $= b$.

Mean

To find a confidence interval for the mean, we shall find a pivotal quantity for estimating $a + cb$, where c is a known constant. Using the distributional results in (15.6), we find

$$\frac{a + cb - \widehat{a}}{\widehat{b}} \sim \frac{2nc - \chi_2^2}{\chi_{2n-2}^2} = f_{n,c}, \text{ say,} \qquad (15.7)$$

where the chi-square random variables are independent. For $0 < \alpha < .5$, let $f_{n,c;\alpha}$ denote the α quantile of $f_{n,c}$. In terms of this quantile, a $1 - 2\alpha$ confidence interval for $a + cb$ is given by

$$\left(\widehat{a} + f_{n,c;\alpha}\widehat{b}, \ \widehat{a} + f_{n,c;1-\alpha}\widehat{b} \right). \qquad (15.8)$$

Noting that the mean of the exponential distribution is $a + b$, an exact $1 - \alpha$ confidence interval for the mean is (15.8) with $c = 1$. Specifically,

$$\left(\widehat{a} + f_{n,1;\alpha/2}\widehat{b}, \ \widehat{a} + f_{n,1;1-\alpha/2}\widehat{b} \right) \qquad (15.9)$$

is a $1 - \alpha$ confidence interval for the mean. The percentiles of $f_{n,1}$ can be estimated by Monte Carlo simulation, or approximated using the modified normal-based approximations (2.14) and (2.15) as follows.

Let $X = 2n - \chi_2^2$ and $Y = \chi_{2n-2}^2$. Then $\mu_x = \mu_y = 2n - 2$, $r = 1$, $X_\alpha = 2 - \chi_{1-\alpha}^2$, $X_{1-\alpha} = 2n - \chi_{2;\alpha}^2$, $Y_\alpha = \chi_{2n-2;\alpha}^2$ and $Y_{1-\alpha} = \chi_{2n-2;1-\alpha}^2$. Substituting these expressions in (2.16), we find approximate values of $f_{n,1;\alpha}$ and $f_{n,1;1-\alpha}$ for $0 < \alpha < .5$. This approximation is quite satisfactory even for samples of size as small as three.

15.6   Prediction Intervals

To find a prediction interval, let $X \sim$ exponential(a, b) independently of the MLEs $\widehat{a}$ and $\widehat{b}$. Then

$$\frac{X - \widehat{a}}{\widehat{b}} \sim \frac{2nZ - \chi_2^2}{\chi_{2n-2}^2} = R_n, \text{ say,} \qquad (15.10)$$

where Z is the exponential$(0, 1)$ random variable, and Z, χ_2^2 and χ_{2n-2}^2 are mutually independent. If $R_{n;\alpha}$ denote the α quantile of R_n, then

$$(\widehat{a} + R_{n;\alpha}\widehat{b}, \ \widehat{a} + R_{n;1-\alpha}\widehat{b}) \qquad (15.11)$$

is a $1 - 2\alpha$ prediction interval for X. The 100α percentile of R_n, denoted by $R_{n;\alpha}$, is given by [Lawless, 1977]

$$R_{n;\alpha} = \begin{cases} n\left\{ [(1-\alpha)(1+1/n)]^{-\frac{1}{n-1}} - 1 \right\}, & \frac{1}{1+n} \leq \alpha < 1, \\ 1 - [\alpha(1+n)]^{-\frac{1}{n-1}}, & 0 < \alpha < \frac{1}{n+1}. \end{cases} \qquad (15.12)$$

Example 15.1. To illustrate the method for finding confidence interval for the mean of an exponential distribution, we shall use the failure mileage data on 19 military carriers given in Grubbs (1971). The data are reproduced here in Table 15.1. The probability plot by Krishnamoorthy and Mathew (2009, Example 7.3) indicated that the data fit a two-parameter exponential distribution. The MLEs

TABLE 15.1: Failure Mileage of 19 Military Carriers

162	200	271	302	393	508	539	629	706	777
884	1008	1101	1182	1463	1603	1984	2355	2880	

Reproduced with permission from *Technometrics*. Copyright [1971] by the American Statistical Association.

based on the above data are $\widehat{a} = X_{(1)} = 162$ and $\widehat{b} = 835.21$. To compute 95% confidence intervals for the location and scale parameters and for the mean, select the dialog box [StatCalc→Continuous→Exponential→Confidence Intervals ...], enter 19 for [S Size], 162 for [MLE of a], 835.21 for [MLE of b], .95 for [Conf Level], and click on [Conf Interval] to get (703, 1582). In the same dialog box, we can find 95% confidence interval for the location parameter as $(0, 160.82)$, and 95% confidence interval for the scale parameter as $(583.02, 1487.54)$.

15.7 Tolerance Limits and Survival Probability

The p quantile of a two-parameter exponential distribution is given by $a - b\ln(1-p)$. A $1 - \alpha$ upper confidence limit for this quantile is a $(p, 1 - \alpha)$ upper tolerance limit for the exponential(a, b) distribution. A pivotal quantity for estimating the quantile is given by (15.7) with $c = -\ln(1 - p)$. That is,

$$-\frac{\chi_2^2 + 2n\ln(1 - p)}{\chi_{2n-2}^2}$$

is the pivotal quantity for the quantile $a - \ln(1 - p)b$. Let $E_{p;\alpha}$ denote the α quantile of $E_p = \frac{\chi_2^2 + 2n\ln(1-p)}{\chi_{2n-2}^2}$. Then

$$\widehat{a} - E_{p;\alpha}\widehat{b} \qquad (15.13)$$

is a $1 - \alpha$ upper confidence limit for q_p, which in turn is a $(p, 1 - \alpha)$ upper tolerance limit for the exponential(a, b) distribution. Similarly, we see that

$$\widehat{a} - E_{1-p,1-\alpha}\widehat{b} \qquad (15.14)$$

is a $1 - \alpha$ lower confidence limit for $q_{1-p} = a - b\ln(p)$, or equivalently, a $(p, 1 - \alpha)$ lower tolerance limit for the exponential(a, b) distribution. Note that the percentile $E_{p;\alpha}$ can be estimated using Monte Carlo simulation.

The percentiles of E_p can be approximated as follows. Let $c = -\ln(1 - p)$, $X = 2nc - \chi_2^2$, and $Y = \chi_{2n-2}^2$. Then $\mu_x = E(X) = 2nc - 2$, $\mu_y = 2n - 2$, $X_\alpha = 2nc - \chi_{2;1-\alpha}^2$, $X_{1-\alpha} = 2nc - \chi_{2;\alpha}^2$, $Y_\alpha = \chi_{2n-2;\alpha}^2$, and $Y_{1-\alpha} = \chi_{2n-2;1-\alpha}^2$. Substituting these expressions in (2.16), we find approximate values of $f_{n,c;\alpha}$ and

$f_{n,c;1-\alpha}$ for $0 < \alpha < .5$. Simulation studies by Krishnamoorthy and Xia (2014) indicated that this approximation is quite satisfactory even for samples of size as small as three.

For a given sample size n, MLEs, content level p, and coverage level $1-\alpha$, the dialog box [StatCalc→Continuous→Exponential→Tolerance Limits ...] calculates one-sided lower and upper tolerance limits. There are other exact methods available in the literature. However, these methods are computationally complex (e.g., Guenther et al. 1976).

Survival Probability

A confidence limit for a survival probability can be deduced from the tolerance limit in the preceding section. Let t denote the specified time at which we like to estimate the survival probability

$$S_t = P(X > t|a,b) = 1 - F(x|a,b) = \exp\left(-\frac{t-a}{b}\right).$$

The value of p for which the $(p, 1-\alpha)$ lower tolerance limit is equal to t is the $1-\alpha$ lower confidence limit for $P(X > t)$. That is, p is determined so that

$$\widehat{a} - E_{1-p,1-\alpha}\widehat{b} = t \quad \Leftrightarrow \quad \frac{\widehat{a}-t}{\widehat{b}} = E_{1-p,1-\alpha}.$$

It can be verified that the value of p that satisfies the above equation is the α quantile of

$$\exp\left\{-\frac{1}{2n}A\right\}, \quad \text{where} \quad A = \left(\frac{t-\widehat{a}_0}{\widehat{b}_0}\right)\chi^2_{2n-2} + \chi^2_2,$$

which in turn is a 95% lower confidence limit of $P(X > t)$. Finally, the $1 - 2\alpha$ confidence interval for $P(X > t)$ is expressed as

$$\left(\exp\left\{-\frac{1}{2n}A_{1-\alpha},\right\}, \exp\left\{-\frac{1}{2n}A_\alpha\right\}\right) \tag{15.15}$$

where A_q is the q quantile of A.

For a given $(\widehat{a}_0, \widehat{b}_0)$, Monte Carlo simulation can be used to estimate the percentiles of A. A convenient approximation to the percentiles of A can be obtained using the modified normal approximation (MNA) in (2.14) and (2.15). This approximation yields the αth quantile of A as

$$A_\alpha \simeq \widehat{\eta}_t(2n-2) + 2 - \left[\widehat{\eta}_t^2(2(n-2) - L^*)^2 + (2 - \chi^2_{2;\alpha})\right]^{\frac{1}{2}}, \quad 0 < \alpha < .5, \tag{15.16}$$

where

$$\widehat{\eta}_t = \frac{t-\widehat{a}}{\widehat{b}}, \quad L^* = \begin{cases} \chi^2_{2n-2;\alpha} & \text{if } \widehat{\eta}_t > 0, \\ \chi^2_{2n-2;1-\alpha} & \text{otherwise} \end{cases}.$$

Similarly, the approximate $1 - \alpha$ quantile for A is expressed as

$$A_{1-\alpha} \simeq \widehat{\eta}_t(2n-2) + 2 + \left[\widehat{\eta}_t^2(2(n-2) - U^*)^2 + (2 - \chi^2_{2;1-\alpha})\right]^{\frac{1}{2}}, \quad 0 < \alpha < .5, \tag{15.17}$$

where

$$U^* = \begin{cases} \chi^2_{2n-2;1-\alpha} & \text{if } \widehat{\eta}_t > 0, \\ \chi^2_{2n-2;\alpha} & \text{otherwise} \end{cases}.$$

This approximation is quite satisfactory even for sample of sizes as small as four.

Example 15.2. We shall compute (.90, .95) tolerance limits based on mileage data in Table 15.1. Recall that the MLEs are $\hat{a} = X_{(1)} = 162$ and $\hat{b} = 835.21$. To compute a (0.95, 0.95) lower tolerance limit, select the dialog box [StatCalc→Continuous→Exponential→Tolerance Limits...], enter 19 for [S Size], 162 for [MLE of a], 835.21 for [MLE of b], .90 for [Cont Level], .95 for [Conf Level], and click on [Tol Limits] to get 114.56 (lower) and 3236.3 (upper). This means that at least 90% of military carriers work 114.56 units of miles or more with 95% confidence. We can also say that at most 10% of military carriers work more than 3236.3 units of miles with confidence 95%. These tolerance limits were estimated using 1,000,000 simulation runs.

We once again use the failure mileage data in Table 15.1 to find a 95% lower confidence limit for $P(X > 300)$, where X represents the failure milage of a military carrier. To compute a 95% lower confidence limit using *StatCalc*, select the dialog box [StatCalc→Continuous→Exponential→ Tolerance Limits...], enter 19 for [S Size], 162 for [MLE of a], 835.21 for [MLE of b], .95 for [Conf Level], 300 for [Time t], and click on [Conf Limits] to get .720 (lower) and .882 (upper). This means that at least 72% of military carriers work 300 units of miles or more with confidence 95%. These confidence limits were estimated using 1,000,000 simulation runs.

To find the approximate 95% lower confidence limit (15.17) for $P(X > 300)$, we found $\hat{\eta}_t = 0.1652$, $U^* = 50.9985$, $\chi^2_{2;.95} = 5.9914$, and $A_{.95} = 12.6464$. Using these quantities in (15.17), we find the 95% lower confidence limit for $P(X > 300)$ is

$$\exp\left(-\frac{1}{38}A_{.95}\right) = 0.717,$$

which is the same as the one based on 1,000,000 simulation runs.

15.8   Two-Sample Case

Let $\hat{a}_i$ and $\hat{b}_i$ denote the MLEs based on a sample of size n_i from an exponential(a_i, b_i) distribution, $i = 1, 2$. It follows from (15.6) that

$$\frac{2n_1\hat{b}_1}{b_1} \sim \chi^2_{2n_1-2} \text{ and } \frac{2n_2\hat{b}_2}{b_2} \sim \chi^2_{2n_2-2}. \qquad (15.18)$$

The above pivotal quantities are also independent because the MLEs are based on independent samples.

15.8.1   Confidence Interval for Comparing Two Parameters

Let $\hat{a}_{i0}$ and $\hat{b}_{i0}$ be observed values of $\hat{a}_i$ and $\hat{b}_i$, respectively. Based on the distributional results (15.6), generalized pivotal quantities (GPQs; see Section 2.7.2) for a_i is given by

$$G_{a_i} = \hat{a}_{i0} - \frac{\chi^2_2}{\chi^2_{2n-2}}\hat{b}_{i0}, \quad i = 1, 2. \qquad (15.19)$$

The GPQ for $a_1 - a_2$ can be expressed as

$$G_{a_1} - G_{a_2} = \widehat{a}_{10} - \widehat{a}_{20} - \left(\frac{\widehat{b}_{10}}{n_1 - 1} F_{2,2n_1-2} - \frac{\widehat{b}_{20}}{n_2 - 1} F_{2,2n_2-2} \right),$$

where the F random variables are independent. The lower and upper α quantiles of $G_{a_1} - G_{a_2}$ form a $1 - 2\alpha$ confidence interval for $a_1 - a_2$. Notice that, for a given $(\widehat{a}_{10}, \widehat{b}_{10}\widehat{a}_{20}\widehat{b}_{20})$, the distribution of the above GPQ can be obtained by Monte Carlo simulation.

A closed-form approximate confidence interval for $a_1 - a_2$ can also be obtained by approximating the percentiles of T_{n_1,n_2} using the expressions in (2.14) and (2.15). To find the approximation, note that $E(F_{2,2n_i-2}) = \frac{n_i-1}{n_i-2}$, $i = 1, 2$. Using these expectations in (2.14) and (2.15), and letting $D = \frac{\widehat{b}_{10}}{n_1-2} - \frac{\widehat{b}_{20}}{n_2-2}$, we find

$$T_{n_1,n_2;\alpha} = \begin{cases} D - \sqrt{w_1^2 \left(\frac{n_1-1}{n_1-2} - F_{2,2n_1-2;\alpha} \right)^2 + w_2^2 \left(\frac{n_2-1}{n_2-2} - F_{2,2n_2-2;1-\alpha} \right)^2}, \\ \qquad\qquad\qquad\qquad\qquad\qquad\qquad\qquad\qquad 0 < \alpha < .5, \\[2ex] D + \sqrt{w_1^2 \left(\frac{n_1-1}{n_1-2} - F_{2,2n_1-2;\alpha} \right)^2 + w_2^2 \left(\frac{n_2-1}{n_2-2} - F_{2,2n_2-2;1-\alpha} \right)^2}, \\ \qquad\qquad\qquad\qquad\qquad\qquad\qquad\qquad\qquad 0.5 < \alpha < 1, \end{cases}$$

where $w_i = \widehat{b}_{i0}/(n_i - 1)$, $i = 1, 2$. The approximate $1 - 2\alpha$ confidence interval for $a_1 - a_2$ is given by

$$\left(\widehat{a}_{10} - \widehat{a}_{20} - T_{n_1 n_2;1-\alpha}, \ \widehat{a}_{10} - \widehat{a}_{20} - T_{n_1 n_2;\alpha} \right). \qquad (15.20)$$

15.8.2   Ratio of Scale Parameters

It follows from (15.6) that

$$\frac{b_2}{b_1} \frac{\widehat{b}_1}{\widehat{b}_2} \sim \frac{n_2(n_1 - 1)}{n_1(n_2 - 1)} F_{2n_1-2,2n_2-2},$$

where $F_{m,n}$ denotes the F distribution with numerator df $= m$ and the denominator df $= n$. On the basis of the above distribution, an exact $1 - \alpha$ confidence interval for the ratio b_1/b_2 is given by

$$\left(\frac{n_1(n_2 - 1)\widehat{b}_1}{n_2(n_1 - 1)\widehat{b}_2} F_{2n_2-2,2n_1-2;\alpha/2}, \ \frac{n_1(n_2 - 1)\widehat{b}_1}{n_2(n_1 - 1)\widehat{b}_2} F_{2n_2-2,2n_1-2;1-\alpha/2} \right),$$

where $F_{a,b;p}$ denotes the $100p$th percentile of an F distribution with the numerator df a and the denominator df b.

15.8.3   Confidence Interval for the Difference between Two Means

We shall describe an interval estimation method for the difference between two means, $(a_1 + b_1) - (a_2 + b_2)$, based on the generalized variable approach (Section

2.7.2). Let $\widehat{a}_{i0}$ and $\widehat{b}_{i0}$ be observed values of $\widehat{a}_i$ and $\widehat{b}_i$, respectively. Based on the distributional results (15.6), generalized pivotal quantities (GPQs) for a_i and b_i are

$$G_{a_i} = \widehat{a}_{i0} - \frac{\chi_2^2}{\chi_{2n-2}^2}\widehat{b}_{i0} \text{ and } G_{b_i} = \frac{2n_i\widehat{b}_{i0}}{\chi_{2n-2}^2}, \tag{15.21}$$

respectively.

A generalized pivotal quantity for the mean $\mu_i = a_i + b_i$ can be obtained by substitution as

$$G_{\mu_i} = G_{a_i} + G_{b_i} = \widehat{a}_{i0} + \left(\frac{2n_i - \chi_2^2}{\chi_{2n_i-2}^2}\right)\widehat{b}_{i0},$$

where the chi-square random variables are independent. The lower $\alpha/2$ quantile of G_{μ_i} and the upper $\alpha/2$ quantile of G_{μ_i} form a $1 - \alpha$ confidence interval for μ_i, which is the same as the exact confidence interval in (15.9). A GPQ for $\mu_1 - \mu_2$ is obtained as

$$
\begin{aligned}
G_{\mu_1} - G_{\mu_2} &= \widehat{a}_{10} - \widehat{a}_{20} + \widehat{b}_{10}\left(\frac{2n_1 - \chi_2^2}{\chi_{2n_1-2}^2}\right) - \widehat{b}_{20}\left(\frac{2n_2 - \chi_2^2}{\chi_{2n_2-2}^2}\right) \\
&= \widehat{a}_{10} - \widehat{a}_{20} + \widehat{b}_{10}f_{n_1,1} - \widehat{b}_{20}f_{n_2,1}.
\end{aligned}
\tag{15.22}
$$

For a given $(\widehat{a}_{10}, \widehat{b}_{10}, \widehat{a}_{20}, \widehat{b}_{20})$, the distribution of $G_{\mu_1} - G_{\mu_2}$ does not depend on any unknown parameters, and so they can be estimated by Monte Carlo simulation. The lower $\alpha/2$ quantile and the upper $\alpha/2$ quantile of $G_{\mu_1} - G_{\mu_2}$ form a $1 - \alpha$ confidence interval for $\mu_1 - \mu_2$.

Approximate Confidence Intervals for the Mean and Mean Difference

The percentiles of $G_{\mu_1} - G_{\mu_2}$ can also be approximated using the modified normal-based approximation as follows. It follows from (15.22) that to find a confidence interval for $\mu_1 - \mu_2$, it is enough to find percentiles of $Q_{n_1,n_2} = \widehat{b}_{10}f_{n_1,1} - \widehat{b}_{20}f_{n_2,1}$, which can be approximated using the approximation in (2.16) as follows. Let $u_i = E(f_{n_i,1}) = (n_i - 1)/(n_i - 2)$, $i = 1, 2$. Then

$$Q_{n_1,n_2;\alpha} \simeq \widehat{b}_{10}u_1 - \widehat{b}_{20}u_2 - \sqrt{\widehat{b}_{10}^2\left(u_1 - f_{n_1,1;\alpha}\right)^2 + \widehat{b}_{20}^2\left(u_2 - f_{n_2,1;1-\alpha}\right)^2}, \quad 0 < \alpha < .5, \tag{15.23}$$

and

$$Q_{n_1,n_2;1-\alpha} \simeq \widehat{b}_{10}u_1 - \widehat{b}_{20}u_2 + \sqrt{\widehat{b}_{10}^2\left(u_1 - f_{n_1,1;1-\alpha}\right)^2 + \widehat{b}_{20}^2\left(u_2 - f_{n_2,1;\alpha}\right)^2}, \quad .5 < \alpha < 1, \tag{15.24}$$

where $f_{n_i,\alpha}$ is as defined in (15.9). Using the above approximations, we find an approximate $1 - 2\alpha$ confidence interval for $\mu_1 - \mu_2$ as

$$\left(\widehat{a}_1 - \widehat{a}_2 + Q_{n_1,n_2;\alpha}, \widehat{a}_1 - \widehat{a}_2 + Q_{n_1,n_2;1-\alpha}\right). \tag{15.25}$$

Even though the above confidence interval is obtained by first approximating the percentiles of $f_{n_i,1}$ and then approximating the percentiles of Q_{n_1,n_2}, it is satisfactory in terms of coverage probabilities.

Approximate percentiles of $f_{n,1}$ can be obtained using the MNA for the percentiles of the ratio of independent random variables in (2.16). To find these approximate percentiles of f_n, let $X = 2n - \chi_2^2$ and $Y = \chi_{2n-2}^2$. Then $\mu_x = \mu_y = 2n - 2$,

$X_\alpha = 2 - \chi^2_{1-\alpha}$, $X_{1-\alpha} = 2n - \chi^2_{2;\alpha}$, $Y_\alpha = \chi^2_{2n-2;\alpha}$ and $Y_{1-\alpha} = \chi^2_{2n-2;1-\alpha}$. Substituting these expressions in (2.16), we find approximate values of $f_{n;\alpha}$ and $f_{n;1-\alpha}$ for $0 < \alpha < .5$. This approximation is quite satisfactory even for samples of size four [Krishnamoorthy and Xia, 2014].

15.9   Properties and Results

Properties

1. Memoryless Property: For a given $t > 0$ and $s > 0$,

$$|P(X > t + s | X > s) = P(X > t),$$

where X is the exponential random variable with pdf (15.3).

2. Let $X_1, \ldots, X_n$ be independent exponential$(0, b)$ random variables. Then

$$\sum_{i=1}^{n} X_i \sim \text{gamma}(n, b).$$

3. Let $X_1, \ldots, X_n$ be a sample from an exponential$(0, b)$ distribution. Then, the smallest order statistic $X_{(1)} = \min\{X_1, ..., X_n\}$ has the exponential$(0, b/n)$ distribution.

Relation to Other Distributions

1. Pareto: If X follows a Pareto distribution with pdf $\lambda \sigma^\lambda / x^{\lambda+1}$, $x > \sigma$, $\sigma > 0$, $\lambda > 0$, then $Y = \ln(X)$ has the exponential(a, b) distribution with $a = \ln(\sigma)$ and $b = 1/\lambda$.

2. Power Distribution: If X follows a power distribution with pdf $\lambda x^{\lambda-1}/\sigma^\lambda$, $0 < x < \lambda$, $\sigma > 0$, then $Y = \ln(1/X)$ has the exponential(a, b) distribution with $a = \ln(1/\sigma)$ and $b = 1/\lambda$.

3. Weibull: See Section 25.9.

4. Extreme Value Distribution: See 26.9.

5. Geometric: Let X be a geometric random variable with success probability p. Then

$$P(X \le k|p) = P(Y \le k+1),$$

where Y is an exponential random variable with mean $b^* = (-\ln(1 - p))^{-1}$ [Prochaska, 1973].

15.10   Random Number Generation

```
Input:  a = location parameter
        b = scale parameter
Output: x is a random number from the exponential(a, b) distribution

Generate u from uniform(0, 1)
Set x = a - b*ln(u)
```

16

Gamma Distribution

16.1 Description

The gamma distribution can be viewed as a generalization of the exponential distribution with mean $1/\lambda$, $\lambda > 0$. An exponential random variable with mean $1/\lambda$ represents the waiting time until the first event to occur, where events are generated by a Poisson process with mean λ, while the gamma random variable X represents the waiting time until the ath event to occur. Notice that $X = \sum_i^a Y_i$, where $Y_1, \ldots, Y_n$ are independent exponential random variables with mean $1/\lambda$. The probability density function of X is given by

$$f(x|a,b) = \frac{1}{\Gamma(a)b^a} e^{-x/b} x^{a-1}, \quad x > 0,\ a > 0,\ b > 0, \qquad (16.1)$$

where $b = 1/\lambda$. The distribution defined by (16.1) is called the gamma distribution with shape parameter a and the scale parameter b. It should be noted that (16.1) is a valid probability density function (pdf) for any $a > 0$ and $b > 0$. The cumulative distribution function is given by

$$F(x|a,b) = \frac{1}{\Gamma(a)b^a} \int_0^x e^{-t/b} t^{a-1} dt, \quad x > 0,\ a > 0,\ b > 0.$$

The three-parameter gamma distribution has the pdf

$$f(x|a,b,c) = \frac{1}{\Gamma(a)b^a} e^{-(x-c)/b} (x-c)^{a-1}, \quad a > 0,\ b > 0,\ x > c,$$

where c is the location parameter. The standard form of the gamma distribution (when $b = 1$ and $c = 0$) has the pdf

$$f(x|a,b) = \frac{1}{\Gamma(a)} e^{-x} x^{a-1}, \quad x > 0,\ a > 0, \qquad (16.2)$$

and cumulative distribution function

$$F(x|a) = \frac{1}{\Gamma(a)} \int_0^x e^{-t} t^{a-1} dt. \qquad (16.3)$$

The cdf in (16.3) is often referred to as the *incomplete gamma function*.

   The gamma distribution with a positive integer shape parameter a is called the

Erlang Distribution. If a is a positive integer, then

$$
\begin{aligned}
P(X \le x|a,b) &= P(\text{waiting time until the } a\text{th event is at most } x \text{ units of time}) \\
&= P(\text{observing at least } a \text{ events in } x \text{ units of time when the} \\
&\quad\ \text{mean waiting time per event is } b) \\
&= P(\text{observing at least } a \text{ events in a Poisson process when} \\
&\quad\ \text{the mean number of events is } x/b) \\
&= \sum_{k=a}^{\infty} \frac{e^{-x/b}(x/b)^k}{k!},
\end{aligned}
$$

which is $P(Y \ge a)$, where $Y \sim \text{Poisson}(x/b)$. Thus, if a is a positive integer, then the gamma cdf can be easily evaluated.

The gamma probability density plots in Figure 16.1 indicate that the degree of asymmetry of the gamma distribution diminishes as a increases. For large a, $(X - a)/\sqrt{a}$ is approximately distributed as the standard normal random variable.

16.2   Moments

Mean:	ab
Variance:	ab^2
Mode:	$b(a - 1), \ a > 1$
Coefficient of Variation:	$1/\sqrt{a}$
Coefficient of Skewness:	$2/\sqrt{a}$
Coefficient of Kurtosis:	$3 + 6/a$
Moment Generating Function:	$(1 - bt)^{-a}, \ t < \frac{1}{b}$
Moments about the Origin:	$\dfrac{\Gamma(a+k)b^k}{\Gamma(a)} = b^k \prod_{i=1}^{k} (a + i - 1), \ k = 1, 2, \ldots$

16.3   Probabilities, Percentiles, and Moments

The dialog box [StatCalc→Continuous→ Gamma] computes probabilities, percentiles, moments, and also the parameters when other values are given.

To compute probabilities: Enter the values of the shape parameter a, scale parameter b, and the observed value x; click on [P(X <= x)]. When $a = 2$, $b = 3$, and $x = 5.3$, $P(X \le 5.3) = 0.527172$ and $P(X > 5.3) = 0.472828$.

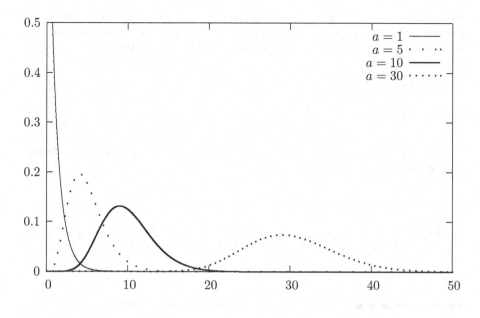

FIGURE 16.1: Gamma probability density functions

To compute percentiles: Enter the values of a, b, and the cumulative probability; click [x]. When $a = 2$, $b = 3$ and the cumulative probability $= 0.05$, the 5th percentile is 1.06608. That is, $P(X \leq 1.06608) = 0.05$.

To compute other parameters: Enter the values of the probability, one of the parameters, and a positive value for x; click on the parameter that is missing. For example, when $b = 3$, $x = 5.3$ and $P(X \leq x) = 0.9$, the value of the shape parameter $a = 0.704973$.

To compute moments: Enter the values of a and b; click [M].

Example 16.1. The distribution of 50-year summer rainfall (in inches) in a certain part of India is approximately gamma with $a = 3.0$ and $b = 2.0$.

a. Find the percentage of summer rainfalls that exceed 6 inches.

b. Find an interval that will contain 95% of the summer rainfall totals.

Solution: Let X denote the total summer rainfall in a year.

a. Select the dialog box [StatCalc→Continuous→ Gamma] from *StatCalc*, enter 3 for a, 2 for b, and 6 for observed x; click [P(X <= x)] to get $P(X > 6) = 0.42319$. That is, about 42% of the summer rainfall totals exceed 6 inches.

b. To find a right endpoint, enter 3 for a, 2 for b, and 0.975 for cumulative probability; click [x] to get 16.71. To find a lower endpoint, enter 0.025 for the cumulative probability and click [x] to get 0.73. Thus, 95% of the summer rainfall totals are between 0.73 and 16.71 inches.

Example 16.2. Customers enter a fast food restaurant, according to a Poisson process, on average 4 for every 3-minute period during the peak hours 11 a.m. – 1 p.m. Let X denote the waiting time in minutes until arrival of the 60th customer.

a. Find $E(X)$.

b. Find $P(X > 50)$.

Solution: The mean number of customers per minute is 4/3. Therefore, mean waiting time in minutes is $b = 3/4$.

a. $E(X) = ab = 60$ x $3/4 = 45$ min.

b. To find the probability using [StatCalc→Continuous→ Gamma], enter 60 for a, $3/4 = 0.75$ for b, and 50 for x; click [P(X <= x)] to get $P(X > 50) = 0.19123$.

16.4 Applications

The gamma distribution arises in situations where one is concerned about the waiting time for a finite number of independent events to occur, assuming that events occur at a constant rate and chances that more than one event occurs in a small interval of time are negligible. This distribution has applications in reliability and queuing theory. Examples include the distribution of failure times of components, the distribution of times between calibration of instruments that need re-calibration after a certain number of uses, and the distribution of waiting times of k customers who will arrive at a store. The gamma distribution can also be used to model the amounts of daily rainfall in a region. For example, the data on daily rainfall in Sydney, Australia, (October 17–November 7; years 1859–1952) were modeled by a gamma distribution. A gamma distribution was postulated because precipitation occurs only when water particles can form around dust of sufficient mass, and the waiting time for such accumulation of dust is similar to the waiting time aspect implicit in the gamma distribution (Das, 1955). Stephenson et al. (1999) showed that the gamma and Weibull distributions provide good fits to the wet-day rainfall distribution in India.

In exposure/pollution data analysis, gamma models are used as alternatives to lognormal models. Maxim et al. (2006) have observed that the gamma distribution is a possible distribution for concentrations of carbon/coke fibers in plants that produce green or calcined petroleum coke. Gibbons (1994), Bhaumik and Gibbons (2006), Krishnamoorthy, Mathew and Mukherjee (2008) and Bhaumik et al. (2009) noted that gamma distributions are potentially useful for applications in many fields, including environmental monitoring, groundwater monitoring, industrial hygiene, genetic research, and industrial quality control. For specific applications in groundwater monitoring, see examples in the sequel.

16.5 Estimation of Parameters: Three-parameter Case

Let $X_1, \ldots, X_n$ be a sample from a gamma distribution with the shape parameter a, scale parameter b, and the location parameter c. Let $\bar{X}$ denote the sample mean.

Maximum Likelihood Estimators

The MLEs of a, b and c are the solutions of the equations

$$\sum_{i=1}^{n} \ln(X_i - c) - n \ln b - n\psi(a) = 0$$

$$\sum_{i=1}^{n} (X_i - c) - nab = 0$$

$$\sum_{i=1}^{n}(X_i - c)^{-1} + n[b(a-1)]^{-1} = 0, \tag{16.4}$$

where ψ is the digamma function (see Section 1.8). These equations may yield reliable solutions if a is expected to be at least 2.5.

If the location parameter c is known, the MLEs of a and b are the solutions of the equations

$$\frac{1}{n} \sum_{i=1}^{n} \ln(X_j - c) - \ln(\bar{X} - c) - \psi(a) + \ln a = 0 \quad \text{and} \quad ab = \bar{X}.$$

If a is also known, then $\bar{X}/a$ is the UMVUE of b.

Moment Estimators

Moment estimators are given by

$$\hat{a} = \frac{4m_2^3}{m_3^2}, \quad \hat{b} = \frac{m_3}{2m_2} \quad \text{and} \quad \hat{c} = \bar{X} - 2\frac{m_2^2}{m_3},$$

where

$$m_k = \frac{1}{n} \sum_{i=1}^{n} (X_i - \bar{X})^k, \quad k = 1, 2, \ldots$$

is the kth sample central moment.

16.6 One-Sample Inference

Let $X_1, \ldots, X_n$ be a sample from a gamma(a, b) distribution. Define

$$\bar{X} = \frac{1}{n} \sum_{i=1}^{n} X_i \quad \text{and} \quad \tilde{G} = \left(\prod_{i=1}^{n} X_i \right)^{\frac{1}{n}}. \tag{16.5}$$

Note that $\bar{X}$ is the usual sample mean, and $\tilde{G}$ is the sample geometric mean. In the sequel, we shall describe inferential methods based on $\bar{X}$ and $\tilde{G}$ for some one-sample problems involving two-parameter gamma distribution.

Maximum Likelihood Estimators

The log-likelihood function is expressed as

$$l(a, b|\bar{X}, \tilde{G}) = -n \ln \Gamma(a) - na \ln b - n\bar{X}/b + (a-1)n\ln \tilde{G}. \qquad (16.6)$$

The maximum likelihood estimate (MLE) $\hat{a}$ is the solution of the equation

$$\ln(a) - \psi(a) = \ln(\bar{X}/\tilde{G}), \qquad (16.7)$$

where ψ is the digamma function. Letting $s = \ln(\bar{X}/\tilde{G})$, an approximation to $\hat{a}$ is given by

$$\hat{a} \simeq \frac{3 - s + \sqrt{(s-3)^2 + 24s}}{12s}. \qquad (16.8)$$

Using the above approximate MLE as the initial value a_0, the MLE can be evaluated by the Newton–Raphson iterative scheme

$$a_1 = a_0 - \frac{\ln a_0 - \psi(a_0) - s}{1/a_0 - \psi'(a_0)},$$

where $\psi'(x) = \frac{\partial \psi(x)}{\partial x}$ is the trigamma function. The MLE of b is $\hat{b} = \bar{X}/\hat{a}$. Note that the MLE $\hat{a}$ is implicitly a function of $\bar{X}/\tilde{G}$, and so it is invariant under a scale transformation of the samples, and $\hat{b} = \bar{X}/\hat{a}$ is scale equivariant.

16.6.1 One-Sample Tests

Signed Likelihood Ratio Test (SLRT) for the Shape Parameter

Consider testing

$$H_0 : a \leq a_0 \quad \text{vs.} \quad H_a : a > a_0, \qquad (16.9)$$

where a_0 is a specified value. It is easy to verify that, for a fixed a_0, the MLE of b is given by $\hat{b}_{a_0} = \bar{X}/a_0$. The SLRT statistic is expressed as

$$
\begin{aligned}
R(a_0) &= \text{sign}(\hat{a} - a_0)\left\{2[\ln l(\hat{a}, \hat{b}) - \ln l(a_0, \hat{b}_{a_0})]\right\}^{1/2} \\
&= \text{sign}(\hat{a} - a_0)\sqrt{2n}\left[\ln \frac{\Gamma(a_0)}{\Gamma(\hat{a})} + (\hat{a} - a_0)\left[\ln\left(\tilde{G}/\bar{X}\right) - 1\right]\right. \\
&\quad + \left. (\hat{a}\ln\hat{a} - a_0 \ln a_0)\right]^{1/2},
\end{aligned}
$$

where $\hat{a}$ and $\hat{b}$ are the MLEs.

As the testing problem is invariant under scale transformation $X \to cX$, where c is a positive constant, the distribution of $R(a_0)$ depends only on a_0. Therefore, the null distribution can be evaluated empirically by Monte Carlo simulation. In particular, for an observed value of the SLRT statistic $R_0(a_0)$, the null hypothesis in (16.9) is rejected if the p-value $P(R(a_0) > R_0(a_0)) < \alpha$, where $0 < \alpha < .5$

is a specified level of significance. Note that this p-value can be estimated using simulated samples from a gamma$(a_0, 1)$ distribution. So this test is exact, except for the simulation error.

SLRT for the Scale Parameter

Consider testing

$$H_0 : b \leq b_0 \quad \text{vs.} \quad H_a : b > b_0, \tag{16.10}$$

where b_0 is a specified value. For a fixed b, the MLE $\widehat{a}_b$ of a is the solution of the equation

$$\psi(a) - \ln \frac{\widetilde{G}}{b} = 0. \tag{16.11}$$

For calculation details, see Krishnamoorthy and Novelo (2014). The SLRT statistic for testing $b = b_0$ is expressed as

$$
\begin{aligned}
R(b_0) &= \operatorname{sign}(\widehat{b} - b_0) \left\{ 2[\ln l(\widehat{a}, \widehat{b}|\bar{X}, \widetilde{G}) - \ln l(\widehat{a}_{b_0}, b_0|\bar{X}, \widetilde{G})] \right\}^{1/2} \\
&= \operatorname{sign}(\widehat{b} - b_0) \sqrt{2n} \left\{ \ln \frac{\Gamma(\widehat{a}_{b_0})}{\Gamma(\widehat{a})} + \widehat{a} \left(\ln \frac{\widetilde{G}}{\widehat{b}} - 1 \right) \right. \\
&\quad \left. - \widehat{a}_{b_0} \ln \frac{\widetilde{G}}{b_0} + \frac{\bar{X}}{b_0} \right\}^{1/2}. \tag{16.12}
\end{aligned}
$$

Extensive simulation studies by Krishnamoorthy and Novelo (2014) indicated that the distribution of the SLRT statistic does not depend on any parameters. So the p-value of the SLRT for testing (16.10) is given by $P(R^*(1) > R_0(b_0)|a = 1, b = 1)$, where $R_0(b_0)$ is an observed value of $R(b_0)$, and $R^*(1)$ is the test statistic in (16.12) based on a random sample of size n from a gamma(1,1) distribution. Note that, for an observed SLRT statistic $R_0(b_0)$, this p-value can be estimated by Monte Carlo simulation.

Modified LRT for the Mean

The SLRT statistic for testing $\mu = \mu_0$ is given by

$$R(\mu_0) = \operatorname{sign}(\widehat{\mu} - \mu_0) \left\{ 2[l(\widehat{a}, \widehat{b}|\bar{X}, \widetilde{G}) - l(\widehat{a}_{\mu_0}, \mu_0|\bar{X}, \widetilde{G})] \right\}^{1/2}, \tag{16.13}$$

where $l(a, b|\bar{X}, \widetilde{G})$ is the log-likelihood function in (16.6),

$$l(a, \mu|\bar{X}, \widetilde{G}) = -n \ln \Gamma(a) - na \ln(a/\mu) - na\bar{X}/\mu + (a - 1)n \ln \widetilde{G},$$

and $\widehat{a}_{\mu_0}$ is the MLE of a at $\mu = \mu_0$. This constrained MLE $\widehat{a}_{\mu_0}$ is obtained as the root of the equation

$$\ln a - \psi(a) = \ln \frac{\mu_0}{\widetilde{G}} + \frac{\bar{X}}{\mu_0} - 1. \tag{16.14}$$

As the above equation is similar to (16.7), a Newton–Raphson iterative scheme is readily obtained. The modified LRT by Fraser et al. (1997) is given by

$$MLRT(\mu_0) = R(\mu_0) - \frac{1}{R(\mu_0)} \ln \left(\frac{R(\mu_0)}{Q} \right), \tag{16.15}$$

where $R(\mu_0)$ is defined in (16.13), and

$$Q = \sqrt{n\widehat{a}}\,(\widehat{\mu}/\mu_0 - 1)\,(\psi'(\widehat{a}) - 1/\widehat{a})^{\frac{1}{2}}/(\psi'(\widehat{a}_{\mu_0}) - 1/\widehat{a}_{\mu_0})^{\frac{1}{2}},$$

and $\psi'(x)$ is the trigamma function. This MLRT has third-order accuracy in the sense that the standard normal approximation to the distribution of MLRT(μ_0) is accurate up to $O(n^{-3/2})$. For testing

$$H_0 : \mu \le \mu_0 \quad \text{vs.} \quad H_a : \mu > \mu_0, \tag{16.16}$$

the MLRT rejects the null hypothesis if $MLRT(\mu_0) > z_{1-\alpha}$. For a two-sided alternative hypothesis, the MLRT rejects the null hypothesis if $MLRT(\mu_0) > z_{1-\alpha/2}$.

For a given $(n, \bar{X}, \ln(\widetilde{G}))$, the dialog box [StatCalc→Continuous→ Gamma→Test, CI...] uses the above testing methods to compute p-values for testing the shape parameter, scale parameter, and the mean.

Example 16.3. The data in Table 16.1 represent vinyl chloride concentrations (in μg/L) collected from clean upgradient monitoring wells. The data are taken from Bhaumik and Gibbons (2006), who have used the data to find prediction limits. A quantile–quantile plot in Figure 16.2 clearly indicates that a gamma model fits these data well. The mean $\bar{X} = 1.8794$ and the geometric mean $\widetilde{G} = 1.0959$. The ln(GM)

TABLE 16.1: Vinyl Chloride Concentrations in Monitoring Wells

5.1	2.4	.4	.5	2.5	.1	6.8	1.2	.5	.6	5.3	2.3	1.8
1.2	1.3	1.1	.9	3.2	1.	.9	.4	.6	8	.4	2.7	.2
2	.2	.5	.8	2	2.9	.1	4					

$= .0915856$. The MLEs are $\widehat{a} = 1.063$ and $\widehat{b} = 1.769$. Note that the sample size $n = 34$. To compute the p-value for testing $H_0 : a \le .5$ vs. $H_a : a > .5$, select the dialog box [StatCalc→Continuous→ Gamma→Test, CI...], enter 34 for the sample size, 1.8794 for the sample mean, 0.0915856 for [Sam log(GM)], and .8 for [H0: a = a0]; click on [p-value] to get the SLRT statistic 3.198 with the p-value of 0.0014. The p-value indicates that the shape parameter of the concentration distribution is greater than 0.5.

Suppose it is desired to test $H_0 : b \ge 50$ vs. $H_a : b < 50$. To find the p-value, enter 50 for [H0: b = b0] and click on [p-value] to get the SLRT statistic -7.08115 and the p-value .001. Thus, the scale parameter of the vinyl chloride concentration distribution is less than 50.

Example 16.4. In this example, we shall use the lifetime data on air conditioning equipments given in Proschan (1963). Keating et al. (1990) used the data for illustrating a gamma model-based inference. The lifetime data in operating hours for plane number 7909 with 13 Boeing 720 aircrafts are as given in Table 16.2. The MLEs are $\widehat{a} = 1.6710$ and $\widehat{b} = 49.981$. The sample mean $\bar{X} = 83.517$, the geometric mean $\widetilde{X} = 60.154$, and $\ln(GM) = \ln(60.154) = 4.0969$. The sample size is $n = 29$.

For this example, it is desired to test $H_0 : a \le 1$ vs. $H_a : a > 1$. To calculate the p-value using *StatCalc*, select the dialog box [StatCalc→Continuous→ Gamma→Test, CI...], enter 29 for the sample size, 83.5172 for the [Sample AM], 4.0969 for [Sam log(GM)], 1 for [H0: a = a0], and click [p-values] to get the SLRT statistic 1.9897

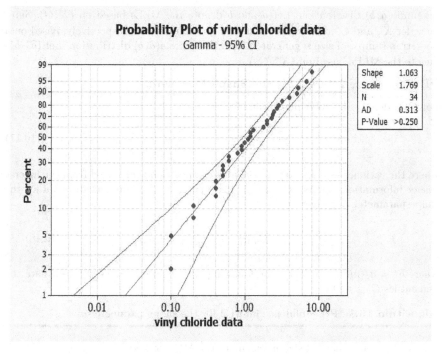

FIGURE 16.2: Gamma Q–Q plot for vinyl chloride concentrations

TABLE 16.2: Lifetime Data on Air Conditioning Equipment

90	10	60	186	61	49	14	24	56	20	79	84
44	59	29	118	25	156	310	76	26	44	23	62
130	208	70	101	208							

with the p-value .039. As this p-value is less than .05, we can conclude that the shape parameter of the lifetime distribution is greater than 1.

To test $H_0 : \mu \geq 110$ vs. $H_a : \mu < 110$, enter 110 for [H0: M = M0] in the dialog box noted in the preceding paragraph, and click on [p-values] to get the MLRT statistic -1.6984 and the p-value is $\Phi(-1.6984) = .045$, where Φ is the standard normal distribution. On the basis of the p-value, we can conclude that the mean concentration is less than 110.

16.6.2   One-Sample Confidence Intervals

The tests in the preceding sections are not conducive to find confidence intervals for the parameters or for the mean. So we shall describe simple parametric bootstrap (PB) confidence intervals for the parameters and the mean. Let $\bar{X}$ and $\tilde{G}$ denote the mean and geometric mean, respectively, based on a sample of size n from a gamma(a, b) distribution. Let $\widehat{a}$ and $\widehat{b}$ denote the MLEs based on $(\bar{X}, \widetilde{G})$. Similarly, let $\bar{X}^*$ and $\widetilde{G}^*$ denote the mean and geometric mean, respectively, based on a bootstrap sample of size n generated from the gamma$(\widehat{a}, \widehat{b})$ distribution. Let $(\widehat{a}^*, \widehat{b}^*)$ denote the MLEs based on $(\bar{X}^*, \widetilde{G}^*)$.

PB Confidence Interval for the Shape Parameter

Let $Q_{a;\alpha}$ denote the 100α percentile of

$$Q_a = \frac{\widehat{a}^* - \widehat{a}}{\widehat{\sigma}_{\widehat{a}^*}} = \frac{\widehat{a}^* - \widehat{a}}{\{\widehat{a}^*/[n(\widehat{a}^*\psi'(\widehat{a}^*) - 1)]\}^{1/2}}, \qquad (16.17)$$

where the variance estimate $\widehat{\sigma}_{\widehat{a}^*}^2$ in the above expression is obtained from the inverse Fisher information matrix. The $100(1 - 2\alpha)$ percent PB confidence interval for the shape parameter a is given by

$$\left(\widehat{a} - Q_{a;\alpha}\widehat{\sigma}_{\widehat{a}}, \ \widehat{a} - Q_{a;1-\alpha}\widehat{\sigma}_{\widehat{a}}\right), \qquad (16.18)$$

where $\widehat{\sigma}_{\widehat{a}}^2 = \widehat{a}/[n(\widehat{a}\psi'(\widehat{a}) - 1)]$. The following algorithm can be used to estimate the percentiles $Q_{a;\alpha}$ and $Q_{a;1-\alpha}$.

Algorithm 16.1. PB confidence interval for the shape parameter a

1. For a given sample of size n, calculate the MLEs $\widehat{a}$ and $\widehat{b}$.
2. Generate a bootstrap sample of size n from gamma$(\widehat{a}, \widehat{b})$ distribution, and calculate the MLEs $\widehat{a}^*$ and $\widehat{b}^*$ based on the bootstrap sample.
3. Set $Q_a = \frac{\widehat{a}^* - \widehat{a}}{\{\widehat{a}^*/[n(\widehat{a}^* \cdot \psi'(\widehat{a}^*) - 1)]\}^{1/2}}$
4. Repeat steps 2 and 3 for a large number of times, say, 100,000.
5. The 100α lower percentile and the 100α upper percentile of Q_as are estimates of $Q_{a;\alpha}$ and $Q_{a;1-\alpha}$, respectively.

PB Confidence Interval for the Scale Parameter

To find the PB confidence interval for the scale parameter b, we note that

$$Q_b = \frac{\widehat{b}^* - \widehat{b}}{\widehat{\sigma}_{\widehat{b}^*}} = \frac{\widehat{b}^* - \widehat{b}}{\left\{\widehat{b}^{*2}\psi'(\widehat{a}^*)/[n(\widehat{a}^*\psi'(\widehat{a}^*) - 1)]\right\}^{1/2}}, \qquad (16.19)$$

where the variance estimate of $\widehat{b}^*$ is obtained from the Fisher information matrix. Letting $\widehat{\sigma}_{\widehat{b}}^2 = \widehat{b}^2\psi'(\widehat{a})/[n(\widehat{a}\psi'(\widehat{a}) - 1)]$, the PB confidence interval for b is given by

$$\left(\widehat{b} - Q_{b;1-\alpha}\widehat{\sigma}_{\widehat{b}}, \ \widehat{b} - Q_{b;\alpha}\widehat{\sigma}_{\widehat{b}}\right), \qquad (16.20)$$

where $Q_{b;\alpha}$ is the 100α percentile of Q_b. The above PB confidence interval can be estimated using an algorithm similar to Algorithm 16.1.

PB Confidence Interval for the Mean

Recall that the mean of a gamma distribution is given by $\mu = ab$, and so the MLE of μ is $\widehat{\mu} = \widehat{ab} = \bar{X}$. The variance estimate

$$\widehat{\sigma}_{\bar{X}}^2 = \frac{\widehat{ab^2}}{n} = \frac{\bar{X}^2}{n\widehat{a}}.$$

The PB pivotal is given by

$$Q_\mu = \frac{\widehat{\mu}^* - \widehat{\mu}}{\widehat{\sigma}_{\bar{X}^*}} = \frac{(\bar{X}^* - \bar{X})}{\bar{X}^*/\sqrt{n\widehat{a}^*}}, \tag{16.21}$$

where the MLEs $\widehat{a}^*$ and $\bar{X}^*$ are based on a bootstrap sample from gamma$(\widehat{a}, \widehat{b})$ distribution. The $100(1 - 2\alpha)$ percent PB confidence interval for μ is given by

$$\left(\bar{X} - Q_{\mu;1-\alpha} \frac{\bar{X}}{\sqrt{n\widehat{a}}}, \ \bar{X} - Q_{\mu;\alpha} \frac{\bar{X}}{\sqrt{n\widehat{a}}} \right), \tag{16.22}$$

where $Q_{\mu;\alpha}$ is the 100α percentile of Q_μ defined in (16.21).

For a given $(n, \bar{X}, \ln(\widetilde{G}))$, the dialog box [StatCalc→Continuous→ Gamma→Test, CI...] computes the above PB confidence intervals the shape parameter, scale parameter, and the mean. These PB confidence intervals are quite satisfactory even for small samples [Krishnamoorthy and Novelo, 2014].

Example 16.5. Let us compute 95% confidence intervals for the mean lifetime of air conditioning equipments based on the data in Table 16.2. Recall that $\bar{X} = 83.5172$, $\ln(GM) = 4.0969$, and the sample size $n = 29$. To compute the 95% PB confidence interval, select the dialog box [StatCalc→Continuous→ Gamma→Test, CI...], enter 29 for the sample size, 83.5172 for the [Sample AM], 4.0969 for [Sam log(GM)], .95 for [Conf Level], and click [CI] to get (63.1, 114.7).

The 95% confidence interval (L, U) based on MLRT test in (16.15) is determined such that

$$MLRT(L) = z_{.975} = 1.96 \ \text{ and } \ MLRT(U) = z_{.025} = -1.96.$$

Using the PB confidence interval as starting values and trial-error, it can be verified that $MLRT(63.38) = 1.96$ and $MLRT(115.16) = 1.96$. Thus, the 95% confidence interval for the mean time to failure on the basis of MLRT in (16.15) is (63.4, 115.2), which is in good agreement with the PB confidence interval.

By clicking appropriate [CI] radio button in the same dialog box, we find 95% PB confidence intervals for the shape parameter as $(.948, 2.492)$ and for the scale parameter as $(32.07, 99.51)$.

Example 16.6. Grice and Bain (1980) discuss the data (Gross and Clark, 1975) on survival times on 20 mice exposed to 240 rads of gamma radiation. The data are reproduced here in Table 16.3. The mean $\bar{X} = 113.45$ and $\ln(GM) = 4.6734$. The MLEs are $\widehat{a} = 8.805$ and $\widehat{b} = 12.885$. To compute the 95% PB confidence interval for the mean survival time, select the dialog box [StatCalc→Continuous→ Gamma→Test, CI...], enter 20 for the sample size, 113.45 for the [Sample AM], 4.6734 for [Sam log(GM)], .95 for [Conf Level], and click [CI] to get (97.0, 134.1).

TABLE 16.3: Survival Time Data

152	152	115	109	137	88	94	77	160	165
125	40	128	123	136	101	62	153	83	69

To find a 95% confidence interval for the shape parameter, click on [CI] (under CI for the Shape Parameter) to get (4.0, 14.30).
To find a 95% confidence interval for the mean based on the MLRT and the PB confidence interval in the preceding paragraph, we find

$$MLRT(97.2) = 1.96 = z_{.975} \quad \text{and} \quad MLRT(134.3) = -1.96 = z_{.025}.$$

Thus, the 95% MLRT confidence interval for the mean is (97.2, 134.3), which is in good agreement with the PB confidence interval (97.0, 134.1).

Remark 16.1. If the shape parameter a is known, then an exact interval estimation method for b is available. In this case, $S = n\bar{X} \sim \text{gamma}(na, b)$, and the endpoints of $1 - \alpha$ confidence interval (b_L, b_U) satisfy

$$P(S \leq S_0 | b_U) = \alpha/2$$

and

$$P(S \geq S_0 | b_L) = \alpha/2.$$

Using the result that the gamma distribution is stochastically increasing in b, the confidence interval can be obtained as

$$(b_L, b_U) = \left(\frac{S_0}{G_{na,1;1-\alpha/2}}, \frac{S_0}{G_{na,1;\alpha/2}} \right),$$

where $G_{na,1;\alpha}$ denotes the α quantile of a gamma distribution with the shape parameter na and the scale parameter 1, is a $1 - \alpha$ confidence interval for b [Guenther 1969 and 1971].

16.6.3   Prediction Intervals, Tolerance Intervals, and Survival Probability

Approximate prediction intervals, tolerance intervals and confidence intervals for survival probability can be obtained using the cube root transformation. Specifically, if $X \sim \text{gamma}(a, b)$ distribution, then $X^{1/3}$ is approximately normally distributed (Wilson-Hilferty, 1931). As a result, tolerance limits, prediction limits, and confidence limits for a survival probability involving gamma distributions are easily obtained by applying normal-based methods for cube root transformed samples, and then taking third power. Krishnamoorthy et al. (2008) proposed this cube root transformed approach, and their simulation studies indicated that the approximate results are very satisfactory even for samples as small as four.
    Let $X_1, ..., X_n$ be a sample from a gamma(a, b) distribution. Let

$$Y_i = X_i^{\frac{1}{3}}, \ i = 1, ..., n, \ \bar{Y} = \frac{1}{n} \sum_{i=1}^{n} Y_i \text{ and } S_y^2 = \frac{1}{n-1} \sum_{i=1}^{n} (Y_i - \bar{Y})^2.$$

Prediction Intervals

An approximate $1 - \alpha$ prediction interval for a future individual is

$$\left(\bar{Y} \pm t_{n-1;1-\alpha/2} S_y \sqrt{1 + 1/n} \right)^3,$$

where $t_{m;q}$ denotes the qth quantile of a t distribution with df $= m$.

Upper Prediction Limits for at least l of m Observations from Each of r Locations

As noted in Section 11.4.4, construction of an upper prediction limit (UPL) for at least l of m observations from a normal population at each of r locations is needed in ground water quality detection monitoring in the vicinity of hazardous waste management facilities (HWMF), and in process monitoring. Let k_u denote the factor for the normal case determined by the integral equation (11.10). Then

$$\left(\bar{Y} + k_u S_y \right)^3$$

is the desired UPL. Numerical investigation by Krishnamoorthy et al. (2008) indicated that the above approximate UPL is very accurate even for small samples.

Tolerance Intervals

An approximate $(p, 1 - \alpha)$ upper tolerance limit for the sampled gamma population is given by

$$\left(\bar{Y} + cS_y \right)^3, \quad \text{where } c = \frac{1}{\sqrt{n}} t_{n-1;1-\alpha}(z_p \sqrt{n}),$$

where $t_{m;q}(\delta)$ denotes the q quantile of a noncentral t distribution with df $= m$ and the noncentrality parameter δ, and z_p is the pth quantile of a standard normal distribution. Furthermore, $\left(\bar{Y} - cS_y \right)^3$ is an approximate $(p, 1 - \alpha)$ lower tolerance limit.

An approximate two-sided tolerance interval is given by

$$\left(\bar{Y} \pm kS_y \right)^3,$$

where k is the normal tolerance factor as defined in (11.6.1).

Survival Probability

Note that, for a given $t > 0$, $P(X > t) = P(X^{1/3} > t^{1/3})$, where Y is approximately normally distributed. On the basis of this approximation, we can find a confidence interval for the survival probability $P(X > t)$ following (11.7) in Section 11.4.3. Let L and U be determined by the following equations:

$$t_{n-1;1-\alpha}(\sqrt{n}L) = \frac{\sqrt{n}(t^{1/3} - \bar{Y})}{S_y} \text{ and } t_{n-1;\alpha}(\sqrt{n}U) = \frac{\sqrt{n}(t^{1/3} - \bar{Y})}{S_y}. \quad (16.23)$$

An approximate $1 - 2\alpha$ approximate confidence interval for the survival probability $P(X > t)$ is given by

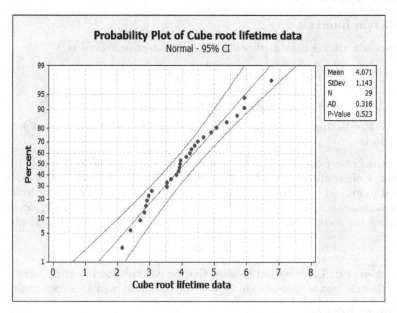

FIGURE 16.3: Normal Q–Q plot for cube root transformed lifetime data on air conditioning equipment

$$(1 - \Phi(U), \quad 1 - \Phi(L)).$$

Example 16.7. Let us calculate (.90, .95) lower tolerance limit for lifetime data on air conditioning equipments in Example 16.4. The Q–Q plot in Figure 16.3 clearly indicates that a normal model for cube root transformed data is tenable. The mean and the standard deviation of the cube root transformed sample are

$$\bar{Y} = 4.0710 \quad \text{and} \quad S_y = 1.1425.$$

Tolerance limits and prediction limits can be calculated using [StatCalc →Continuous→Gamma→Tolerance ...]. In this dialog box, just enter 29 for [Sample Size n], 0.9 for [Proportion p], .95 for [Coverage Prob g], 4.071 for [x-bar*] and 1.1425 for [s*]; click on [1-sided] to get 8.344.

Suppose it is desired to find 95% lower confidence limit for $P(X > 15)$. Select the dialog box [StatCalc→Continuous→Normal→Coefficients ...], enter 29 for [S Size], 4.0710 for the [Sample Mean], 1.1425 for [Sample SD], .95 for [Conf Level], $15^{1/3} = 2.4662$ for [Value of t], and click [CI for $P(X > t)$] to get .8111. This means that at least 81% of air conditioning equipment will last 15 units of time with confidence 95%.

Example 16.8. Let us calculate tolerance limits and prediction limits for the vinyl chloride concentration data in Table 16.1. The mean of the cube root transformed data is $\bar{Y} = 1.1022$ and $S_y = 0.3999$. To compute (.95, .95) upper tolerance limit, select the dialog box ([StatCalc→Continuous→Normal→Tolerance ...]), enter 34 for [Sample Size n], .95 for [Proportion p], .95 for [Coverage Prob g], 1.1022 for [x-bar*], and 0.3999 for [s*]; click on [1-sided] to get 7.674. This means that at least 95% of the concentrations are less than 7.674 with confidence 95%.

We shall now calculate 95% UPL so that at least one of three future measurements from each of five locations fall below the UPL. To find the desired factor, choose the dialog box in the preceding paragraph, enter the sample size, the mean, and SD as in the preceding paragraph and enter (34, 5, 3, 1) for (n, r, m, l) and .95 for [Conf Level]; click on [1-sided] to get 3.008.

16.7 Comparison of Several Gamma Distributions

Let $(\bar{X}_i, \widetilde{G}_i)$ denote the (arithmetic mean, geometric mean) based on a sample of size n_i from gamma(a_i, b_i) distribution, $i = 1, ..., k$. In the following we shall describe the likelihood ratio statistics for comparing several parameters, means, and homogeneity of several gamma distributions.

LRT for Equality of Shape Parameters

Consider testing

$$H_0 : a_1 = ... = a_k \quad \text{vs.} \quad H_a : a_i \neq a_j \text{ for some } i \neq j. \tag{16.24}$$

Following (16.6), the log-likelihood function under H_0 is expressed as

$$\sum_{i=1}^{k} l(a, b_i) = -\sum_{i=1}^{k} n_i \left(\ln \Gamma(a) + a \ln(b_i) + \frac{\bar{X}_i}{b_i} - (a - 1) \ln \widetilde{G}_i \right), \tag{16.25}$$

where a is the unknown common shape parameter. It can be readily checked that the constrained MLE of the common shape parameter, denoted by $\hat{a}_c$, that maximizes (16.25) is the solution of the equation

$$\ln a - \psi(a) = \sum_{i=1}^{n} w_i \ln \frac{\bar{X}_i}{\widetilde{G}_i}, \tag{16.26}$$

where $w_i = n_i / \sum_{j=1}^{k} n_j$. Noting that the above equation is similar to (16.7), the root can be found using the Newton–Raphson method with the starting value as defined in (16.8) with $s = \sum_{i=1}^{n} w_i \ln \frac{\bar{X}_i}{\widetilde{G}_i}$. The constrained MLEs $\hat{b}_{ic} = \bar{X}_i / \hat{a}_c$, $i = 1, ..., k$.

The LRT statistic for testing (16.24) is given by

$$
\begin{aligned}
\Lambda_a &= 2 \left\{ \sum_{i=1}^{k} l(\hat{a}_i, \hat{b}_i) - \sum_{i=1}^{k} l(\hat{a}_c, \hat{b}_{ic}) \right\} \\
&= 2 \sum_{i=1}^{k} n_i \left(\ln \frac{\Gamma(\hat{a}_c)}{\Gamma(\hat{a}_i)} - (\ln \hat{a}_c^{\hat{a}_c} - \ln \hat{a}_i^{\hat{a}_i}) \right. \\
&\quad + \left. (\hat{a}_c - \hat{a}_i) \left[\ln \frac{\bar{X}_i}{\widetilde{G}_i} + 1 \right] \right), \tag{16.27}
\end{aligned}
$$

where $l(a, b)$ is given in (16.6), $(\widehat{a}_i, \widehat{b}_i)$ is the MLE of (a_i, b_i) based on $(\bar{X}_i, \widetilde{G}_i)$, $i = 1, ..., k$, and $\widehat{a}_c$ is constrained MLE of a satisfying equation (16.26).

Since the LRT statistic Λ_a is invariant under the transformation $X_{ij} \rightarrow c_i X_{ij}$, $j = 1, ..., n_i$, $i = 1, ..., k$, and $c_i > 0$ for all i. The null distribution of Λ_a may depend on the unknown common shape parameter under H_0 in (16.24). However, extensive simulation studies by Krishnamoorthy, Lee and Wang (2015) indicate that the null distribution of Λ_a does not depend on the common unknown shape parameter, so it appears that the LRT is exact.

The necessary percentiles to carry out the test or p-values can be estimated by Monte Carlo simulation based on independent samples, each of size n_i, generated from gamma$(1, 1)$ distribution (or from the standard exponential distribution) as described in the following algorithm.

Algorithm 16.2. Calculation of the p-value of the LRT for equality of shape parameters

---

For a given set of sample sizes $(n_1, ..., n_k)$:

1. Generate independent samples

$$U_{ij} \sim \text{uniform}(0, 1), \ j = 1, ..., n_i, \ i = 1, ..., k$$

   Set $X_{ij} = -\ln(U_{ij})$

2. Compute $\bar{X}_i = \frac{1}{n_i} \sum_{j=1}^{n_i} X_{ij}$ and $\ln \widetilde{G}_i = \frac{1}{n} \sum_{j=1}^{n_i} \ln(X_{ij})$, $i = 1, ..., k$

3. Find the MLEs $(\widehat{a}_i, \widehat{b}_i)$ and the constrained MLEs $(\widehat{a}_c, \widehat{b}_{ic})$ based on $(n_i, \bar{X}_i, \ln \widetilde{G}_i)$, $i = 1, ..., k$

4. Compute the LRT statistic Λ_a using (16.27)

5. Repeat steps 1–4 for a large number of times, say, 100,000

6. The $100(1-\alpha)$ percentile of the 100,000 LRT statistics is a Monte Carlo estimate of the $100(1 - \alpha)$ percentile of the null distribution of Λ_a.

The LRT rejects the null hypothesis in (16.24) when an observed value Λ_a^0 of Λ_a is greater than the above $100(1 - \alpha)$ percentile. For an observed value Λ_a^0 of Λ_a, the percentage of 100,000 LRT statistics that is greater than Λ_a^0 is an estimate of the p-value. For a vector n of sample sizes, vector $\bar{X} = \mathbf{xb}$, and vector $\mathbf{lgm} = \ln \widetilde{G}$, R function 16.2 calculates the p-value of the LRT for equality of shape parameters.

Example 16.9. Schickdanz and Krause (1970) fitted normal, lognormal, and gamma distributions to weekly rainfall data from Springfield, Ill., during the seasons of summer, fall and winter for 1960–64. They found that the gamma distribution is the best fit among these three distributions. The sample sizes along with MLEs are given for each season in Table 16.4.

To test equality of the shape parameters of the rainfall distributions during these three seasons, we calculated the constrained MLEs $\widehat{a}_c = .8430$, $\widehat{b}_{1c} = 1.0723$, $\widehat{b}_{2c} = .9057$, and $\widehat{b}_{3c} = .4370$. The LRT statistic in (16.27) is 1.2673 with the p-value of .540. This p-value clearly indicates that the shape parameters are not significantly different.

TABLE 16.4: Weekly Rainfall Data with Sample Statistics and MLEs

Sample Statistics	Summer $n_1 = 58$	Fall $n_2 = 51$	Winter $n_3 = 57$
$\bar{X}$	.9040	.7635	.3684
$\ln \tilde{G}$	-.8471	-1.0417	-1.5850
MLEs $(\hat{a}, \hat{b})$	(.7959, 1.1358)	(.7725, .9884)	(.9860, .3736)

The above calculation can be carried out using the R function 16.2 as follows.

```
> n = c(58,51,57)
> xb = c(.9040, .7635, .3684)
> lgm = c(-.8471,-1.0417,-1.5850)
> test.gamma.shapes(100000,n,xb,lgm)
[1] 1.267333 0.540 # (LRT statistic, p-value)
```

LRT for Equality of Scale Parameters

Consider testing

$$H_0 : b_1 = ... = b_k \quad \text{vs.} \quad H_a : b_i \neq b_j \text{ for some } i \neq j. \tag{16.28}$$

The log-likelihood function under H_0 can be expressed as

$$\sum_{i=1}^{k} l(a_i, b) = -\sum_{i=1}^{k} n_i \left(\ln \Gamma(a_i) + a_i \ln(b) + \frac{\bar{X}_i}{b} - (a_i - 1) \ln(\tilde{G}_i) \right), \tag{16.29}$$

where b is the unknown common scale parameter under H_0. The constrained MLEs $\hat{a}_{ic}$s of a_is are the solutions of the equations

$$n_i \ln \left(\sum_{j=1}^{k} n_j a_j \right) - n_i \psi(a_i) - n_i \ln \left(\sum_{j=1}^{k} n_j \bar{X}_j \right) + n_i \ln(\tilde{G}_i) = 0, \quad i = 1, ..., k \tag{16.30}$$

and

$$\hat{b}_c = \frac{\sum_{i=1}^{k} n_i \bar{X}_i}{\sum_{i=1}^{k} n_i \hat{a}_{ic}}.$$

For calculation details of $\hat{a}_{ic}$s, see Krishnamoorthy, Lee and Wang (2015). The LRT statistic for testing the equality of scale parameters is given by

$$\begin{aligned} \Lambda_b &= 2 \left\{ \sum_{i=1}^{k} l(\hat{a}_i, \hat{b}_i) - \sum_{i=1}^{k} l(\hat{a}_i, \hat{b}_c) \right\} \\ &= 2 \sum_{i=1}^{k} n_i \left(\ln \frac{\Gamma(\hat{a}_{ic})}{\Gamma(\hat{a}_i)} + \ln \left(\frac{\hat{b}_c^{\hat{a}_{ic}}}{\hat{b}_i^{\hat{a}_i}} \right) + (\hat{a}_{ic} - \hat{a}_i) \left(1 - \ln(\tilde{G}_i) \right) \right), \tag{16.31} \end{aligned}$$

where $l(a, b)$ is given in (16.6), $(\hat{a}_i, \hat{b}_i)$ is the MLE of (a_i, b_i) based on $(\bar{X}_i, \tilde{G}_i)$,

$i = 1, ..., k$, and $\hat{a}_{ic}$s are the constrained MLEs of a_is satisfying equation (16.30), and $\hat{b}_c = \sum_{i=1}^{k} n_i \bar{X}_i / \sum_{i=1}^{k} n_i \hat{a}_{ic}$.

On the basis of extensive simulation studies, Krishnamoorthy, Lee and Wang (2015) noted that the null distribution of Λ_b does not depend on any parameters. Furthermore, the null distribution of Λ_b and that of Λ_a are the same, and both distributions depend only on sample sizes. So Algorithm 16.2 can be used to find the percentiles or the p-value of an observed value of the LRT statistic Λ_b.

For given sample sizes, $(\bar{X}_1, ..., \bar{X}_k)$ and $(\ln \tilde{G}_1, ..., \ln \tilde{G}_k)$, R function 16.3 calculates the p-value of the LRT for equality of the shape parameters or of the LRT for equality of scale parameters.

Example 16.10. To test the equality of the scale parameters of the rainfall distributions in Example 16.9, we calculated the constrained MLEs as $\hat{b}_c = .9480$, $\hat{a}_{1c} = .8474$, $\hat{b}_{2c} = .6396$, and $\hat{b}_{3c} = .8345$. The LRT statistic for testing the equality of the scale parameters is 13.46 with the p-value .001, which indicates that the scale parameters of the rainfall distributions are significantly different. These quantities can be calculated using R function 16.3 as follows.

```
> n = c(58,51,57)
> xb = c(.9040, .7635, .3684)
> lgm = c(-.8471,-1.0417,-1.5850)
> test.gamma.scales(100000,n,xb,lgm)
[1] 13.45636  0.00144
```

LRT for Equality of Means

Let $(\bar{X}_i, \tilde{G}_i)$ denote the (arithmetic mean, geometric mean) based on a sample of size n_i from a gamma$(a_i b_i)$ distribution, $i = 1, ..., k$. Let $\mu_i = a_i b_i$, $i = 1, ..., k$, and consider testing

$$H_0 : \mu_1 = ... = \mu_k \quad \text{vs.} \quad H_a : \mu_i \neq \mu_j \text{ for some } i \neq j.$$

Denoting the unknown common mean under H_0 by μ, the log-likelihood function under H_0 can be expressed as

$$\sum_{i=1}^{k} l(a_i, \mu) = -\sum_{i=1}^{k} n_i \left(\ln \Gamma(a_i) + a_i \ln \frac{\mu}{a_i} + \bar{X}_i \frac{a_i}{\mu} - (a_i - 1) \ln \tilde{G}_i \right). \quad (16.32)$$

The MLEs of a_is, under H_0, are the solutions of the equations

$$\ln a_i - \psi(a_i) = \ln \frac{\mu}{\tilde{G}_i} + \frac{\bar{X}_i}{\mu} - 1, \quad (16.33)$$

where

$$\mu = \frac{\sum_{i=1}^{k} n_i a_i \bar{X}_i}{\sum_{i=1}^{k} n_i a_i}.$$

Denoting the MLEs satisfying the above equations (16.33) by $\hat{a}_{ic}$, the MLE of μ is given by the above expression with a_i replaced by $\hat{a}_{ic}$.

The LRT statistic is given by

$$\Lambda_\mu = 2\left\{\sum_{i=1}^{k} l(\widehat{a}_i, \widehat{b}_i) - \sum_{i=1}^{k} l(\widehat{a}_{ic}, \widehat{\mu}_c)\right\}$$

$$= 2\sum_{i=1}^{k} n_i \left(\ln\frac{\Gamma(\widehat{a}_{ic})}{\Gamma(\widehat{a}_i)} + \widehat{a}_{ic}\ln\frac{\widehat{\mu}_c}{\widehat{a}_{ic}} - \widehat{a}_i\ln\frac{\bar{X}_i}{\widehat{a}_i}\right.$$

$$\left. + \widehat{a}_i\left(\frac{\widehat{b}_i}{\widehat{b}_{ic}} - 1\right) - (\widehat{a}_{ic} - \widehat{a}_i)\ln\widetilde{G}_i\right), \qquad (16.34)$$

where $l(a_i, b_i)$ is defined in (16.6) and $l(a_i, \mu)$ is defined in (16.32).

Simulation results by Krishnamoorthy, Lee and Wang (2015) indicate that the null distribution of the LRT statistic Λ_μ depends only on the number of groups k and the sample sizes, not on the common mean μ under H_0 and the shape parameters. Thus, for a given set of sample sizes, Monte Carlo simulation can be used to estimate the p-value or the percentile of Λ_μ.

For given sample sizes, sample means, and logarithm of sample geometric means, R function 16.4 calculates the p-value for testing equality of means.

Example 16.11. To test the equality of the means of the rainfall distributions in Example 16.9, the constrained MLEs are computed as $\widehat{\mu}_c = .6694$, $\widehat{a}_{1c} = .7421$, $\widehat{a}_{2c} = .7556$, and $\widehat{a}_{3c} = .8002$. The LRT statistic for testing the equality of the means is 20.17 with the p-value .00005. This p-value provides a strong evidence to conclude that the means are quite different. These quantities can be calculated using R function 16.4 as follows.

```
> n = c(58,51,57)
> xb = c(.9040, .7635, .3684)
> lgm = c(-.8471,-1.0417,-1.5850)
> test.gamma.means(100000,n,xb,lgm)
[1] 20.17182   0.00005
```

LRT for Homogeneity of Several Gamma Distributions

Consider testing

$$H_0 : (a_1, b_1) = ... = (a_k, b_k) \quad \text{vs.} \quad H_a : (a_i, b_i) \neq (a_j, b_j) \text{ for some } i \neq j. \quad (16.35)$$

The log-likelihood function under H_0 is simply the log-likelihood function based on a single sample of size $N = \sum_{i=1}^{k} n_i$ from a gamma(a, b) distribution, and is expressed as

$$l(a, b) = -N\ln\Gamma(a) - Na\ln b - N\frac{\bar{\bar{X}}}{b} + N(a-1)\ln\widetilde{\widetilde{G}}, \qquad (16.36)$$

where $(\bar{\bar{X}}, \widetilde{\widetilde{G}})$ is the (mean, geometric mean) based on all N observations. The LRT statistic for testing H_0 in (16.35) is given by

$$\Lambda_E = 2\left(\sum_{i=1}^{k} l(\widehat{a}_i, \widehat{b}_i) - l(\widehat{a}, \widehat{b})\right), \qquad (16.37)$$

where $l(a, b)$ is given in (16.36), $\widehat{a}_i$ and $\widehat{b}_i$ are the MLEs based on the ith sample.

For large samples, the LRT rejects H_0 in (16.35) if $\Lambda_E > \chi^2_{2k-2;1-\alpha}$. On the basis of extensive simulation studies Krishnamoorthy, Lee and Wang (2015) conjectured that the distribution of the statistic Λ_E does not depend on any unknown parameters. So the percentiles of the null distribution of Λ_E or the p-value can be estimated by Monte Carlo simulation.

Example 16.12. To test if the rainfall distributions in Example 16.9 are identical, the MLEs based on all three samples are computed as $\widehat{a} = .7741$ and $\widehat{b} = .8745$. The LRT statistic for testing the equality of the means is 20.26 with the p-value .0005. This p-value provides a strong evidence to conclude that the rainfall distributions are quite different. The LRT statistic and the p-value can be calculated using R function 16.5 as follows.

```
> n = c(58,51,57)
> xb = c(.9040, .7635, .3684)
> lgm = c(-.8471,-1.0417,-1.5850)
> test.gamma.means(100000,n,xb,lgm)
[1] 20.17182  0.00005
```

One-Sided Tests for Comparing Two Parameters

Solutions to two-sample problems can be readily obtained as a special case of k-sample problems. In order to test one-sided hypotheses in a two-sample problem, the LRT statistics in the preceding sections can be modified as follows. For example, to test

$$H_0 : a_1 \leq a_2 \quad \text{vs.} \quad H_a : a_1 > a_2,$$

the LRT statistic Λ_a in (16.27) can be modified as

$$\Lambda_a^* = \text{sign}(\widehat{a}_1 - \widehat{a}_2)\sqrt{2\Lambda_a},$$

where $\text{sign}(x) = 1$ if $x > 0$, and is -1 if $x < 0$. The distribution of the above statistic also does not depend on any parameters, so Monte Carlo simulation can be used to find the p-value or percentile of Λ_a^*.

The test statistics Λ_b and Λ_μ can be modified similarly to handle a one-sided hypothesis test for equality of the scale parameters and for equality of two means.

Approximate Test for $a_1 - a_2$

The approximate test by Shiue, Bain and Engelhardt (1988) is described as follows. Let $R_i = \ln(\bar{X}_i/\widetilde{G}_i)$, and $\nu_i = (n_1 - 1)\left(1 + 1/(1 + 4.3\widehat{a}_i)^2\right)$, $i = 1, 2$. Then the above test rejects the null hypothesis of equal shape parameter at the level of significance α, if

$$\frac{n_1 R_1/(n_1 - 1)}{n_2 R_2/(n_2 - 1)} > F_{\nu_1,\nu_2;1-\alpha}, \tag{16.38}$$

where $F_{m,n;p}$ denotes the $100p$ percentile of the F distribution with degrees of freedoms (dfs) m and n. Numerical studies by Krishnamoorthy and Novelo (2014) indicated that the SLRT and the above test are very similar in terms of type I error rates and powers. As this test is simple to carry out, it is recommended for applications. *StatCalc* uses this approximate test to calculate the p-value for testing $a_1 = a_2$.

For a given $\{n_1, n_2, \bar{X}_1, \bar{X}_2, \ln \widetilde{G}_1, \ln \widetilde{G}_2\}$, the dialog box [StatCalc→Continuous

→Gamma→Test, CI for Mean Difference ...] calculates the p-values for testing equality of two shape parameters, two scale parameters, or two means.

Example 16.13. Experimental Meteorology Laboratory conducted randomized pyrotechnic seeding experiments on single clouds in south Florida during 1968 and 1970. Overall, 26 seeded and 26 control clouds were compared in the experiment to judge the effect of seeding. The data (in acre-feet per cloud) are given in Table 1 of Simpson (1972). The seeded rain data fit a gamma distribution very well (p-value >.250) whereas the data on control rain barely fit a gamma model (p-value > .057). We shall use the data to illustrate some two-sample methods described in earlier sections.

The calculated statistics for seeded rain are as follows: $\bar{X}_1 = 441.98$, $\ln(\widetilde{G}_1) = 5.134$, $\hat{a}_1 = .6396$, $\hat{b}_1 = 691.05$. For control rain (after replacing an entry of 0 by 1), $\bar{X}_2 = 164.59$, $\ln(\widetilde{G}_2) = 3.990$, $\hat{a}_2 = .5608$, $\hat{b}_2 = 293.51$. The mean difference $\bar{X}_1 - \bar{X}_2 = 277.4$.

Let μ_1 denote the mean amount of seeded rain, and let μ_2 denote the same for the control rain. To test the effect of seeding, one may want to test $H_0 : \mu_1 \le \mu_2$ vs. $H_a : \mu_1 > \mu_2$. To find the p-value, select the dialog box [StatCalc→Continuous→Gamma→Test, CI for Mean Difference ...], enter 26 for both sample sizes, 441.98 and 164.59 for [Sample AM], 5.134 and 3.990 for [Sample log(GM)]; click on [p-value] (under Test for M1 = M2) to get .006. This p-value indicates that there is a seeding effect on rainfall.

To find a 95% PB confidence interval for $\mu_1 - \mu_2$, enter .95 for [Conf Level] and click on [CI] (under CI for M1 - M2) to get (94.1, 579.6). This confidence interval indicates that on the average seeding effect on rainfall exceeded by 92.5 to 579.5 acre-feet.

To illustrate the test for the difference between two shape parameters, let us consider testing $H_0 : a_1 = a_2$ vs. $H_a : a_1 \ne a_2$, where a_1 is the shape parameter for seeded rain, and a_2 is the shape parameter for the control rain. To find the p-value, click on [p-value] (under Test for a1-a2) to get .350 for [Ha: a1 > a2]. The p-value for the two-tailed test is $2 \times .350 = .7$. Thus, the null hypothesis is not rejected. A 95% confidence interval for $a_1 - a_2$ can be obtained by clicking [CI], and is $(-.279, .448)$.

The p-value of the test $H_0 : b_1 \le b_2$ vs. $H_a : b_1 > b_2$ can be found by clicking on [p-value] under [Test for b1 - b2]. For this example, the p-value is .048, which indicates that b_1 is barely greater than b_2.

16.8 Properties and Results

1. An Identity: Let $F(x|a,b)$ and $f(x|a,b)$ denote, respectively, the cdf and pdf of a gamma random variable X with parameters a and b. Then

$$F(x|a, 1) = F(x|a + 1, 1) + f(x|a + 1, 1).$$

2. Additive Property: Let $X_1, \ldots, X_k$ be independent gamma random variables with the same scale parameter but possibly different shape parameters

$a_1, \ldots, a_k$, respectively. Then

$$\sum_{i=1}^{k} X_i \sim \text{gamma}\left(\sum_{i=1}^{k} a_i, b\right).$$

3. Exponential: Let $X_1, \ldots, X_n$ be independent exponential random variables with mean b. Then

$$\sum_{i=1}^{n} X_i \sim \text{gamma}(n, \ b).$$

4. Chi-square: When $a = n/2$ and $b = 2$, the gamma distribution specializes to the chi-square distribution with df $= n$.

5. Normal: If X is a gamma(a) random variable, then

$$X^{\frac{1}{3}} \sim N(\mu, \sigma^2), \quad \text{approximately},$$

where $\mu = a^{\frac{1}{3}} \left(1 - \frac{1}{9a}\right)$ and $\sigma^2 = 1/(9a^{\frac{1}{3}})$ [Wilson and Hilferty, 1931].

6. Beta: See Section 17.6.2.

7. Student's t: If X and Y are independent gamma$(n, 1)$ random variables, then

$$\sqrt{n/2}\left(\frac{X - Y}{\sqrt{XY}}\right) \sim t_{2n}.$$

16.9   Random Number Generation

```
Input:  a = shape parameter gamma(a) distribution
Output: x = gamma(a) random variate
        y = b*x is a random number from gamma(a, b).
```

Algorithm 16.3. Gamma variate generator

```
For a = 1:
Generate u from uniform(0, 1) return x = -ln(u)
```

The following algorithm for $a > 1$ is due to Schmeiser and Lal (1980). When $0 < a < 1$, $X = $ gamma(a) variate can be generated using relation that $X = U^{1/a}Z$, where Z is a gamma$(a + 1)$ random variate.

```
        Set f(x) = exp(x3*ln(x/x3) + x3 - x)
        x3 = a-1
        d = sqrt(x3)
        k =1
        x1 = x2 = f2 = 0
        If d >= x3, go to 2
        x2 = x3 - d
        k = 1- x3/x2
```

```
          x1 = x2 + 1/k
          f2 = f(x2)

2         Set x4 = x3 + d
              r = 1 - x3/x4
              x5 = x4 + 1/r
              f4 = f(x4)
              p1 = x4 - x2
              p2 = p1 - f2/k
              p3 = p2 + f4/r

3         Generate u, v from uniform(0, 1)
          Set u = u*p3
          If u > p1 go to 4
          Set x = x2 + u
          If x > x3 and v <= f4 + (x4 - x)*(1 - f4)/(x4 - x3), return x
          If x < x3 and v <= f2 + (x - x2)*(1 - f2)/(x3 - x2), return x
          go to 6

4         If u > p2, go to 5
          Set u = (u - p1)/(p2 - p1)
          x = x2 - ln(u)/k
          If x < 0, go to 3
          Set v = v*f2*u
          If v <= f2*(x - x1)/(x2 - x1) return x
          go to 6
5         Set u = (u - p2)/(p3 - p2)
          x = x4 - ln(u)/r
          v = v*f4*u
          If v <= f4*(x5 - x)/(x5 - x4) return x

6         If ln(v) <= x3*ln(x/x3) + x3 - x, return x
          else go to 3
```

x is a random number from the gamma(a, 1) distribution.

16.10   Computational Method for Probabilities

To compute $P(X \leq x)$ when $a > 0$ and $b = 1$:
The Pearson series for the cdf is given by

$$P(X \leq x) = \exp(-x)x^a \sum_{i=0}^{\infty} \frac{1}{\Gamma(a+1+i)}x^i. \qquad (16.39)$$

The cdf can also be computed using the continued fraction:

$$P(X > x) = \frac{\exp(-x)x^a}{\Gamma(a)} \left(\frac{1}{x+1-a-} \frac{1 \cdot (1-a)}{x+3-a-} \frac{2 \cdot (2-a)}{x+5-a-} \cdots \right). \qquad (16.40)$$

To compute $\Gamma(a+1)$, use the relation $\Gamma(a+1) = a\Gamma(a)$ [Press et al., 1992].

The series (16.39) converges faster for $x < a + 1$, while the continued fraction (16.40) converges faster for $x \geq a + 1$. A method of evaluating continued fraction is given in Kennedy and Gentle (1980, p. 76).

The following R function computes the gamma distribution function.

R function 16.1. Calculation of the gamma cdf[a]

```
gamcdf = function(x, a){
one = 1.0; maxitr = 1000; err = 1.0e-12
# alng(x) = logarithmic gamma function; R function 1.1
com  = exp(a*log(x)-alng(a)-x)
a0 = a
term = one/a;   su = one/a
for(i in 1:maxitr){
a0 = a0 + one
term = term*x/a0;
su = su + term;
if (abs(term) < su*err){break}
}
return(su*com)
}
```

[a]Electronic version of this R function can be found in HBSDA.r, located in *StatCalc* directory.

16.11   R Programs

The following R functions are also provided in the file "HBSDA.r" in *StatCalc* installation directory.

R function 16.2. Calculation of the p-value of the LRT for equality of shape parameters

```
test.gamma.shapes = function(nr, n, xb, lgm){
k = length(n); stats = seq(1:nr)
s = log(xb)-lgm
stat0 = lrt.stat.shapes(n, s)
for(i in 1:nr){
for(j in 1:k){
x = -log(runif(n[j]))
xb[j] = mean(x); lgm[j] = mean(log(x))
s[j] = log(xb[j])-lgm[j]
}
stats[i] = lrt.stat.shapes(n, s)
}
pval = sum(stats > stat0)/nr
return(c(stat0, pval))
}
# LRT statistic Lambda_a

lrt.stat.shapes = function(n, s){
k = length(n); sn = sum(n)
w = n/sn; ah = seq(1:k)
for(i in 1:k){
ah[i] = gam.mles(n[i], s[i])}
sti = sum(w*s)
a0 = (3-sti+sqrt((sti-3)**2+24*sti))/12/sti
l = 1
repeat{
a1 = a0-(log(a0)-digamma(a0)-sti)/(1/a0-trigamma(a0))
if(abs(a1-a0) <= 1.0e-7 | l >= 300){break}
a0 = a1
l = l+1}
ahc = a1
lrts = sum(n*(lgamma(ahc)-lgamma(ah)))-sum(n*(ahc*log(ahc)+
    -ah*log(ah)))+sum(n*(ahc-ah)*(s+1))
return(2*lrts)
}
```

```
# MLE of the shape parameter

gam.mles = function(n, s){
a0 = (3.0-s+sqrt((s-3.0)^2+24.0*s))/12.0/s
l = 1
repeat{
ans = (log(a0)-digamma(a0)-s)
a1 = a0-ans/(1.0/a0-trigamma(a0))
if(abs(ans) <= 1.0e-7 | l >= 30){break}
a0 = a1
l = l+1}
ah = a1
return(ah)
}
```

R function 16.3.   Calculation of the p-value of the LRT for equality of scale parameters

```
test.gamma.scales = function(nr, n, xb, lgm){
k = length(n); stats = seq(1:nr)
s = log(xb)-lgm
stat0 = lrt.stat.scales(n, xb, lgm)
for(i in 1:nr){
for(j in 1:k){
x = -log(runif(n[j]))
xb[j] = mean(x); lgm[j] = mean(log(x))
s[j] = log(xb[j])-lgm[j]
}
stats[i] = lrt.stat.shapes(n, s)
}
pval = sum(stats > stat0)/nr
return(c(stat0, pval))
}
```

```
# LRT statistic Lambda_b

lrt.stat.scales = function(n, xb, lgm){
k = length(n); ah = seq(1:k); bh = seq(1:k)
s = log(xb)-lgm
for(i in 1:k){
mle = gam.mles(n[i],s[i]) # see R function 15.1
ah[i] = mle[1]; bh[i] = xb[i]/ah[i]}
mlc = cons.mles.scales(n, xb, lgm)
ahc = mlc[1:k]; bhc = mlc[k+1]
stat = sum(n*(lgamma(ahc)-lgamma(ah)))+sum(n*(ahc*log(bhc)+
    -ah*log(bh)))+sum(n*(ahc-ah)*(1-lgm))
stat = 2*stat
return(stat)
}
```

```
# Calculation of the MLEs under the null hypothesis

cons.mles.scales = function(n, xb, lgm){
k = length(n); fm = matrix(0,k,k)
s = log(sum(n*xb))-lgm
a0 = (3.0-s+sqrt((s-3.0)^2+24.0*s))/12.0/s
l = 1
repeat{
sumna = sum(n*a0)
f = n*(log(sumna)-digamma(a0)-s)
for(i in 1:k){
for(j in 1:k){
if(i == j){
fm[i,i] = n[i]**2/sum(n*a0)-n[i]*trigamma(a0[i])}
else{
fm[i,j] = n[i]*n[j]/sumna}
}}
b0 = fm%*%a0-f
a1 = solve(fm,b0)
err = max(abs(f))
if(err <= 1.0e-5 | l >= 30){break}
a0 = a1
l = l + 1
}
ahc = a1; bhc = sum(n*xb)/sum(n*ahc)
return(c(ahc,bhc))
}
```

R function 16.4. Calculation of the p-value of the LRT for equality of gamma means

```
test.gamma.means = function(nr, n, xb, lgm){
k = length(n); stat = seq(1:nr)
stat0 = lrt.stat.means(n, xb, lgm)
pval = 0
for(j in 1:nr){
for(i in 1:k){
x = -log(runif(n[i]))
xb[i] = mean(x); lgm[i] = mean(log(x))
}
stat[j] = lrt.stat.means(n, xb, lgm)
}
pval = sum(stat > stat0)/nr; return(c(stat0, pval))
}
```

```
# LRT statistic Lambda_mu

lrt.stat.means = function(n, xb, lgm){
k = length(n)
ah = seq(1:k)
bh = seq(1:k)
mlc = cons.mles.means(n, xb, lgm)
uhc = mlc[1]
ahc = mlc[2:(k+1)]
bhc = uhc/ahc
ah = mlc[(k+2):(2*k+1)]
bh = mlc[(2*k+2):(3*k+1)]
stat = sum(n*(lgamma(ahc)-lgamma(ah)))+sum(n*(ahc*log(bhc)+
     - ah*log(bh)))+sum(n*ah*(bh/bhc-1))-sum(n*(ahc-ah)*lgm)
stat = 2*stat
return(stat)
}
```

```
# Constrained MLEs under the null hypothesis of equal mean

cons.mles.means = function(n, xb, lgm){
k = length(n); ah = seq(1:k); bh = seq(1:k)
sumn = sum(n); gm = matrix(0,k,k)
for(i in 1:k){
ah[i] = gam.mles(n[i], log(xb[i])-lgm[i])
bh[i] = xb[i]/ah[i]}
u0 = sum(n*ah*xb)/sum(n*ah)
s = log(u0)-lgm+xb/u0-1
a0 = (3.0-s+sqrt((s-3.0)^2+24.0*s))/12.0/s; a0s = a0;
sumna = sum(n*a0)
u0 = sum(n*a0*xb)/sumna; u0s = u0
l = 1
repeat{
g = n*(log(a0)-digamma(a0)-log(u0)+lgm-xb/u0+1)
for(i in 1:k){
for(j in 1:k){
if(i == j){
gm[i,i] = n[i]*(1/a0[i]-trigamma(a0[i])+
         +n[i]*(xb[i]-u0)**2/u0**2/sumna)}
else{
gm[i,j] = n[i]*n[j]*(xb[i]-u0)*(xb[j]-u0)/u0**2/sumna}
}}
b0 = gm%*%a0-g
mls = solve(gm,b0)
if(min(mls) < 0){
return(c(u0s,a0s, ah, bh))}
err = max((mls-a0)**2)
if(err <= 1.0e-7 | l > 300){break}
a0 = mls
sumna = sum(n*a0)
u0 = sum(n*xb*a0)/sumna
l = l + 1
}
ahc = a0; uhc = u0
return(c(uhc, ahc, ah, bh))
}
```

```
# Approximate MLEs under the null hypothesis of equal mean

cons.mles.apprx = function(n, xb, lgm){
k = length(n); ah = seq(1:k); bh = seq(1:k)
sumn = sum(n);
for(i in 1:k){
ah[i] = gam.mles(n[i], log(xb[i])-lgm[i])
bh[i] = xb[i]/ah[i]}
u0 = sum(n*ah*xb)/sum(n*ah)
s = log(u0)-lgm+xb/u0-1
a0 = (3.0-s+sqrt((s-3.0)^2+24.0*s))/12.0/s
u0 = sum(n*a0*xb)/sum(n*a0)
s = log(u0)-lgm+xb/u0-1
a0 = (3.0-s+sqrt((s-3.0)^2+24.0*s))/12.0/s
u0 = sum(n*a0*xb)/sum(n*a0)
return(c(u0,a0,ah,bh))
}
```

R function 16.5. Calculation of the p-value of the LRT for homogeneity of several gamma distributions

```
test.gamma.equal = function(nr, n, xb, lgm){
k = length(n); stat = seq(1:nr)
stat0 = lrt.stat.equal(n, xb, lgm)
for(j in 1:nr){
for(i in 1:k){
x = -log(runif(n[i]))
xb[i] = mean(x); lgm[i] = mean(log(x))
}
stat[j] = lrt.stat.equal(n,xb,lgm)
}
pval = sum(stat > stat0)/nr
return(c(stat0, pval))
}
```

```
# LRT statistic Lambda_E

lrt.stat.equal = function(n, xb, lgm){
k = length(n); s = log(xb)-lgm
N = sum(n); xbb = sum(n*xb)/N; ah = seq(1:k); bh = seq(1:k)
GM = sum(n*lgm)/N; S = log(xbb)-GM
AH = gam.mles(N, S)
BH = xbb/AH
for(i in 1:k){
ah[i] = gam.mles(n[i], s[i])
bh[i] = xb[i]/ah[i]}
tr1 = -N*lgamma(AH)-N*AH*log(BH)-N*xbb/BH+N*(AH-1)*GM
tr2 = -sum(n*lgamma(ah))-sum(n*ah*log(bh))-sum(n*ah)+
      +sum(n*(ah-1)*lgm)
slrt = 2*(tr2-tr1)
return(c(AH,BH,slrt))
}
```

17

Beta Distribution

17.1 Description

The probability density function (pdf) of a beta random variable with shape parameters a and b is given by

$$f(x|a,b) = \frac{1}{B(a,b)} x^{a-1}(1-x)^{b-1}, \quad 0 < x < 1, \ a > 0, \ b > 0,$$

where the beta function

$$B(a,b) = \frac{\Gamma(a)\Gamma(b)}{\Gamma(a+b)}.$$

We denote the above beta distribution by beta(a, b). A situation where the beta distribution arises is given below.

Consider a Poisson process with arrival rate of λ events per unit time. Let W_k denote the waiting time until the kth arrival of an event and W_s denote the waiting time until the sth arrival, $s > k$. Then, W_k and $W_s - W_k$ are independent gamma random variables with

$$W_k \sim \text{gamma}\left(k, \frac{1}{\lambda}\right) \quad \text{and} \quad W_s - W_k \sim \text{gamma}\left(s - k, \frac{1}{\lambda}\right).$$

The proportion of the time taken by the first k arrivals in the time needed for the first s arrivals is

$$\frac{W_k}{W_s} = \frac{W_k}{W_k + (W_s - W_k)} \sim \text{beta}(k, s - k).$$

The beta density plots are given for various values of a and b in Figure 17.1. We observe from the plots that the beta density is U shaped when $a < 1$ and $b < 1$, symmetric about 0.5 when $a = b > 1$, J shaped when $(a-1)(b-1) < 0$, and unimodal for other values of a and b. For equally large values of a and b, the cumulative probabilities of a beta distribution can be approximated by a normal distribution.

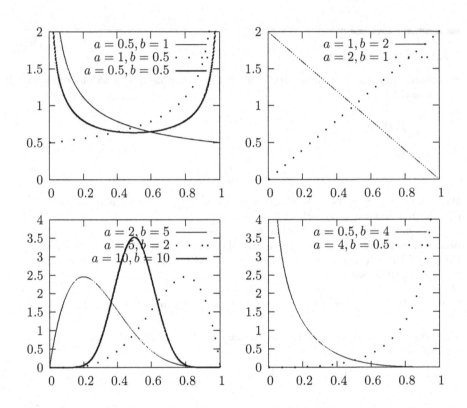

FIGURE 17.1: Beta probability density functions

17.2 Moments

Mean:	$\frac{a}{a+b}$
Variance:	$\frac{ab}{(a+b)^2(a+b+1)}$
Mode:	$\frac{a-1}{a+b-2}, \quad a > 1, \ b > 1$
Mean Deviation:	$\frac{\Gamma(a+b)}{\Gamma(a)\Gamma(b)} \frac{2a^a b^b}{(a+b)^{(a+b+1)}}$
Coefficient of Skewness:	$\frac{2(b-a)(a+b+1)^{1/2}}{(a+b+2)(ab)^{1/2}}$
Coefficient of Variation:	$\frac{\sqrt{b}}{\sqrt{a(a+b+1)}}$
Coefficient of Kurtosis:	$\frac{3(a+b+1)[2(a+b)^2+ab(a+b-6)]}{ab(a+b+2)(a+b+3)}$
Characteristic Function:	$\frac{\Gamma(a+b)}{\Gamma(a)} \sum_{k=0}^{\infty} \frac{\Gamma(a+k)(it)^2}{\Gamma(a+b+k)\Gamma(k+1)}$
Moments about the Origin:	$E(X^k) = \prod_{i=0}^{k-1} \frac{a+i}{a+b+i}, \quad k = 1, 2, ...$

17.3 Probabilities, Percentiles, and Moments

The dialog box [StatCalc→Continuous→Beta] computes the distribution function, percentiles and moments of a beta distribution.

To compute probabilities: Enter the values of the parameters a and b, and the value of x; click [P(X <= x)]. For example, hen $a = 2$, $b = 3$, and $x = 0.4$, $P(X \leq 0.4) = 0.5248$ and $P(X > 0.4) = 0.4752$.

To compute percentiles: Enter the values of a, b, and the cumulative probability; click [x]. For example, when $a = 2$, $b = 3$, and the cumulative probability $= 0.40$, the 40th percentile is 0.329167. That is, $P(X \leq 0.329167) = 0.40$.

To compute other parameters: Enter the values of one of the parameters, cumulative probability, and the value of x; click on the missing parameter. For example, hen $b = 3$, $x = 0.8$, and the cumulative probability $= 0.40$, the value of a is 12.959.

To compute moments: Enter the values of a and b and click [M].

17.4   Inferences

Let $X_1, \ldots, X_n$ be a sample from a beta distribution with shape parameters a and b. Let

$$\bar{X} = \frac{1}{n} \sum_{i=1}^{n} X_i \text{ and } S^2 = \frac{1}{n-1} \sum_{i=1}^{n} (X_i - \bar{X})^2.$$

Moment Estimators

$$\widehat{a} = \bar{X} \left[\frac{\bar{X}(1 - \bar{X})}{S^2} - 1 \right] \text{ and } \widehat{b} = \frac{(1 - \bar{X})\widehat{a}}{\bar{X}}.$$

Maximum Likelihood Estimators

MLEs are the solution of the equations

$$\psi(\widehat{a}) - \psi(\widehat{a} + \widehat{b}) = \frac{1}{n} \sum_{i=1}^{n} \ln(X_i)$$

$$\psi(\widehat{b}) - \psi(\widehat{a} + \widehat{b}) = \frac{1}{n} \sum_{i=1}^{n} \ln(1 - X_i),$$

where $\psi(x)$ is the digamma function given in Section 2.11. Moment estimators can be used as initial values to solve the above equations numerically.

17.5   Applications with an Example

As mentioned in earlier chapters, the beta distribution is related to many other distributions such as Student's t, F, noncentral F, binomial, and negative binomial distributions. Therefore, cumulative probabilities and percentiles of these distributions can be obtained from those of beta distributions. For example, as mentioned in Sections 4.4.3 and 8.6, percentiles of beta distributions can be used to construct exact confidence limits for binomial and negative binomial success probabilities. In Bayesian analysis, the beta distribution is considered as a conjugate prior distribution for the binomial success probability p. Beta distributions are often used to model data consisting of proportions. Applications of beta distributions in risk analysis are mentioned in Johnson (1997).

Chia and Hutchinson (1991) used a beta distribution to fit the frequency distribution of daily cloud durations, where cloud duration is defined as the fraction of daylight hours not receiving bright sunshine. They used data collected from 11 Australian locations to construct empirical frequency distributions of 132 (11 stations by 12 months) daily cloud durations. Sulaiman et al. (1999) fitted Malaysian sunshine data covering a 10-year period to a beta distribution. Nicas (1994) pointed out that beta distributions offer greater flexibility than lognormal distributions in modeling respirator penetration values over the physically plausible interval [0,1]. An approach for dynamically computing the retirement probability and the retirement rate when the age manpower follows a beta distribution is given in Shivanagaraju et al. (1998). The coefficient of kurtosis of the beta distribution has been used as a good indicator of the condition of a gear (Oguamanam et al., 1995). Schwarzenberg-Czerny (1997) showed that the phase dispersion minimization statistic (a popular method for searching for nonsinusoidal pulsations) follows a beta distribution. In the following, we give an illustrative example.

Example 17.1. National Climatic Center (North Carolina, USA) reported the following data in Table 16.1 on percentage of day during that sunshine occurred in Atlanta, Georgia, November 1–30, 1974. Daniel (1990) considered these data to demonstrate the application of a run test for testing randomness. Let us check if a beta model fits the data.

TABLE 17.1: Percentage of Sunshine Period in a Day in November 1974

85	85	99	70	17	74	100	28	100	100	31	86	100	0	100
100	45	7	12	54	87	100	100	88	50	100	100	100	48	0

To fit a beta distribution, we first compute the mean and variance of the data:

$$\bar{x} = 0.6887 \quad \text{and} \quad s^2 = 0.1276.$$

Using the computed mean and variance, we compute the moment estimators (see Section 17.4) as

$$\hat{a} = 0.4687 \quad \text{and} \quad \hat{b} = 0.2116.$$

The observed quantiles q_j (that is, the ordered proportions) for the data are given in the second column of Table 16.2. The estimated shape parameters can be used to compute the beta quantiles so that they can be compared with the corresponding observed quantiles. For example, when the observed quantile is 0.31 (at $j = 7$), the corresponding beta quantile Q_j can be computed as

$$Q_j = \text{beta}^{-1}(0.21667; \hat{a}, \hat{b}) = 0.30308,$$

where $\text{beta}^{-1}(p; \hat{a}, \hat{b})$ denotes the $100p$th percentile of the beta distribution with shape parameters $\hat{a}$ and $\hat{b}$. Comparison between the sample quantiles and the corresponding beta quantiles (see the Q–Q plot in Figure 16.1) indicates that the data set is well fitted by the beta($\hat{a}$, $\hat{b}$)distribution. Using this fitted beta distribution, we can estimate the probability that the sunshine period exceeds a given proportion in a November day in Atlanta. For example, the estimated probability that at least 70% of a November day will have sunshine is given by $P(X \geq 0.7) = 0.61546$, where X is the beta(0.4687, 0.2116) random variable.

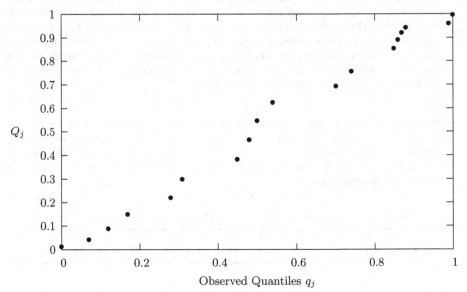

Figure 16.1 Beta Q–Q plots of the sunshine data

TABLE 17.2: Observed and Beta Quantiles for Sunshine Data

j	Observed Quantiles q_j	Cumulative Probability $(j-0.5)/30$	Beta Quantiles Q_j	j	Observed Quantiles q_j	Cumulative Probability $(j-0.5)/30$	Beta Quantiles Q_j
1	0			12	0.7	0.383333	0.69772
2	0	0.05	0.01586	13	0.74	0.416667	0.75946
3	0.7	0.083333	0.04639	14	0.85	....	....
4	0.12	0.116667	0.09263	15	0.85	0.483333	0.85746
5	0.17	0.15	0.15278	16	0.86	0.516667	0.89415
6	0.28	0.183333	0.22404	17	0.87	0.55	0.92344
7	0.31	0.216667	0.30308	18	0.88	0.583333	0.94623
8	0.45	0.25	0.38625	19	0.99	0.616667	0.96346
9	0.48	0.283333	0.47009	20	1	...	...
10	0.5	0.316667	0.55153	...	...	...	...
11	0.54	0.35	0.62802	30	1	0.983333	1

17.6 Properties and Results

17.6.1 An Identity and Recurrence Relations

1. Let $F(x|a,b)$ denote the cumulative distribution of a beta(a, b) random variable; that is $F(x|a,b) = P(X \le x|a,b)$.

   a. $F(x|a,b) = 1 - F(1 - x|b,a)$.

   b. $F(x|a,b) = xF(x|a-1,b) + (1-x)F(x|a,b-1)$, $a > 1, b > 1$.

   c. $F(x|a,b) = [F(x|a+1,b) - (1-x)F(x|a+1,b-1)]/x$, $b > 1$.

   d. $F(x|a,b) = [aF(x|a+1,b) + bF(x|a,b+1)]/(a+b)$.

   e. $F(x|a,b) = \frac{\Gamma(a+b)}{\Gamma(a+1)\Gamma(b)}x^a(1-x)^{b-1} + F(x|a+1,b-1)$, $b > 1$.

   f. $F(x|a,b) = \frac{\Gamma(a+b)}{\Gamma(a+1)\Gamma(b)}x^a(1-x)^b + F(x|a+1,b)$.

   g. $F(x|a,a) = \frac{1}{2}F(1 - 4(x - 0.5)^2|a, 0.5)$, $x \le 0.5$.

   [Abramowitz and Stegun 1965, p. 944]

17.6.2 Relation to Other Distributions

1. If X is a beta random variable, then $Y = (-\ln(X))^{\frac{1}{3}}$ has an approximate normal distribution.

2. Chi-square Distribution: Let X and Y be independent chi-square random variables with degrees of freedom (df) m and n, respectively. Then

$$\frac{X}{X+Y} \sim \text{beta}(m/2, n/2) \text{ distribution.}$$

3. Student's t Distribution: Let t be a Student's t random variable with df $= n$. Then

$$P(|t| \le x) = P(Y \le x^2/(n + x^2)) \text{ for } x > 0,$$

where Y is a beta$(1/2, n/2)$ random variable.

4. Uniform Distribution: The beta(a, b) distribution specializes to the uniform$(0,1)$ distribution when $a = 1$ and $b = 1$.

5. Let $X_1,\ldots,X_n$ be independent uniform$(0,1)$ random variables, and let $X_{(k)}$ denote the kth order statistic. Then, $X_{(k)}$ follows a beta$(k, n-k+1)$ distribution.

6. F Distribution: Let X be a beta$(m/2, n/2)$ random variable. Then

$$\frac{nX}{m(1 - X)} \sim F_{m,n} \text{ distribution.}$$

7. Binomial: Let X be a binomial(n, p) random variable. Then, for a given k,

$$P(X \ge k|n, p) = P(Y \le p),$$

where Y is a beta$(k, n - k + 1)$ random variable. Furthermore,

$$P(X \le k|n, p) = P(W \ge p),$$

where W is a beta$(k+1, n - k)$ random variable.

8. Negative Binomial: Let X be a negative binomial(r, p) random variable.

$$P(X \le k|r, p) = P(W \le p),$$

where W is a beta random variable with parameters r and $k + 1$.

9. Gamma: Let X and Y be independent gamma random variables with the same scale parameter b, but possibly different shape parameters a_1 and a_2. Then

$$\frac{X}{X + Y} \sim \text{beta}(a_1, a_2).$$

17.7   Random Number Generation

The following algorithm generates beta(a, b) variates. It uses the approach by Jöhnk (1964) when $\min\{a, b\} < 1$ and Algorithm 2P of Schmeiser and Shalaby (1980) otherwise.

Algorithm 17.1. [1]

```
Input:
    a, b = the shape parameters

Output:
    x is a random variate from beta(a, b) distribution

        if a > 1 and b > 1, goto 1
2       Generate u1 and u2 from uniform(0, 1)
        Set s1 = u1**(1./a)
            s2 = u2**(1./b)
            s = s1 + s2
            x = s1/s
        if(s <= 1.0) return x
        goto 2

1       Set aa = a - 1.0
            bb = b - 1.0
            r = aa + bb
            s = r*ln(r)
            x1 = 0.0; x2 = 0.0
            x3 = aa/r
            x4 = 1.0; x5 = 1.0
            f2 = 0.0; f4 = 0.0
        if(r <= 1.0) goto 4
        d = sqrt(aa*bb/(r-1.0))/r
        if(d >= x3) goto 3
```

---

[1]Reproduced with permission from the American Statistical Association.

```
        x2 = x3 - d
        x1 = x2 - (x2*(1.0-x2))/(aa-r*x2)
        f2 = exp(aa*ln(x2/aa)+bb*ln((1.0-x2)/bb)+s)

3       if(x3+d >= 1.0) goto 4
        x4 = x3 + d
        x5 = x4 - (x4*(1.0-x4)/(aa-r*x4))
        f4 = exp(aa*ln(x4/aa) + bb*ln((1.0-x4)/bb)+s)

4       p1 = x3 - x2
        p2 = (x4 - x3) + p1
        p3 = f2*x2/2.0 + p2
        p4 = f4*(1.0-x4)/2.0+ p3

5       Generate u from uniform(0,1)
        Set u = u*p4
        Generate w from uniform(0,1)
        if(u > p1) goto 7
        x = x2 + w*(x3-x2)
        v = u/p1
        if(v <= f2 + w*(1.0-f2)) return x
        goto 10

7       if(u > p2) goto 8
        x = x3 + w*(x4 - x3)
        v = (u - p1)/(p2 - p1)
        if(v <= 1.0 - (1.0-f4)/w) return x
        goto 10

8       Generate w2 from uniform(0,1)
        if(w2 > w) w = w2
        if(u > p3) goto 9
        x = w*x2
        v = (u-p2)/(p3-p2)*w*f2
        if(x <= x1) goto 10
        if(v <= f2*(x-x1)/(x2-x1)) return x
        goto 10

9       x = 1.0 - w*(1.0-x4)
        v = ((u-p3)/(p4-p3))*(1.0-x)*f4/(1.0-x4)
        if(x >= x5) goto 10
        if(v <= f4*(x5-x)/(x5-x4)) return x

10      ca = ln(v)
        if(ca >= -2.0*r*(x-x3)**2) goto 5
        if(ca <= aa*ln(x/aa)+bb*ln((1.0-x)/bb) + s) return x
        goto 5
```

For other equally efficient algorithms, see Cheng (1978).

17.8   Computation of the Distribution Function

The recurrence relation

$$F(x|a,b) = \frac{\Gamma(a+b)}{\Gamma(a+1)\Gamma(b)}x^a(1-x)^b + F(x|a+1,b)$$

can be used to evaluate the cdf at a given x, a and b. The above relation produces the series

$$
\begin{aligned}
F(x|a,b) &= \frac{x^a(1-x)^b}{\text{Beta}(a+1,b)}\left(\frac{1}{a+b} + \frac{x}{a+1} + \frac{(a+b+1)x^2}{(a+1)(a+2)}\right.\\
&+ \left.\frac{(a+b+1)(a+b+2)}{(a+1)(a+2)(a+3)}x^3 + \ldots\right).
\end{aligned}
\tag{17.1}
$$

If $x > 0.5$, then, to speed up the convergence, compute first $F(1-x|b,a)$, and then use the relation that $F(x|a,b) = 1 - F(1-x|b,a)$ to evaluate $F(x|a,b)$.

Majumder and Bhattacharjee (1973a, Algorithm AS 63) proposed a slightly faster approach than the above method. Their algorithm uses a combination of the recurrence relations 1(e) and 1(f) in Section 16.6.1, depending on the parameter configurations and the value of x at which the cdf is evaluated. For computing percentiles of a beta distribution, see Majumder and Bhattacharjee (1973b, Algorithm AS 64).

The following R function evaluates the cdf of a beta(a,b) distribution, and is based on the above method.

R function 17.1.   Calculation of the cdf of a beta(a,b) distribution

```
betacdf = function(x, a, b){
one = 1.0; errtol = 1.0e-12; zero = 0.0; maxitr = 1000
if(x > 0.5){
xx = one-x; aa = b; bb = a
check = TRUE}
else{
xx = x; aa = a; bb = b
check = FALSE}
# alng(x) = logarithmic gamma function; R function 1.1
bet = alng(aa+bb+one)-alng(aa+one)-alng(bb)
su = zero
term = xx/(aa+one)
i = 1
```

```
repeat{
su = su + term
if(term <= errtol | i >= maxitr){break}
term = term*(aa+bb+i)*xx/(aa+i+one)
i = i + 1
}
ans = (su + one/(aa+bb))*exp(bet+aa*log(xx)+bb*log(one-xx))
if(check) ans = one-ans
return(ans)
}
```

18

Noncentral Chi-square Distribution

18.1 Description

The probability density function (pdf) of a noncentral chi-square random variable with the degrees of freedom n and the noncentrality parameter δ is given by

$$f(x|n,\delta) = \sum_{k=0}^{\infty} \frac{\exp\left(-\frac{\delta}{2}\right)\left(\frac{\delta}{2}\right)^k}{k!} \frac{\exp\left(-\frac{x}{2}\right) x^{\frac{n+2k}{2}-1}}{2^{\frac{n+2k}{2}}\Gamma\left(\frac{n+2k}{2}\right)}, \qquad (18.1)$$

where $x > 0$, $n > 0$, and $\delta > 0$. This random variable is usually denoted by $\chi_n^2(\delta)$. It is clear from the density function (18.1) that conditionally given K, $\chi_n^2(\delta)$ is distributed as χ_{n+2K}^2, where K is a Poisson random variable with mean $\delta/2$. Thus, the cumulative distribution of $\chi_n^2(\delta)$ can be written as

$$P(\chi_n^2(\delta) \le x|n,\delta) = \sum_{k=0}^{\infty} \frac{\exp\left(-\frac{\delta}{2}\right)\left(\frac{\delta}{2}\right)^k}{k!} P(\chi_{n+2k}^2 \le x). \qquad (18.2)$$

The plots of the noncentral chi-square pdfs in Figure 18.1 show that, for fixed n, $\chi_n^2(\delta)$ is stochastically increasing with respect to δ, and for large values of n, the pdf is approximately symmetric about its mean $n + \delta$.

18.2 Moments

Mean:	$n + \delta$
Variance:	$2n + 4\delta$
Coefficient of Variation:	$\frac{\sqrt{(2n+4\delta)}}{(n+\delta)}$
Coefficient of Skewness:	$\frac{(n+3\delta)\sqrt{8}}{(n+2\delta)^{3/2}}$
Coefficient of Kurtosis:	$3 + \frac{12(n+4\delta)}{(n+2\delta)^2}$

Moment Generating Function:	$(1 - 2t)^{-n/2} \exp[t\delta/(1 - 2t)]$
Moments about the Origin:	$E(X^k) = 2^k \Gamma(n/2 + k) \sum\limits_{j=0}^{\infty} \binom{k}{j} \frac{(\delta/2)^j}{\Gamma(n/2+j)}$,
	$k = 1, 2, \ldots$
	[Johnson and Kotz 1970, p. 135]

18.3 Probabilities, Percentiles, and Moments

The dialog box [StatCalc→Continuous→NC Chi-sqr] computes the distribution function, percentiles, moments, and other parameters of a noncentral chi-square distribution.

To compute probabilities: Enter the values of the degrees of freedom (df), noncentrality parameter, and the value of x; click [P(X <= x)]. For example, when df = 13.0, noncentrality parameter = 2.2 and the observed value $x = 12.3$,

$$P(X \leq 12.3) = 0.346216 \quad \text{and} \quad P(X > 12.3) = 0.653784.$$

To compute percentiles: Enter the values of the df, noncentrality parameter, and the cumulative probability; click [x]. For example, when df = 13.0, noncentrality parameter = 2.2, and the cumulative probability = 0.95, the 95th percentile is 26.0113. That is,
$$P(X \leq 26.0113) = 0.95.$$

To compute other parameters: Enter the values of one of the parameters, the cumulative probability, and click on the missing parameter. For example, when df = 13.0, the cumulative probability = 0.95, and $x = 25.0$, the value of the noncentrality parameter is 1.57552.

To compute moments: Enter the values of the df and the noncentrality parameter; click [M].

18.4 Applications

The noncentral chi-square distribution is useful in computing the power of the goodness-of-fit test based on the usual chi-square statistic (see Section 1.4.2)

$$Q = \sum_{i=1}^{k} \frac{(O_i - E_i)^2}{E_i},$$

where O_i is the observed frequency in the ith cell, $E_i = N p_{i0}$ is the expected frequency in the ith cell, p_{i0} is the specified (under the null hypothesis) probability

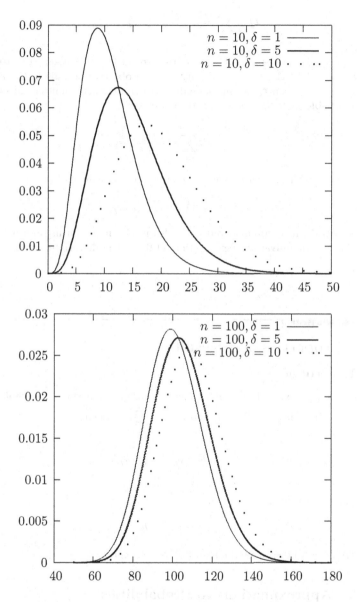

FIGURE 18.1: Noncentral chi-square probability density functions

that an observation falls in the ith cell, $i = 1, \cdots, k$, and N = total number of observations. The null hypothesis will be rejected if

$$Q = \sum_{i=1}^{k} \frac{(O_i - E_i)^2}{E_i} > \chi^2_{k-1, 1-\alpha},$$

where $\chi^2_{k-1, 1-\alpha}$ denotes the $100(1 - \alpha)$th percentile of a chi-square distribution with df $= k - 1$. If the true probability that an observation falls in the ith cell is p_i, $i = 1, \cdots, k$, then Q is approximately distributed as a noncentral chi-square random variable with the noncentrality parameter

$$\delta = N \sum_{i=1}^{k} \frac{(p_i - p_{i0})^2}{p_{i0}},$$

and df $= k - 1$. Thus, an approximate power function is given by

$$P\left(\chi^2_{k-1}(\delta) > \chi^2_{k-1, 1-\alpha}\right).$$

The noncentral chi-square distribution is also useful in computing approximate tolerance factors for univariate (see Section 11.6) and multivariate normal tolerance factor.

18.5   Properties and Results

18.5.1   Properties

1. Let $X_1, \ldots, X_n$ be independent normal random variables with $X_i \sim N(\mu_i, 1)$, $i = 1, 2, ..., n$, and let $\delta = \sum_{i}^{n} \mu_i^2$. Then

$$\sum_{i=1}^{n} X_i^2 \sim \chi^2_n(\delta).$$

2. For any real valued function h,

$$E[h(\chi^2_n(\delta))] = E[E(h(\chi^2_{n+2K})|K)],$$

where K is a Poisson random variable with mean $\delta/2$.

18.5.2   Approximations to Probabilities

Let $a = n + \delta$ and $b = \delta/(n + \delta)$.

1. Let Y be a chi-square random variable with df $= a/(1+b)$. Then

$$P(\chi^2_n(\delta) \le x) \simeq P\left(Y \le \frac{x}{1+b}\right).$$

2. Let Z denote the standard normal random variable. Then

    a. $P(\chi_n^2(\delta) \leq x) \simeq P\left(Z \leq \dfrac{\left(\frac{x}{a}\right)^{1/3} - \left[1 - \frac{2}{9}\left(\frac{1+b}{a}\right)\right]}{\sqrt{\frac{2}{9}\left(\frac{1+b}{a}\right)}}\right).$

    b. $P(\chi_n^2(\delta) \leq x) \simeq P\left(Z \leq \sqrt{\frac{2x}{1+b}} - \sqrt{\frac{2a}{1+b} - 1}\right).$

18.5.3 Approximations to Percentiles

Let $\chi_{n,p}^2(\delta)$ denote the $100p$th percentile of the noncentral chi-square distribution with df $= n$, and noncentrality parameter δ. Define $a = n + \delta$ and $b = \delta/(n+\delta)$

1. Patnaik's (1949) Approximation:

$$\chi_{n,p}^2(\delta) \simeq c\chi_{f,p}^2,$$

    where $c = 1 + b$, and $\chi_{f,p}^2$ denotes the $100p$th percentile of the central chi-square distribution with df $f = a/(1+b)$.

2. Normal Approximations: Let z_p denote the $100p$th percentile of the standard normal distribution.

    a. $\chi_{n,p}^2(\delta) \simeq \frac{1+b}{2}\left(z_p + \sqrt{\frac{2a}{1+b} - 1}\right)^2.$

    b. $\chi_{n,p}^2(\delta) \simeq a\left[z_p\sqrt{\frac{2}{9}\left(\frac{1+b}{a}\right)} - \frac{2}{9}\left(\frac{1+b}{a}\right) + 1\right]^3.$

18.6 Random Number Generation

The following exact method can be used to generate random numbers when the degrees of freedom $n \geq 1$. The following algorithm is based on the additive property of the noncentral chi-square distribution given in Section 17.5.1.

Algorithm 18.1. Noncentral chi-square variate generator

For a given n and δ:
Set u = sqrt(δ)
Generate z_1 from N(u, 1)
Generate y from gamma($(n-1)/2, 2$)
return x = $z_1^2 + y$
x is a random variate from $\chi_n^2(\delta)$ distribution.

18.7   Evaluating the Distribution Function

The following computational method is due to Benton and Krishnamoorthy (2003), and is based on the following infinite series expression for the cdf.

$$P(\chi_n^2(\delta) \le x) = \sum_{i=0}^{\infty} P(X = i) I_{x/2}(n/2 + i), \qquad (18.3)$$

where X is a Poisson random variable with mean $\delta/2$, and

$$I_y(a) = \frac{1}{\Gamma(a)} \int_0^y e^{-t} t^{a-1} dt, \quad a > 0, \ y > 0, \qquad (18.4)$$

is the incomplete gamma function. To compute (18.3), evaluate first the kth term, where k is the integer part of $\delta/2$, and then compute the other Poisson probabilities and incomplete gamma functions recursively using forward and backward recursions. To compute Poisson probabilities, use the relations

$$P(X = k+1) = \frac{\delta/2}{k+1} P(X = k), \quad k = 0, 1, \dots$$

and

$$P(X = k-1) = \frac{k}{\delta/2} P(x = k), \quad k = 1, 2, \dots.$$

To compute the incomplete gamma function, use the relations

$$I_x(a+1) = I_x(a) - \frac{x^a \exp(-x)}{\Gamma(a+1)}, \qquad (18.5)$$

and

$$I_x(a-1) = I_x(a) + \frac{x^{a-1} \exp(-x)}{\Gamma(a)}. \qquad (18.6)$$

Furthermore, the series expansion

$$I_x(a) = \frac{x^a \exp(-x)}{\Gamma(a+1)} \left(1 + \frac{x}{(a+1)} + \frac{x^2}{(a+1)(a+2)} + \cdots \right)$$

can be used to evaluate $I_x(a)$.

When computing the terms using both forward and backward recurrence relations, stop if

$$1 - \sum_{j=k-i}^{k+i} P(X = j)$$

is less than the error tolerance or the number of iterations is greater than a specified integer. While computing using only forward recurrence relation, stop if

$$\left(1 - \sum_{j=0}^{2k+i} P(X = j) \right) I_x(2k + i + 1)$$

is less than the error tolerance or the number of iterations is greater than a specified integer.

The R function 18.1 computes the noncentral chi-square distribution function, and is based on the algorithm given in Benton and Krishnamoorthy (2003).

R function 18.1. Calculation of the noncentral chi-square cdf

```
ncchicdf = function(xx, dfs, lambda){
one = 1.0; half = 0.5; zero = 0.0; maxitr = 1000; errtol = 1.0e-12
if(xx <= zero){return(0)}
x = half*xx; del = half*lambda; k = floor(del)+one
a = half*dfs + k
gamkf = gamcdf(x, a); gamkb = gamkf
if(del == zero){return (gamkf)}
poikf = poipr(k, del); poikb = poikf
# alng(x) = logarithmic gamma function in Section 1.8
xtermf = exp((a-one)*log(x)-x-alng(a)); xtermb = xtermf*x/a
su = poikf * gamkf; remain = one - poikf
i = 0
repeat{
i = i + 1
xtermf = xtermf*x/(a+i-one)
gamkf = gamkf - xtermf; poikf = poikf * del/(k+i)
termf = poikf * gamkf; su = su + termf; error = remain * gamkf
remain = remain - poikf
# Do forward and backward computations "maxitr" times or until
# convergence
if (i > k){
if(error <= errtol | i > maxitr){break}
}
else{
xtermb = xtermb * (a-i+one)/x
gamkb = gamkb + xtermb; poikb = poikb * (k-i+one)/del
termb = gamkb * poikb
su = su + termb; remain = remain - poikb
if(remain <= errtol | i > maxitr){break}
}}
return(su)
}
```

19

Noncentral F Distribution

19.1 Description

Let $\chi_m^2(\delta)$ be a noncentral chi-square random variable with degrees of freedom (df) $= m$, and noncentrality parameter δ, and χ_n^2 be a chi-square random variable with df $= n$. If $\chi_m^2(\delta)$ and χ_n^2 are independent, then the distribution of the ratio

$$F_{m,n}(\delta) = \frac{\chi_m^2(\delta)/m}{\chi_n^2/n}$$

is called the noncentral F distribution with the numerator df $= m$, the denominator df $= n$, and the noncentrality parameter δ.

The cumulative distribution function (cdf) is given by

$$F(x|m,n,\delta) = \sum_{k=0}^{\infty} \frac{\exp(-\frac{\delta}{2})(\frac{\delta}{2})^k}{k!} P\left(F_{m+2k,n} \le \frac{mx}{m+2k}\right),$$
$$m > 0, n > 0, \delta > 0,$$

where $F_{a,b}$ denotes the central F random variable with the numerator df $= a$, and the denominator df $= b$.

The plots of probability density functions (pdfs) of $F_{m,n}(\delta)$ are presented in Figure 19.1 for various values of m, n and δ. It is clear from the plots that the noncentral F distribution is always right skewed.

19.2 Moments

Mean: $\frac{n(m+\delta)}{m(n-2)}$, $n > 2$

Variance: $\frac{2n^2[(m+\delta)^2+(m+2\delta)(n-2)]}{m^2(n-2)^2(n-4)}$, $n > 4$

$E(F_{m,n}^k)$: $\frac{\Gamma[(n-2k)/2]\,\Gamma[(m+2k)/2]\,n^k}{\Gamma(n/2)m^k} \sum_{j=0}^{k} \binom{k}{j} \frac{(\delta/2)^j}{\Gamma[(m+2j)/2]}$, $n > 2k$

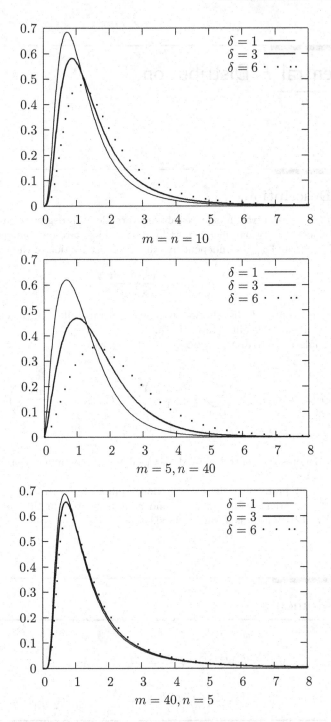

FIGURE 19.1: Noncentral *F* probability density functions

19.3 Probabilities, Percentiles, and Moments

The dialog box [StatCalc→Continuous→NC F] computes cumulative probabilities, percentiles, moments, and other parameters of an $F_{m,n}(\delta)$ distribution.

To compute probabilities: Enter the values of the numerator df, denominator df, noncentrality parameter, and x; click [P(X <= x)]. For example, when numerator df = 4.0, denominator df = 32.0, noncentrality parameter = 2.2, and $x = 2$, $P(X \le 2) = 0.702751$ and $P(X > 2) = 0.297249$.

To compute percentiles: Enter the values of the df, noncentrality parameter, and the cumulative probability; click [x]. For example, when numerator df = 4.0, denominator df = 32.0, noncentrality parameter = 2.2, and the cumulative probability = 0.90, the 90th percentile is 3.22243. That is, $P(X \le 3.22243) = 0.90$.

To compute moments: Enter the values of the numerator df, denominator df and the noncentrality parameter; click [M].

StatCalc also computes one of the dfs or the noncentrality parameter for given other values. For example, when numerator df = 5, denominator df = 12, $x = 2$ and $P(X \le x) = 0.7$, the value of the noncentrality parameter is 2.24162.

19.4 Applications

The noncentral F distribution is useful to compute the powers of a test based on the central F statistic. Examples include analysis of variance and tests based on the Hotelling T^2 statistics. Let us consider the power function of the Hotelling T^2 test for testing about a multivariate normal mean vector.

Let $X_1, \ldots, X_n$ be a sample from an m-variate normal population with mean vector μ and covariance matrix Σ. Define

$$\bar{X} = \frac{1}{n} \sum_{i=1}^{n} X_i \quad \text{and} \quad S = \frac{1}{n-1} \sum_{i=1}^{n} (X_i - \bar{X})(X_i - \bar{X})'.$$

The Hotelling T^2 statistic for testing $H_0 : \mu = \mu_0$ vs. $H_a : \mu \ne \mu_0$ is given by

$$T^2 = n \left(\bar{X} - \mu_0 \right)' S^{-1} \left(\bar{X} - \mu_0 \right).$$

Under H_0, $T^2 \sim \frac{(n-1)m}{n-m} F_{m,n-m}$. Under H_a,

$$T^2 \sim \frac{(n-1)m}{n-m} F_{m,n-m}(\delta),$$

where $F_{m,n-m}(\delta)$ denotes the noncentral F random variable with the numerator df = m, denominator df = $n - m$, and the noncentrality parameter $\delta = n(\mu - \mu_0)' \Sigma^{-1}(\mu - \mu_0)$, and μ is true mean vector. The power of the T^2 test is given by

$$P\left(F_{m,n-m}(\delta) > F_{m,n-m,1-\alpha} \right),$$

where $F_{m,n-m,1-\alpha}$ denotes the $100(1 - \alpha)$th percentile of the F distribution with the numerator df $= m$ and denominator df $= n - m$.

The noncentral F distribution also arises in multiple use confidence estimation in a multivariate calibration problem [Mathew and Zha, 1996].

19.5   Properties and Results

19.5.1   Properties

1.
$$\frac{mF_{m,n}(\delta)}{n + mF_{m,n}(\delta)} \sim \text{noncentral beta} \left(\frac{m}{2}, \frac{n}{2}, \delta \right).$$

2. Let $F(x; m, n, \delta)$ denote the cdf of $F_{m,n}(\delta)$. Then

   a. for a fixed m, n, x, $F(x; m, n, \delta)$ is a nonincreasing function of δ;

   b. for a fixed δ, n, x, $F(x; m, n, \delta)$ is a nondecreasing function of m.

19.5.2   Approximations

1. For a large n, $F_{m,n}(\delta)$ is distributed as $\chi_m^2(\delta)/m$.

2. For a large m, $F_{m,n}(\delta)$ is distributed as $(1 + \delta/m)\chi_n^2(\delta)$.

3. For large values of m and n,

$$\frac{F_{m,n}(\delta) - \frac{n(m+\delta)}{m(n-2)}}{\frac{n}{m} \left[\frac{2}{(n-2)(n-4)} \left(\frac{(m+\delta)^2}{n-2} + m + 2\delta \right) \right]^{1/2}} \sim N(0,\ 1) \text{ approximately.}$$

4. Let $m^* = \frac{(m+\delta)^2}{m+2\delta}$. Then

$$\frac{m}{m+\delta} F_{m,n}(\delta) \sim F_{m*,n} \text{ approximately.}$$

5.

$$\frac{\left(\frac{mF_{m,n}(\delta)}{m+\delta} \right)^{1/3} \left(1 - \frac{2}{9n} \right) - \left(1 - \frac{2(m+2\delta)}{9(m+\delta)^2} \right)}{\left[\frac{2(m+2\delta)}{9(m+\delta)^2} + \frac{2}{9n} \left(\frac{mF_{m,n}(\delta)}{m+\delta} \right)^{2/3} \right]^{1/2}} \sim N(0,\ 1) \text{ approximately.}$$

[Abramowitz and Stegun, 1965]

19.6   Random Number Generation

The following algorithm is based on the definition of the noncentral F distribution given in Section 19.1.

Algorithm 19.1. Noncentral F variate generator

---

1. Generate x from the noncentral chi-square distribution with df $= m$ and non-centrality parameter δ (See Section 18.6).
2. Generate y from the central chi-square distribution with df $= n$.
3. return $F = nx/(my)$.

F is a noncentral $F_{m,n}(\delta)$ random number.

19.7 Evaluating the Distribution Function

The following approach is similar to the one for computing the noncentral χ^2 in Section 18.7, and is based on the method for computing the tail probabilities of a noncentral beta distribution given in Chattamvelli and Shanmugham (1997). The distribution function of $F_{m,n}(\delta)$ can be expressed as

$$P(X \le x | m, n, \delta) = \sum_{i=0}^{\infty} \frac{\exp(-\delta/2)(\delta/2)^i}{i!} I_y(m/2 + i, \, n/2), \qquad (19.1)$$

where $y = mx/(mx + n)$, and

$$I_y(a, \, b) = \frac{\Gamma(a+b)}{\Gamma(a)\Gamma(b)} \int_0^y t^{a-1}(1 - t)^{b-1} dt$$

is the incomplete beta function. Let Z denote the Poisson random variable with mean $\delta/2$. To compute the cumulative distribution function, compute first the kth term in the series (19.1), where k is the integral part of $\delta/2$, and then compute other terms recursively. For Poisson probabilities one can use the forward recurrence relation

$$P(X = k+1 | \lambda) = \frac{\lambda}{k+1} p(X = k | \lambda), \quad k = 0, 1, 2, \ldots,$$

and backward recurrence relation

$$P(X = k - 1 | \lambda) = \frac{k}{\lambda} P(X = k | \lambda), \quad k = 1, 2, \ldots. \qquad (19.2)$$

R function 19.1. Calculation of the beta cdf

"alng" (R function 2.1), "poipr" (R function 6.2) and "betacdf" (R function 17.1) are required.

```
ncfcdf = function(x, dfn, dfd, del){
half = 0.5; zero = 0.0; errtol = 1.0e-14;
if(x <= zero){return(0)}
d = half*del;  k = as.integer(d)
y = dfn*x/(dfn*x+dfd); b = half*dfd; a = half*dfn+k
# betacdf(x, a, b) = beta distribution function
fkf = betacdf(y,a,b);
if(d == zero){return(fkf)}
# poipr(k,d) = Poisson pmf given in Section 5.13
pkf = poipr(k,d); fkb = fkf; pkb = pkf
# Logarithmic gamma function alng(x) in Section 1.8 is required
xtermf = exp(alng(a+b-1)-alng(a)-alng(b)+(a-1)*log(y)+ b*log(1-y))
xtermb = xtermf*y*(a+b-1)/a
cdf = fkf*pkf
sumpois = 1 - pkf
if(k == 0){
i = 1
cdf = back.sum(i,xtermf,a,b,y,k,d,fkf,pkf,cdf,sumpois,errtol)}
else{
for(i in 1:k){
xtermf = xtermf*y*(a+b+(i-1)-1)/(a+i-1)
fkf = fkf - xtermf
pkf = pkf * d/(k+i)
termf = fkf*pkf
xtermb = xtermb *(a-i+1)/(y*(a+b-i))
fkb = fkb + xtermb
pkb = (k-i+1)*pkb/d
termb = fkb*pkb
term = termf + termb
cdf = cdf + term
sumpois = sumpois-pkf-pkb
if (sumpois <= errtol){break}}
i = k + 1
cdf = back.sum(i,xtermf,a,b,y,k,d,fkf,pkf,cdf,sumpois,errtol)}
return(cdf)
}
back.sum = function(i,xtermf,a,b,y,k,d,fkf,pkf,cdf,sumpois,errtol){
repeat{
xtermf = xtermf*y*(a+b+(i-1)-1)/(a+i-1)
fkf = fkf - xtermf
pkf = pkf*d/(k+i)
termf = fkf*pkf
cdf = cdf + termf
sumpois = sumpois-pkf
if(sumpois <= errtol) {return(cdf)}
i = i+1}
}
```

To compute incomplete beta function, use forward recurrence relation

$$I_x(a + 1, \ b) = I_x(a, \ b) - \frac{\Gamma(a+b)}{\Gamma(a)\Gamma(b)} x^a (1 - x)^b,$$

and backward recurrence relation

$$I_x(a - 1, \ b) = I_x(a, \ b) + \frac{\Gamma(a+b-1)}{\Gamma(a)\Gamma(b)} x^{a-1} (1 - x)^b. \tag{19.3}$$

While computing the terms using both forward and backward recursions, stop if

$$1 - \sum_{j=k-i}^{k+i} P(X = j)$$

is less than the error tolerance or the number of iterations is greater than a specified integer; otherwise, stop if

$$\left(1 - \sum_{j=0}^{2k+i} P(X = j) \right) I_x(m/2 + 2k + i, \ n/2)$$

is less than the error tolerance, or the number of iterations is greater than a specified integer.

The R function 19.1 evaluates the cdf of the noncentral F distribution function with numerator df $=$ dfn, denominator df $=$ dfd and the noncentrality parameter "del."

20

Noncentral t Distribution

20.1 Description

Let X be a normal random variable with mean δ and variance 1 and S^2 be a chi-square random variable with degrees of freedom (df) n. If X and S^2 are independent, then the distribution of the ratio $\sqrt{n}X/S$ is called the noncentral t distribution with the degrees of freedom n and the noncentrality parameter δ. The probability density function is given by

$$f(x|n,\delta) \;=\; \frac{n^{n/2}\exp(-\delta^2/2)}{\sqrt{\pi}\,\Gamma(n/2)(n+x^2)^{(n+1)/2}} \sum_{i=0}^{\infty} \frac{\Gamma[(n+i+1)/2]}{i!}\left(\frac{x\delta\sqrt{2}}{\sqrt{n+x^2}}\right)^i,$$

$$-\infty < x < \infty, \quad -\infty < \delta < \infty,$$

where $\left(\frac{x\delta\sqrt{2}}{\sqrt{n+x^2}}\right)^0$ should be interpreted as 1 for all values of x and δ, including 0. The above noncentral t random variable is denoted by $t_n(\delta)$.

The noncentral t distribution specializes to Student's t distribution when $\delta = 0$. We also observe from the plots of pdfs in Figure 20.1 that the noncentral t random variable is stochastically increasing with respect to δ. That is, for $\delta_2 > \delta_1$,

$$P\left(t_n(\delta_2) > x\right) > P\left(t_n(\delta_1) > x\right) \quad \text{for every } x.$$

20.2 Moments

Mean: $\quad \mu_1 = \dfrac{\Gamma[(n-1)/2]\sqrt{n/2}}{\Gamma(n/2)}\delta$

Variance: $\quad \mu_2 = \dfrac{n}{n-2}(1+\delta^2) - \left(\dfrac{\Gamma[(n-1)/2]}{\Gamma(n/2)}\right)^2 (n/2)\delta^2$

Moments about the Origin: $E(X^k) = \frac{\Gamma[(n-k)/2]n^{k/2}}{2^{k/2}\Gamma(n/2)} u_k,$

where $u_{2k-1} = \sum\limits_{i=1}^{k} \frac{(2k-1)!\delta^{2i-1}}{(2i-1)!(k-i)!2^{k-i}}, \quad k = 1, 2, \ldots$

and $u_{2k} = \sum\limits_{i=0}^{k} \frac{(2k)!\delta^{2i}}{(2i)!(k-i)!2^{k-i}}, \quad k = 1, 2, \ldots$

[Bain, 1969]

Coefficient of Skewness: $\dfrac{\mu_1 \frac{n(2n-3+\delta^2)}{(n-2)(n-3)} - 2\mu_2}{\mu_2^{3/2}}$

Coefficient of Kurtosis: $\dfrac{\frac{n^2}{(n-2)(n-4)}(3+6\delta^2+\delta^4) - (\mu_1)^2 \left[\frac{n[(n+1)\delta^2+3(3n-5)]}{(n-2)(n-3)} - 3\mu_2\right]}{\mu_2^2}.$

[Johnson and Kotz, 1970, p. 204]

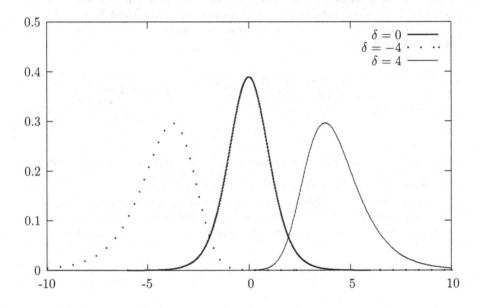

FIGURE 20.1: Noncentral t probability density functions

20.3   Probabilities, Percentiles, and Moments

The dialog box [StatCalc→Continuous→NC t] computes the distribution function, percentiles, moments, and noncentrality parameter.

To compute probabilities: Enter the values of the df, noncentrality parameter, and x; click [P(X <= x)]. For example, when df = 13.0, noncentrality parameter = 2.2, and $x = 2.2$, $P(X \leq 2.2) = 0.483817$ and $P(X > 2.2) = 0.516183$.

To compute percentiles: Enter the values of the df, noncentrality parameter, and the cumulative probability; click [x]. For example, when df = 13.0, noncentrality parameter = 2.2, and the cumulative probability = 0.90, the 90th percentile is 3.87082. That is, $P(X \leq 3.87082) = 0.90$.

To compute other parameters: Enter the values of one of the parameters, the cumulative probability and x. Click on the missing parameter. For example, when df = 13.0, the cumulative probability = 0.40, and $x = 2$, the value of noncentrality parameter is 2.23209.

To compute moments: Enter the values of the df, and the noncentrality parameter; click [M].

20.4 Applications

The noncentral t distribution arises as a power function of a test if the test procedure is based on a central t distribution. More specifically, powers of the t-test for a normal mean and of the two-sample t-test (Sections 11.4.1 and 11.5.2) can be computed using noncentral t distributions. The percentiles of noncentral t distributions are used to compute the one-sided tolerance factors for a normal population (Section 10.6) and tolerance limits for the one-way random effects model (Section 10.6.5). This distribution also arises in multiple-use hypothesis testing about the explanatory variable in calibration problems [Krishnamoorthy, Kulkarni and Mathew (2001), and Benton, Krishnamoorthy and Mathew (2003)].

20.5 Properties and Results

20.5.1 Properties

1. The noncentral distribution $t_n(\delta)$ specializes to the t distribution with df = n when $\delta = 0$.

2. $P(t_n(\delta) \leq 0) = P(Z \leq -\delta)$, where Z is the standard normal random variable.

3. $P(t_n(\delta) \leq t) = P(t_n(-\delta) \geq -t)$.

4. **a.** $P(0 < t_n(\delta) < t) = \sum\limits_{j=0}^{\infty} \frac{\exp(-\delta^2/2)(\delta^2/2)^{j/2}}{\Gamma(j/2+1)} P\left(Y_j \leq \frac{t^2}{n+t^2}\right),$

    b. $P(|t_n(\delta)| < t) = \sum\limits_{j=0}^{\infty} \frac{\exp(-\delta^2/2)(\delta^2/2)^{j}}{j!} P\left(Y_j \leq \frac{t^2}{n+t^2}\right),$

    where Y_j denotes the beta($(j+1)/2, n/2$) random variable, $j = 1, 2, \ldots$ [Craig, 1941, and Guenther, 1978].

5.

$$P(0 < t_n(\delta) < t) = \sum_{j=0}^{\infty} \frac{\exp(-\delta^2/2)(\delta^2/2)^j}{j!} P\left(Y_{1j} \leq \frac{t^2}{n+t^2}\right)$$

$$+ \frac{\delta}{2\sqrt{2}} \sum_{j=0}^{\infty} \frac{\exp(-\delta^2/2)(\delta^2/2)^j}{\Gamma(j+3/2)} P\left(Y_{2j} \leq \frac{t^2}{n+t^2}\right),$$

where Y_{1j} denotes the beta$((j+1)/2, n/2)$ random variable and Y_{2j} denotes the beta$(j+1, n/2)$ random variable, $j = 1, 2, \ldots$ [Guenther, 1978].

6. Relation to the Sample Correlation Coefficient: Let R denote the correlation coefficient of a random sample of $n + 2$ observations from a bivariate normal population. Then, letting

$$\rho = \delta\sqrt{2/(2n+1+\delta^2)},$$

the following function of R,

$$\frac{R}{\sqrt{1-R^2}}\sqrt{\frac{n(2n+1)}{2n+1+\delta^2}} \sim t_n(\delta) \text{ approximately. [Harley, 1957]}$$

20.5.2   An Approximation

Let $X = t_n(\delta)$. Then

$$Z = \frac{X\left(1 - \frac{1}{4n}\right) - \delta}{\left(1 + \frac{X^2}{2n}\right)^{1/2}} \sim N(0,1) \text{ approximately.}$$

[Abramowitz and Stegun 1965, p 949]

20.6   Random Number Generation

The following algorithm for generating $t_n(\delta)$ variates is based on the definition given in Section 20.1.

Algorithm 20.1. Noncentral t variate generator

Generate z from N(0, 1)
Set w = z + δ
Generate y from gamma(n/2, 2)
return x = w*sqrt(n)/sqrt(y)

20.7   Evaluating the Distribution Function

The following method is due to Benton and Krishnamoorthy (2003). Letting $x = \frac{t^2}{n+t^2}$, the distribution function can be expressed as

$$
\begin{aligned}
P(t_n(\delta) \le t) &= \Phi(-\delta) + P(0 < t_n(\delta) \le t) \\
&= \Phi(-\delta) + \frac{1}{2}\sum_{i=0}^{\infty}\left[P_i I_x(i + 1/2,\ n/2) + \frac{\delta}{\sqrt{2}} Q_i I_x(i + 1,\ n/2) \right],
\end{aligned}
\tag{20.1}
$$

where Φ is the standard normal distribution, $I_x(a,b)$ is the incomplete beta function given by

$$
I_x(a,\ b) = \frac{\Gamma(a + b)}{\Gamma(a)\Gamma(b)} \int_0^x y^{a-1}(1 - y)^{b-1} dy,
$$

$P_i = \exp(-\delta^2/2)(\delta^2/2)^i/i!$ and $Q_i = \exp(-\delta^2/2)(\delta^2/2)^i/\Gamma(i + 3/2), i = 0, 1, 2, ...$

To compute the cdf, first compute the kth term in the series expansion (20.1), where k is the integer part of $\delta^2/2$, and then compute the other terms using forward and backward recursions:

$$
P_{i+1} = \frac{\delta^2}{2(i + 1)} P_i, \qquad P_{i-1} = \frac{2i}{\delta^2} P_i, \qquad Q_{i+1} = \frac{\delta^2}{2i + 3} Q_i, \qquad Q_{i-1} = \frac{2i + 1}{\delta^2} Q_i
$$

$$
I_x(a + 1,\ b) = I_x(a,\ b) - \frac{\Gamma(a + b)}{\Gamma(a + 1)\Gamma(b)} x^a(1 - x)^b,
$$

and

$$
I_x(a - 1,\ b) = I_x(a,\ b) + \frac{\Gamma(a + b - 1)}{\Gamma(a)\Gamma(b)} x^{a-1}(1 - x)^b.
$$

Let E_m denote the remainder of the infinite series in (17.7.1) after the mth term. It can be shown that

$$
|E_m| \le \frac{1}{2}(1 + |\delta|/2) I_x(m + 3/2,\ n/2)\left(1 - \sum_{i=0}^{m} P_i\right)
\tag{20.2}
$$

[See Lenth, 1989, and Benton and Krishnamoorthy, 2003].

Forward and backward iterations can be stopped when $1 - \sum_{j=k-i}^{k+i} P_j$ is less than the error tolerance or when the number of iterations exceeds a specified integer. Otherwise, forward computation of (20.1) can be stopped once the error bound (20.2) is less than a specified error tolerance or the number of iterations exceeds a specified integer. The following R function computes the noncentral t cdf.

R function 20.1. Calculation of the noncentral t cdf

```
nctcdf = function(t, dfs, delta){
zero = 0.0; one = 1.0; half = 0.5
errtol = 1.0e-12; maxitr = 1000
if (t < zero){
x = -t
del = -delta
indx = TRUE}
else{
x = t; del = delta
indx = FALSE}
# normcdf(x) is the normal cdf in Section 10.10
ans = normcdf(-del)
if( x == zero){return(ans)}
y = x*x/(dfs+x*x)
dels = half*del*del
k = floor(dels)+1; a = k+half; c = k+one; b = half*dfs
# alng(x) is the logarithmic gamma function in Section 1.8
pkf = exp(-dels+k*log(dels)-alng(k+one))
pkb = pkf
qkf = exp(-dels+k*log(dels)-alng(k+one+half))
qkb = qkf
# betadf(y, a, b) is the beta cdf in Section 16.6
pbetaf = pbeta(y, a, b); pbetab = pbetaf
qbetaf = pbeta(y, c, b); qbetab = qbetaf
pgamf = exp(alng(a+b-one)-alng(a)-alng(b)+(a-one)*log(y)+
            + b*log(one-y))
pgamb = pgamf*y*(a+b-one)/a
qgamf = exp(alng(c+b-one)-alng(c)-alng(b)+(c-one)*log(y)+
            + b*log(one-y))
qgamb = qgamf*y*(c+b-one)/c
rempois = one - pkf
delosq2 = del/1.4142135623731
sum = pkf*pbetaf+delosq2*qkf*qbetaf
cons = half*(one + half*abs(delta))
i = 0
repeat{
i = i + 1
    pgamf = pgamf*y*(a+b+i-2.0)/(a+i-one)
    pbetaf = pbetaf - pgamf
    pkf = pkf*dels/(k+i)
    ptermf = pkf*pbetaf
    qgamf = qgamf*y*(c+b+i-2.0)/(c+i-one)
    qbetaf = qbetaf - qgamf
    qkf = qkf*dels/(k+i-one+1.5)
    qtermf = qkf*qbetaf
    term = ptermf + delosq2*qtermf
```

```
        sum = sum + term
        error = rempois*cons*pbetaf
        rempois = rempois - pkf
#   Do forward and backward computations k times or until convergence
        if (i > k){
            if(error <= errtol | i > maxitr){break}
            }
        else{
            pgamb = pgamb*(a-i+one)/(y*(a+b-i))
            pbetab = pbetab + pgamb
            pkb = (k-i+one)*pkb/dels
            ptermb = pkb*pbetab
            qgamb = qgamb*(c-i+one)/(y*(c+b-i))
            qbetab = qbetab + qgamb
            qkb = (k-i+one+half)*qkb/dels
            qtermb = qkb*qbetab
            term =  ptermb + delosq2*qtermb
            sum = sum + term
            rempois = rempois - pkb
            if (rempois <= errtol | i >= maxitr){break}
         }}
        tnd = half*sum + ans
        if(indx) tnd = one - tnd
        return(tnd)
    }
```

21

Laplace Distribution

21.1 Description

The distribution with the probability density function (pdf)

$$f(x|a,b) = \frac{1}{2b}\exp\left[-\frac{|x-a|}{b}\right],\tag{21.1}$$

$$-\infty < x < \infty,\ -\infty < a < \infty,\ b > 0,$$

where a is the location parameter and b is the scale parameter, is called the Laplace(a,b) distribution. The cumulative distribution function (cdf) is given by

$$F(x|a,b) = \begin{cases} 1 - \frac{1}{2}\exp\left[\frac{a-x}{b}\right] & \text{for } x \geq a, \\ \frac{1}{2}\exp\left[\frac{x-a}{b}\right] & \text{for } x < a. \end{cases}\tag{21.2}$$

The Laplace distribution is also referred to as the *double exponential* distribution (see Figure 21.1). For any given probability p, the inverse distribution is given by

$$F^{-1}(p|a,b) = \begin{cases} a + b\ln(2p) & \text{for } 0 < p \leq 0.5, \\ a - b\ln(2(1-p)) & \text{for } 0.5 < p < 1. \end{cases}\tag{21.3}$$

21.2 Moments

Mean:	a
Median:	a
Mode:	a
Variance:	$2b^2$
Mean Deviation:	b
Coefficient of Variation:	$\frac{b\sqrt{2}}{a}$

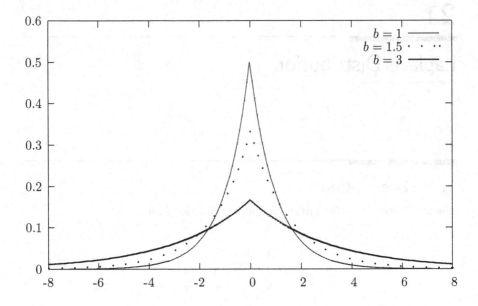

FIGURE 21.1: Laplace probability density functions

Coefficient of Skewness:	0
Coefficient of Kurtosis:	6

$$\text{Moments about the Mean:} \quad E(X-a)^k = \begin{cases} 0 & \text{for } k = 1,3,5,\ldots \\ k!b^k & \text{for } k = 2,4,6,\ldots \end{cases}$$

21.3 Probabilities, Percentiles, and Moments

For given values of a and b, the dialog box [*StatCalc*→Continuous→Laplace] computes the cdf, percentiles, moments, and other parameters of the Laplace(a,b) distribution.

To compute probabilities: Enter the values of the parameters a, b, and the value of x; click [P(X <= x)]. For example, when $a = 3$, $b = 4$, and $x = 4.5$,

$$P(X \le 4.5) = 0.656355 \quad \text{and} \quad P(X > 4.5) = 0.343645.$$

To compute percentiles: Enter the values of a, b, and the cumulative probability; click [x]. For example, when $a = 3$, $b = 4$, and the cumulative probability $= 0.95$, the 95th percentile is 12.2103. That is, $P(X \le 12.2103) = 0.95$.

To compute parameters: Enter value of one of the parameters, cumulative probability, and x; click on the missing parameter. For example, when $a = 3$, cumulative
probability $= 0.7$, and $x = 3.2$, the value of b is 0.391523.

To compute moments: Enter the values of a and b and click [M].

21.4   Maximum Likelihood Estimators

Let $X_1, \ldots, X_n$ be a sample of independent observations from a Laplace distribution
with the pdf (21.1). Let $X_{(1)} < X_{(2)} < \ldots < X_{(n)}$ be the order statistics based on
the sample.

If the sample size n is odd, then the sample median $\hat{a} = X_{((n+1)/2)}$ is the maximum likelihood estimate (MLE) of a. If n is even, then the MLE of a is any number
between $X_{(n/2)}$ and $X_{(n/2+1)}$. The MLE of b is given by

$$\hat{b} = \frac{1}{n}\sum_{i=1}^{n}|X_i - \hat{a}| \text{ (if a is unknown)} \text{ and } \hat{b} = \frac{1}{n}\sum_{i=1}^{n}|X_i - a| \text{ (if a is known)}.$$

Censored Case

Let $X_{(r+1)}, \ldots, X_{(n)}$ be the largest order statistics for a sample of size n from a
Laplace distribution. The MLEs can be obtained as a special case of a general result
in Childs and Balakrishnan (1997), and they are as follows.

$$\hat{a} = \begin{cases} X^* - \hat{b}\ln\left(\frac{n}{2(n-r)}\right), & \text{if } r \geq \frac{n}{2}, \\ X_{\left(\frac{n+1}{2}\right)}, & \text{if } r \leq \frac{n}{2} - 1 \text{ and } n \text{ is odd}, \\ \frac{X_{\left(\frac{n}{2}\right)}+X_{\left(\frac{n+1}{2}\right)}}{2} & \text{if } r \leq \frac{n}{2} - 1, \text{ and } n \text{ is even}, \end{cases} \quad (21.4)$$

and

$$\hat{b} = \begin{cases} \frac{1}{n-r}\left[\sum_{i=r+2}^{n} X_{(i)} - (n-r-1)X^*\right], & \text{if } r \geq \frac{n}{2}, \\ \frac{1}{n-r}\left[\sum_{i=(n+1)/2+1}^{n} X_{(i)} - \sum_{i=r+2}^{(n-1)/2} X_{(i)} - (r+1)X^*\right], \\ \qquad \text{if } r \leq \frac{n}{2} - 1 \text{ and } n \text{ is odd}, \\ \frac{1}{n-r}\left[\sum_{i=n/2+1}^{n} X_{(i)} - \sum_{i=r+2}^{n/2} X_{(i)} - (r+1)X^*\right], \\ \qquad \text{if } r \leq \frac{n}{2} - 1 \text{ and } n \text{ is even}, \end{cases} \quad (21.5)$$

where X^* is $X_{(r+1)}$ if the samples are type II censored, and is the censoring value
X_0 if the samples are type I censored.

The R function 21.1 computes the MLEs for a left censored sample. If the sample
is right censored, then set $y = -x$ and $y_0 = -x_0$, and calculates the MLEs $\hat{a}$ and $\hat{b}$;
change the sign of $\hat{a}$.

Example 21.1. Consider the following sample of size 30 of which 10 observations
are type II left censored.

3.606 3.633 3.642 3.669 3.847 3.975 4.064 4.291 4.475 4.612
4.781 4.861 4.945 5.430 5.476 5.487 5.529 5.599 5.641 6.943

Here $n = 30$, $r = 10$, x is the vector containing all the above 20 data, $x_0 = 3.606$, and type $= 2$. We find the MLEs by calling the R function 21.1.

```
x = c(3.606, 3.633, 3.642, 3.669, 3.847, 3.975, 4.064, 4.291, 4.475,
      4.612, 4.781, 4.861, 4.945, 5.430, 5.476, 5.487, 5.529, 5.599,
      5.641, 6.943)
n= 30; x0 = 3.606; r = 10; type = 2
mles.laplace.cens(n, x, x0, r, type)
[1] 3.84700    0.43635
```

R function 21.1. Calculation of the MLEs based on a left censored sample

```
# n = sample size; x = vector of uncensored observations
# r = number of censored observations;
# x0 = mission time if type I censored or the smallest observation
#      in x if type II censored
# type = 1 if type I censored or 2 if type II censored

mles.laplace.cens = function(n, x, x0, r, type){
nminusr = length(x)
x = sort(x); y = seq(1:n);
if(r == 0){
ah = median(x)
bh = sum(abs(x-ah))/nminusr
return(c(ah,bh))
}
if(type == 2){
x0 = x[r+1]
}
y[1:r]=x0; y[(r+1):n] = x;
if(r >= n/2){
bh = (sum(y[(r+2):n])-(n-r-1)*x0)/(n-r)
}
else if(r <= n/2-1 & n%%2 != 0){
bh = (sum(y[((n+1)/2+1):n])-sum(y[(r+2):((n-1)/2)])-(r+1)*x0)/(n-r)
}
else if(r <= (n/2-1) & n%%2 == 0){
bh = (sum(y[(n/2+1):n])-sum(y[(r+2):(n/2)])-(r+1)*x0)/(n-r)
}
if(r >= n/2){
ah = x0 - bh*log(.5*n/(n-r))
}
```

```
if (r <= (n/2-1) & n%%2 !=0){
ah = y[(n+1)/2]
}
else if (r <= (n/2-1) & n%%2 ==0){
ah = (y[n/2]+y[(n+1)/2])/2
}
return(c(ah,bh))
}
```

Assume now that the above sample is type II right censored. That is, observations above the largest value 6.943 are censored. To find the MLEs, set $x_0 = 6.943$ and type $= 2$, and call the function

```
mles.laplace.cens(n, -x, -x0, r, type)
[1] -5.48700  0.38405
```

The MLEs are $\widehat{a} = 5.48700$ and $\widehat{b} = .38405$.

21.5 Confidence Intervals and Prediction Intervals

Since the MLEs are location-scale equivariant, confidence intervals for the parameters can be obtained using the results of Section 2.7.1. Note that the mean of a Laplace(a, b) distribution is the location parameter a, and a $1-\alpha$ confidence interval for a is given by

$$\left(\widehat{a} - k_u \widehat{b}, \ \widehat{a} - k_l \widehat{b}\right),\tag{21.6}$$

where k_l and k_u are determined by

$$P\left(k_l \leq \frac{\widehat{a}^*}{\widehat{b}^*} \leq k_u\right) = 1 - \alpha,$$

$\widehat{a}^*$ and $\widehat{b}^*$ are the MLEs based on a random sample from a Laplace$(0, 1)$ distribution. Note that k_l and k_u can be estimated using simulated samples from a Laplace$(0, 1)$ distribution.

A $1 - \alpha$ confidence interval for b is given by

$$\left(c_l \widehat{b}, \ c_u \widehat{b}\right),\tag{21.7}$$

where c_u and c_l are determined by

$$P\left(c_l \leq \frac{1}{\widehat{b}^*} \leq c_u\right) = 1 - \alpha,$$

and $\widehat{b}^*$ is the MLE based on a random sample from a Laplace$(0, 1)$ distribution. Note that c_l and c_u can be estimated using simulated samples from a Laplace$(0, 1)$ distribution.

Remark 21.1. The above confidence intervals are exact when samples are type II censored or uncensored except for simulation errors. They can be used as approximate confidence intervals when the samples are type I censored. *StatCalc* calculates the above confidence intervals for the parameters based on a censored on an uncensored sample.

Prediction Intervals

Suppose it is desired to find an upper prediction limit (UPL) based on a sample (uncensored) of size n so that it includes at least l of m observations from each of r locations. As the MLEs are equivariant, the required factor k does not depend on any unknown parameters, and it depends only on (n, R, m, l) and the confidence level. So Monte Carlo simulation can be used to find such factor. For a given $(n, r, m, l, \text{conf level})$, the dialog box [Continuous→Laplace→Prediction...] calculates the factor so that

$$\widehat{a} + k\widehat{b}$$

is the desired UPL.

A pivotal based approach can be used to find a prediction interval for a future observation. Specifically, let $\widehat{a}$ and $\widehat{b}$ be the MLEs based on sample (censored or uncensored), and let X follows the sampled Laplace distribution. Then

$$\frac{X - \widehat{a}}{\widehat{b}}$$

is a pivotal quantity, whose empirical distribution can be evaluated using generated samples from a Laplace(0, 1) distribution. Let k_l and k_u be determined so that

$$P\left(-k_l \le \frac{X^* - \widehat{a}^*}{\widehat{b}^*} \le k_u\right) = 1 - \alpha,$$

where $X^* \sim$ Laplace(0, 1) independently of $\widehat{a}^*$ and $\widehat{b}^*$, which are the MLEs based on a sample from a Laplace(0, 1) distribution. Then

$$\left(\widehat{a} - k_l\widehat{b}, \ \widehat{a} + k_u\widehat{b}\right) \tag{21.8}$$

is a $1 - \alpha$ prediction interval for X.

21.6   Tolerance Intervals

One-Sided Tolerance Limits

Recall that one-sided tolerance limits are one-sided confidence limits for appropriate quantiles of the population. The pth quantile (see (21.3)) is of the form

$$a + q_p b, \quad \text{with} \quad q_p = \begin{cases} \ln(2p), & 0 < p \le .5, \\ -\ln(2(1-p)), & .5 < p < 1. \end{cases} \tag{21.9}$$

As this is a location-scale equivariant function of a and b, a pivotal based approach is readily obtained from Section 2.7.1. Specifically, it follows from Result 2.7.1 that

$$\frac{\widehat{a} - (a + q_p b)}{\widehat{b}} \sim \frac{\widehat{a}^* - q_p}{\widehat{b}^*},$$

where $\widehat{a}^*$ and $\widehat{b}^*$ are the MLEs based on a sample from a Laplace(0, 1) distribution. Let c_l denote the α quantile of $(\widehat{a}^* - q_p)/\widehat{b}^*$. Then

$$\widehat{a} - c_l\widehat{b} \qquad (21.10)$$

is a $(p, 1-\alpha)$ upper tolerance limit for the Laplace(a, b) distribution. For $0 < p < .5$, let c_u denote the $1 - \alpha$ quantile of $(\widehat{a}^* - q_p)/\widehat{b}^*$. Then

$$\widehat{a} - c_u\widehat{b} \qquad (21.11)$$

is a $(p, 1-\alpha)$ lower tolerance limit for the Laplace(a, b) distribution. For the uncensored case, $cu = -cl$.

For a given $(n, r, p, 1 - \alpha)$, StatCalc calculates factors k_l and k_u so that $\widehat{a} - k_l\widehat{b}$ is a $(p, 1 - \alpha)$ lower tolerance limit, and $\widehat{a} + k_u\widehat{b}$ is a $(p, 1 - \alpha)$ upper tolerance limit. These tolerance limits are exact, except for simulation errors.

Tolerance Intervals

The $(p, 1 - \alpha)$ tolerance factor k for constructing a Laplace tolerance interval is determined by

$$P_{\widehat{a}^*, \widehat{b}^*}\left\{F_Z(\widehat{a}^* + k\widehat{b}^*) - F_Z(\widehat{a}^* - k\widehat{b}^*) \geq p\right\} = 1 - \alpha, \qquad (21.12)$$

where

$$F_Z(z) = \begin{cases} 1 - \frac{1}{2}e^{-z}, & z > 0, \\ \frac{1}{2}e^z, & z < 0. \end{cases} \qquad (21.13)$$

Let $v(\widehat{a}^*, p)$ be the root (with respect to r) of the equation

$$F_Z(\widehat{a}^* + r) - F_Z(\widehat{a}^* - r) = p. \qquad (21.14)$$

Then the factor k is the $1 - \alpha$ quantile of $v(\widehat{a}^*, p)/\widehat{b}^*$. See Krishnamoorthy and Xie (2012) for more details.

Algorithm 21.1. Calculation for tolerance factor satisfying (21.12)

---

1. Generate a sample of size n from a Laplace distribution with $a = 0$ and $b = 1$.
2. Discard the smallest r samples, and compute the MLEs (or equivariant estimators) of a and b based on the largest $n - r$ samples, say, $z_1, ..., z_{n-r}$; denote these estimators by $\widehat{a}^*$ and $\widehat{b}^*$.
3. For a given p, and using $\widehat{a}^*$ computed in step 2, find the root $v(\widehat{a}^*, p)$ of equation (21.14).
4. Set $Q = \frac{v(\widehat{a}^*, p)}{\widehat{b}^*}$.
5. Repeat the steps 1–4 for a large number of times, say, N.
6. The $100(1-\alpha)$ percentile of $Q_1, ..., Q_N$ is a Monte Carlo estimate of the tolerance factor k that satisfies (21.12).

StatCalc uses the above algorithm to estimate the factor k.

21.7   Applications with Examples

Because the distribution of differences between two independent exponential variates with mean b is Laplace $(0, b)$, a Laplace distribution can be used to model the difference between the waiting times of two events generated by independent random processes. The Laplace distribution can also be used to describe breaking strength data. Korteoja et al. (1998) studied tensile strength distributions of four paper samples and concluded that among extreme value, Weibull and Laplace distributions, a Laplace distribution fits the data best. Sahli et al. (1997) proposed a one-sided acceptance sampling by variables when the underlying distribution is Laplace. In the following, we see an example where the differences in flood stages are modeled by a Laplace distribution.

Example 21.2. The data in Table 20.1 represent the differences in flood stages for two stations on the Fox River in Wisconsin for 33 different years. The data were first considered by Gumbel and Mustafi (1967), and later Bain and Engelhardt (1973) justified the Laplace distribution for modeling the data. Kappenman (1977) used the data for constructing one-sided tolerance limits.

To fit a Laplace distribution for the observed differences of flood stages, we estimate

$$\widehat{a} = 10.13 \ \ \text{and} \ \ \widehat{b} = 3.36$$

by the maximum likelihood estimates (see Section 21.4). Using these estimates, the population quantiles are estimated as described in Section 1.4.1. For example, to find the population quantile corresponding to the sample quantile 1.96, select [Continuous→Laplace] from *StatCalc*, enter 10.13 for a, 3.36 for b and 0.045 for [P(X <= x)]; click on [x] to get 2.04. The Q–Q plot of the observed differences and the Laplace(10.13, 3.36) quantiles is given in Figure 20.2. The Q–Q plot shows that the sample quantiles (the observed differences) and the population quantiles are in good agreement. Thus, we conclude that the Laplace(10.13, 3.36) distribution adequately fits the data on flood stage differences.

The fitted distribution can be used to estimate the probabilities. For example, the percentage of differences in flood stages exceed 12.4 is estimated by

$$P(X > 12.4 | a = 10.13, b = 3.36) = 0.267631.$$

That is, about 27% of differences in flood stages exceed 12.4.

To find a 95% confidence interval for the mean a, select the dialog box [Cont..→Laplace→CI...] from *StatCalc*, enter 33 for [Sample Size], 0 for [No. Censored], .95 for [Conf Level], and click on [CI for a] to get (8.748, 11.517). To find a 90% confidence interval for the scale parameter b, just click on [CI for b] to get (2.486, 4.971).

To find (.90, .95) one-sided tolerance limits, select the dialog box [Cont..→Laplace →Tolerance...] from *StatCalc*, enter 33 for [Sample Size], 0 for [No. Censored], .90 for [Proportion p], .95 for [Conf Level], and click on [1-sided] to get the factor 2.34. That is,

$$10.13 - 2.34 \times 3.36 = 2.268$$

is a (.90, .95) lower tolerance limit, and

$$10.13 + 2.34 \times 3.36 = 17.992$$

is a $(.90, .95)$ upper tolerance limit.

To find a $(.90, .95)$ two-sided tolerance interval, just click on [2-sided] to get 3.233. This means that $10.13 \pm 3.233 \times 3.36$ is $(.90, .95)$ tolerance interval.

TABLE 21.1: Differences in Flood Stages

j	Observed Differences	$\frac{j-0.5}{33}$	Population Quantiles	j	Observed Differences	$\frac{j-0.5}{33}$	Population Quantiles
1	1.96	–	–	18	10.24	0.530	10.34
2	1.96	0.045	2.04	19	10.25	0.561	10.56
3	3.60	0.076	3.80	20	10.43	0.591	10.80
4	3.80	0.106	4.92	21	11.45	0.621	11.06
5	4.79	0.136	5.76	22	11.48	0.652	11.34
6	5.66	0.167	6.44	23	11.75	0.682	11.65
7	5.76	0.197	7.00	24	11.81	0.712	11.99
8	5.78	0.227	7.48	25	12.34	0.742	12.36
9	6.27	0.258	7.90	26	12.78	0.773	12.78
10	6.30	0.288	8.27	27	13.06	0.803	13.26
11	6.76	0.318	8.61	28	13.29	0.833	13.82
12	7.65	0.348	8.92	29	13.98	0.864	14.50
13	7.84	0.379	9.20	30	14.18	0.894	15.34
14	7.99	0.409	9.46	31	14.40	0.924	16.47
15	8.51	0.439	9.70	32	16.22	0.955	18.19
16	9.18	0.470	9.92	33	17.06	0.985	21.88
17	10.13	0.500	10.13				

Example 21.3. (*Laplace tolerance intervals*). The following data in Table 21.2 are breaking strengths of 100 yarns reported in Duncan (1974). Puig and Stephens (2000) showed that a Laplace distribution fits the samples well. We shall use these samples to construct a $(.90, .95)$ tolerance interval for the breaking strength of yarns. The MLEs based on all 100 measurements are

TABLE 21.2: Breaking Strength of 100 Yarns

62	66	78	79	80	84	84	85	85	86	86	87	88	88	89
89	91	91	91	91	92	92	92	92	93	94	94	94	95	95
95	96	96	96	96	96	97	97	97	97	97	97	98	98	98
98	98	98	98	99	99	99	99	99	100	100	100	100	100	101
101	101	101	102	102	102	102	102	102	102	103	103	103	104	104
104	104	104	104	104	105	105	106	107	107	109	110	111	111	111
111	114	115	117	122	132	132	137	137	138					

$$\widehat{a} = 99 \quad \text{and} \quad \widehat{b} = 8.33.$$

To find a 95% confidence interval for the mean a, select the dialog box [Cont..→Laplace→CI...] from *StatCalc*, enter 100 for [Sample Size], 0 for [No. Censored], .95 for [Conf Level], and click on [CI for a] to get $(97.2, \quad 100.8)$. To find

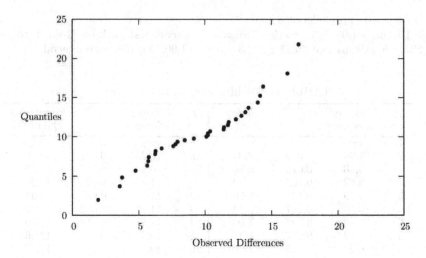

FIGURE 21.2: Q–Q plot of differences in flood stages data

a 90% confidence interval for the scale parameter b, just click on [CI for b] to get
(6.91, 10.30).

The (.90, .95) tolerance factor is computed as 2.76 (using *StatCalc*; see the preceding example). So the tolerance interval for breaking strength is $99 \pm 2.76 \times 8.33 =$
(76, 122). This means that at least 90% of the yarns have breaking strength between
76 and 122 with confidence 95%.

To illustrate the methods for the censored, suppose that the measuring device
can not measure the strength of a yarn if it is below 90. In this case, only $n - r = 84$
yarn strengths are recorded in a sample of size $n = 100$, and the MLEs are $\hat{\mu} = 99$
and $\hat{\sigma} = 8.45$. The (.90, .95) tolerance factor when $n = 100$ and $r = 16$ is 2.81, and
the tolerance interval is (75.26, 122.74). Note that the tolerance interval based on the
censored sample is wider, but still close to the one based on all 100 measurements.
The factor for computing (.90, .95) one-sided lower tolerance limit, when $n = 100$
and $k = 16$, is 2.01, and the limit is $99 - 2.01 \times 8.45 = 82.02$.

To illustrate the methods for type I right censored samples, let us assume that
the right censoring value is 103.5. Note that 27 values are censored with $X^* = 103.5$.
To compute the MLEs, the x-vector is x = c(66, 78, ... ,103,103), which includes 73
uncensored observations. To calculate the MLEs, call the R function 21.1:

```
mle.laplace.cens(100,-x,-103.5,27,1)
[1] -99    7.7123
```

Thus, the MLEs are $\hat{a} = 99$ and $\hat{b} = 7.7123$. To find a (.90, .95)
tolerance interval, the required factor was calculated using the dialog box
[StatCalc→Cont..→Laplace→Tolerance ...] as 2.85, and the (.90, .95) tolerance interval is $99 \pm 2.85 \times 7.7123 = (77, 121)$.

21.8 Relation to Other Distributions

1. Exponential: If X follows a Laplace(a, b) distribution, then $|X - a|/b$ follows an exponential distribution with mean 1. That is, if $Y = |X - a|/b$, then the pdf of Y is $\exp(-y)$, $y > 0$.

2. Chi-square: $|X - a|$ is distributed as $(b/2)\chi_2^2$.

3. Chi-square: If $X_1, \ldots, X_n$ are independent Laplace(a, b) random variables, then

$$\frac{2}{b} \sum_{i=1}^{n} |X_i - a| \sim \chi_{2n}^2.$$

4. F Distribution: If X_1 and X_2 are independent Laplace(a, b) random variables, then

$$\frac{|X_1 - a|}{|X_2 - a|} \sim F_{2,2}.$$

5. Normal: If Z_1, Z_2, Z_3, and Z_4 are independent standard normal random variables, then

$$Z_1 Z_2 - Z_3 Z_4 \sim \text{Laplace}(0, \ 2).$$

6. Exponential: If Y_1 and Y_2 are independent exponential random variables with mean b, then

$$Y_1 - Y_2 \sim \text{Laplace}(0, \ b).$$

7. Uniform: If U_1 and U_2 are uniform(0,1) random variables, then

$$\ln(U_1/U_2) \sim \text{Laplace}(0, \ 1).$$

21.9 Random Number Generation

Algorithm 21.2. Laplace variate generator

For a given a and b:
Generate u from uniform$(0, 1)$
If $u \geq 0.5$, return $x = a - b * \ln(2 * (1 - u))$
else return $x = a + b * \ln(2 * u)$

x is a pseudo random number from the Laplace(a, b) distribution.

22

Logistic Distribution

22.1 Description

The probability density function (pdf) of a logistic distribution with the location parameter a and scale parameter b is given by

$$f(x|a,b) = \frac{1}{b} \frac{\exp\left\{-\left(\frac{x-a}{b}\right)\right\}}{\left[1 + \exp\left\{-\left(\frac{x-a}{b}\right)\right\}\right]^2}, \quad -\infty < x < \infty, \quad -\infty < a < \infty, \quad b > 0.$$

(22.1)

The cumulative distribution function (cdf) is given by

$$F(x|a,b) = \left[1 + \exp\left\{-\left(\frac{x-a}{b}\right)\right\}\right]^{-1}.$$

(22.2)

For $0 < p < 1$, the inverse distribution function is

$$F^{-1}(p|a,b) = a + b\ln[p/(1-p)].$$

(22.3)

The cdf and the inverse distribution function are in simple form, so they easy to calculate. The logistic distribution is symmetric about the location parameter a (see Figure 22.1), and it can be used as a substitute for a normal distribution.

22.2 Moments

Mean:	a
Variance:	$\frac{b^2\pi^2}{3}$
Mode:	a
Median:	a
Mean Deviation:	$2b\ln(2)$
Coefficient of Variation:	$\frac{b\pi}{a\sqrt{3}}$

Coefficient of Skewness:	0
Coefficient of Kurtosis:	4.2
Moment Generating Function:	$E(e^{tY}) = \pi \text{cosec}(t\pi)$, where $Y = (X - a)/b$.
Inverse Distribution Function:	$a + b \ln[p/(1 - p)]$
Survival Function:	$\frac{1}{1+\exp[(x-a)/b]}$
Inverse Survival Function:	$a + b \ln\{(1 - p)/p\}$
Hazard Rate:	$\frac{1}{b[1+\exp[-(x-a)/b]]}$
Hazard Function:	$\ln\{1 + \exp[(x - a)/b]\}$

22.3   Probabilities, Percentiles, and Moments

For given values of a and b, the dialog box [StatCalc→Continuous→Logistic→ Probabilities ...] computes the cdf, percentiles and moments of a Logistic(a, b) distribution.

To compute probabilities: Enter the values of the parameters a, b, and the value of x; click [P(X <= x)]. For example, when $a = 2$, $b = 3$, and the observed value $x = 1.3$, $P(X \leq 1.3) = 0.44193$ and $P(X > 1.3) = 0.55807$.

To compute percentiles: Enter the values a, b, and the cumulative probability; click [x]. For example, when $a = 2$, $b = 3$, and the cumulative probability $= 0.25$, the 25th percentile is -1.29584. That is, $P(X \leq -1.29584) = 0.25$.

To compute other parameters: Enter the values of one of the parameters, cumulative probability and x; click on the missing parameter. For example, when $b = 3$, cumulative probability $= 0.25$ and $x = 2$, the value of a is 5.29584.

To compute moments: Enter the values of a and b and click [M].

22.4   Maximum Likelihood Estimators

Let $X_1, \ldots, X_n$ be a sample of independent observations from a logistic distribution with parameters a and b. Explicit expressions for the maximum likelihood estimates (MLEs) of a and b are not available. Likelihood equations can be solved only nu-

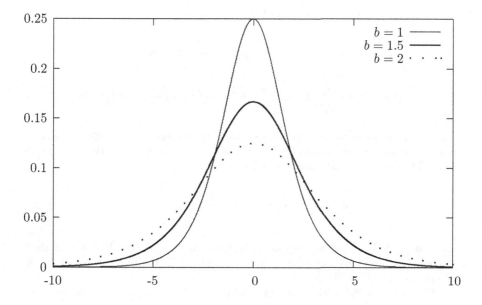

FIGURE 22.1: Logistic probability density functions; $a = 0$

merically, and they are

$$\sum_{i=1}^{n}\left[1 + \exp\left(\frac{X_i - a}{b}\right)\right]^{-1} = \frac{n}{2}$$

$$\sum_{i=1}^{n}\left(\frac{X_i - a}{b}\right)\frac{1 - \exp[(X_i - a)/b]}{1 + \exp[(X_i - a)/b]} = n. \qquad (22.4)$$

The sample mean and standard deviation can be used to estimate a and b. Specifically,

$$\widehat{a} = \frac{1}{n}\sum_{i=1}^{n}X_i \quad \text{and} \quad \widehat{b} = \frac{\sqrt{3}}{\pi}\sqrt{\frac{1}{n-1}\sum_{i=1}^{n}(X_i - \bar{X})^2}.$$

These estimators may be used as initial values to solve the equations in (22.4) iteratively for a and b.

Censored Case

Suppose that in a sample of n observations, r observations are censored, and denote the ordered uncensored observations by $X_1, ..., X_{n-r}$. Define $Z_i = (X_i - a)/b$, $i = 1, ..., n-r$, and $Z_1^* = (X_1 - a)/b$ if the sample is type II censored, and is $(X_0 - a)/b$ if the sample is type I censored with censoring time X_0. Let $h(z) = \exp(-z)/[1 +$

$\exp(-z)$]. The MLEs are the roots of the equations

$$f_1(a,b) = (n-r) - 2\sum_{i=1}^{n-r} h(Z_i) - rh(Z_1^*) = 0,$$

$$f_2(a,b) = \sum_{i=1}^{n-r} z_i - 2\sum_{i=1}^{n-r} Z_i h(Z_i) - (n-r) - rZ_1^* h(Z_1^*) = 0, \qquad (22.5)$$

which can be obtained as a special case from Harter and Moore (1967). Newton–Raphson iterative method can be used to find the MLEs satisfying the above equations. The following equivariant estimators

$$\tilde{a} = \frac{1}{n-r}\sum_{i=1}^{n-r} x_i \ \text{ and } \ \tilde{b} = \frac{1}{\tilde{n}}(x_n - x_{r+1}), \qquad (22.6)$$

where $\tilde{n} = 2\sum_{i=1}^{n-r}\frac{1}{i}$ can be used as initial values for the Newton-Raphson method. Partial derivatives to implement Newton–Raphson method are given in Krishnamoorthy and Xie (2011).

The following R function computes MLEs based on a censored sample of size n with k censored observations.

R function 22.1. Calculation of the MLEs based on a left censored sample

---

```
# n = sample size; k = number of censored observations
# x0 = censoring value (type I) or the smallest uncensored
#       value (type II)
# y = vector containing uncensored values
mles.logistic.cens = function(n, k, x0, y){
nmk = n-k; x = as.vector(n); z= as.vector(nmk);
hz = function(x){
return(exp(-x)/(1+exp(-x)))}
x = c(rep(x0,k),y); one = 1.0
x = sort(x); a0 = sum(x[(k+1):n])/nmk
sk = 2*sum(1/seq(1:nmk)); b0 = (x[n]-x[k+1])/sk
j = 1
repeat{
z = (x[(k+1):n]-a0)/b0; zs1 = (x0-a0)/b0
hz1 = hz(zs1); fz1 = hz1/(one+exp(-zs1))
hzi = hz(z); fzi = hzi/(one+exp(-z))
f1 =  - sum(hzi);  f2 = sum(z-2*z*hzi)
f1a = 2*sum(fzi);  f1b = 2*sum(z*fzi)
f2a = sum(hzi - z*fzi);  f2b =  sum(-0.5*z+z*hzi-z*z*fzi)
f1 = 2*f1 + nmk - k*hz1;  f2 = f2 - nmk-k*zs1*hz1
f1a = -(f1a+k*fz1)/b0;  f1b = -(f1b+k*zs1*fz1)/b0
f2a = (2*f2a-nmk+k*hz1-k*zs1*fz1)/b0
f2b = (2*f2b+k*zs1*hz1-k*zs1**2*fz1)/b0
```

```
detf = f1a*f2b-f1b*f2a
f1ai =  f2b/detf; f1bi = -f2a/detf
f2ai = -f1b/detf; f2bi = f1a/detf
ahat = a0 - (f1ai*f1+f1bi*f2); bhat = b0 - (f2ai*f1+f2bi*f2)
if((abs(f1) <= 1.0e-5 & abs(f2) <= 1.0e-5) | j >= 30){break}
j = j + 1; a0 = ahat; b0 = bhat
}
return(c(ahat,bhat))
}
```

22.5 Confidence Intervals and Prediction Intervals

Recall that the MLEs are location-scale equivariant, so confidence intervals for the parameters can be obtained using the results of Section 2.7.1. Note that the mean of a Logistic(a, b) distribution is the location parameter a, and a $1 - \alpha$ confidence interval for a is given by

$$\left(\widehat{a} - k_u\widehat{b}, \ \widehat{a} - k_l\widehat{b}\right),\tag{22.7}$$

where k_l and k_u are determined by

$$P\left(k_l \leq \frac{\widehat{a}^*}{\widehat{b}^*} \leq k_u\right) = 1 - \alpha,$$

and $\widehat{a}^*$ and $\widehat{b}^*$ are the MLEs based on a random sample from a Logistic($0, 1$) distribution. Note that k_l and k_u can be estimated using simulated samples from a Loistic($0, 1$) distribution.

A $1 - \alpha$ confidence interval for b is given by

$$\left(c_l\widehat{b}, c_u\widehat{b}\right),\tag{22.8}$$

where c_u and c_l are determined by

$$P\left(c_l \leq \frac{1}{\widehat{b}^*} \leq c_u\right) = 1 - \alpha,$$

and $\widehat{b}^*$ is the MLE based on a random sample from a Logistic($0, 1$) distribution. The critical values c_l and c_u can be estimated using simulated samples from a Logistic($0, 1$) distribution.

Prediction Intervals

Let $\widehat{a}$ and $\widehat{b}$ denote the MLEs based on a sample (could be censored) of size n from a Logistic(a, b) distribution. Let X follow the same distribution independently of the sample. A prediction interval for X is given by

$$\left(\widehat{a} - k_l\widehat{b}, \ \widehat{a} + k_u\widehat{b}\right),\tag{22.9}$$

where k_l and k_u is determined by

$$P\left(-k_l \le \frac{X^* - \widehat{a}^*}{\widehat{b}^*} \le k_u\right) = 1 - \alpha,$$

where $X^* \sim \text{Logistic}(0, 1)$ distribution, and $\widehat{a}^*$ and $\widehat{b}^*$ are the MLEs based on a sample from a $\text{Logistic}(0, 1)$ distribution. Also, X^* and $(\widehat{a}^*, \widehat{b}^*)$ are independent. Note that the factors k_l and k_u can be estimated by Monte Carlo simulation by generating independent samples from a $\text{Logistic}(0, 1)$ distribution.

Example 22.1. The data in Table 22.1 represent failure mileage (in units of 1000 miles) of different locomotive controls in a life test involving 96 locomotive controls. The test was terminated after 135,000 miles, and by then 37 controls had failed. This example is discussed in Schmee and Nelson (1977), and also in Lawless (2003, Section 5.3). These authors noted that a lognormal distribution gives a good fit to the data. Lawless has noted that the data also fit a log-logistic model (log-transformed failure data fit a logistic distribution). In this type of situation, a lower tolerance limit is

TABLE 22.1: Failure Mileage (in 1000) of Locomotive Controls

22.5	37.5	46.0	48.5	51.5	53.0	54.5	57.5	66.5	68.0
69.5	76.5	77.0	78.5	80.0	81.5	82.0	83.0	84.0	91.5
93.5	102.5	107.0	108.5	112.5	113.5	116.0	117.0	118.5	119.0
120.0	122.5	123.0	127.5	131.0	132.5	134.0			

desired to assess the reliability of the controls, and to estimate the lifetime at certain mileage.

For the above data, the sample size $n = 96$ and the number of censored data is $r = 59$. Recall that the data were type I right censored with the censoring value 135. So the R function 22.1 for left censored data can be applied to the right censored data after taking negative log transformation. Specifically, the MLEs for the parameters can be computed using the R function 22.1 as follows.

```
#locomotive mileage data
> x =  c(22.5,37.5,46.0,...,132.5 ,134.0)
> y = -log(x); x0 = -log(135)
> n = 96; k = 59 # number of censored observations
> mles.logistic.cens(n, k, x0, y)
[1] -5.0829458  0.3836752
```

Thus the MLEs are $\widehat{a} = 5.083$ and $\widehat{b} = 0.384$.

To compute 95% confidence intervals for the parameters a and b using *StatCalc*, select [StatCalc→Continuous→Logistic→Confidence ...], enter 96 for n, 59 for the number of observations censored k, 5.083 for [MLE of a], .384 for [MLE of b], and .95 for [Conf Level]; click on [Confidence Intervals] to get (4.85, 5.23) for a and (.27, .49) for b.

To find a 90% prediction interval for a failure time of a locomotive control using *StatCalc*, select [StatCalc→Continuous→Logistic→Prediction ...], enter 96 for n, 59 for k, and .95 for [Conf Level]; click on [Factors for ...] to get factors 3.20 (lower) and 3.05 (upper). Thus, the 95% prediction interval is

$$(5.083 - 3.20 \times 0.384, 5.083 + 3.05 \times 0.384) = (3.85, 6.25).$$

By taking exponentiation, we find the prediction interval as (46.993, 518.013). Thus, a locomotive control may survive between 46,993 and 518,013 miles with confidence 95%.

22.6   Tolerance Intervals

One-Sided Tolerance Limits

Recall that one-sided tolerance limits are one-sided confidence limits for appropriate quantiles of the population. The pth quantile (see (22.3)) is given by

$$a + q_p b, \quad \text{with} \quad q_p = \ln\left(\frac{p}{1-p}\right). \tag{22.10}$$

As this is a location-scale equivariant function of a and b, a pivotal based approach is readily obtained from Section 2.7.1. Specifically, it follows from Result 2.7.1 that

$$\frac{\widehat{a} - (a + q_p b)}{\widehat{b}} \sim \frac{\widehat{a}^* - q_p}{\widehat{b}^*},$$

where $X \sim Y$ means that X and Y are identically distributed, and $\widehat{a}^*$ and $\widehat{b}^*$ are the MLEs based on sample of size n from a Logistic$(0,1)$ distribution. Let c_l denote the α quantile of $(\widehat{a}^* - q_{1-p})/\widehat{b}^*$. Then

$$\widehat{a} - c_l \widehat{b} \tag{22.11}$$

is a $(p, 1 - \alpha)$ upper tolerance limit for the Laplace(a, b) distribution. To find a $(p, 1 - \alpha)$ lower tolerance limit, let c_u denote the $1 - \alpha$ quantile of $(\widehat{a}^* - q_{1-p})/\widehat{b}^*$. Then

$$\widehat{a} - c_u \widehat{b} \tag{22.12}$$

is a $(p, 1 - \alpha)$ lower tolerance limit for the Logistic(a, b) distribution. For the uncensored case, $c_u = -c_l$.

For a given sample size n, the number of censored observations r, and the values of $(p, 1 - \alpha)$, *StatCalc* calculates factors $k_l = c_u$ and $k_u = -c_l$ so that $\widehat{a} - k_l \widehat{b}$ is a $(p, 1 - \alpha)$ lower tolerance limit, and $\widehat{a} + k_u \widehat{b}$ is a $(p, 1 - \alpha)$ upper tolerance limit. These tolerance limits are exact, except for simulation errors, provided the samples are type II censored, and can be used as approximate if the samples are type I censored.

Tolerance Intervals

The $(p, 1 - \alpha)$ tolerance factor k for constructing a logistic tolerance interval is determined by

$$P_{\widehat{a}^*,\widehat{b}^*}\left\{F_Z(\widehat{a}^* + k\widehat{b}^*) - F_Z(\widehat{a}^* - k\widehat{b}^*) \geq p\right\} = 1 - \alpha, \qquad (22.13)$$

where

$$F_Z(z) = [1 + \exp(-z)]^{-1}. \qquad (22.14)$$

Let $v(\widehat{a}^*, p)$ be the root (with respect to x) of the equation

$$F_Z(\widehat{a}^* + x) - F_Z(\widehat{a}^* - x) = p. \qquad (22.15)$$

Then, the factor k is the $1 - \alpha$ quantile of $v(\widehat{a}^*, p)/\widehat{b}^*$. See Krishnamoorthy and Xie (2011) for more details.

An algorithm similar to Algorithm 21.1 for calculating Laplace tolerance intervals can be used to compute the factor k satisfying 22.15. The dialog box [StatCalc→Continuous→Logistic→Tolerance ...] computes the factors for finding one-sided and two-sided tolerance intervals for a logistic distribution.

Remark 22.1. The factors for the type II censored case can be used as an approximation for the type I censored case with the following adjustment. Let $k_{n,r,p,1-\alpha}$ denote the tolerance factor for a sample of size n with r type II censored observations, and let $\widehat{P}_{x_0}$ denote the proportion of censored observations in the sample. If $P_{x_0} \leq .20$, then one could use the factor $k_{n,r,p,1-\alpha}$, otherwise use $k_{n,r-1,p,1-\alpha}$. The tolerance intervals with this adjustment seem to be satisfactory as long as $\widehat{P}_{x_0} \leq .70$.

Example 22.2. To find the $(.90, .90)$ two-sided tolerance interval for locomotive control data in Table 22.1, we computed the factor (with $n = 96$ and $r - 1 = 58$) as 3.74. Thus, the tolerance interval is $5.083 \pm 3.74 \times 0.384 = (3.647, 6.519)$. By taking exponentiation, we get $(38.359, 677.900)$. Thus, we are 90% confident that at least 90% of locomotive controls survive 38,359 to 677,900 miles.

22.7 Applications

The logistic distribution can be used as a substitute for a normal distribution. It is also used to analyze data related to stocks. Braselton et al. (1999) considered the day-to-day percent changes of the daily closing values of the S&P 500 index from January 1, 1926 through June 11, 1993. These authors found that a logistic distribution provided the best fit for the data even though the lognormal distribution has been used traditionally to model these daily changes. An application of the logistic distribution in nuclear-medicine is given in Prince et al. (1988). de Visser and van den Berg (1998) studied the size grade distribution of onions using a logistic distribution. The logistic distribution is also used to predict the soil-water retention based on the particle-size distribution of Swedish soil (Rajkai et al., 1996). Scerri and Farrugia (1996) compared the logistic and Weibull distributions for modeling wind speed data. Applicability of a logistic distribution to study citrus rust mite damage on oranges is given in Yang et al. (1995).

22.8   Properties and Results

1. If X is a Logistic(a, b) random variable, then $(X - a)/b \sim$ Logistic$(0, 1)$.

2. If u follows a uniform$(0, 1)$ distribution, then $a + b[\ln(u) - \ln(1 - u)] \sim$ Logistic(a, b).

3. If Y is a standard exponential random variable, then

$$-\ln\left[\frac{e^{-y}}{1 - e^{-y}}\right] \sim \text{Logistic}(0, 1).$$

4. If Y_1 and Y_2 are independent standard exponential random variables, then

$$-\ln\left(\frac{Y_1}{Y_2}\right) \sim \text{Logistic}(0, 1).$$

For more results and properties, see Balakrishnan (1991).

22.9   Random Number Generation

Algorithm 22.1. Logistic variate generator

For a given a and b:
Generate u from uniform$(0, 1)$
return $x = a + b * (\ln(u) - \ln(1 - u))$

23

Lognormal Distribution

23.1 Description

A positive random variable X is lognormally distributed if $\ln(X)$ is normally distributed. The probability density function (pdf) of X is given by

$$f(x|\mu,\sigma) = \frac{1}{\sqrt{2\pi}x\sigma} \exp\left[-\frac{(\ln x - \mu)^2}{2\sigma^2}\right], \quad x > 0, \ \sigma > 0, \ -\infty < \mu < \infty. \quad (23.1)$$

Note that if $Y = \ln(X)$, and Y follows a normal distribution with mean μ and standard deviation σ, then the distribution of X is called lognormal. Since X is actually an antilogarithmic function of a normal random variable, some authors refer to this distribution as antilognormal. We denote this distribution by lognormal(μ, σ^2).

The cumulative distribution function (cdf) of a lognormal(μ, σ^2) distribution is given by

$$
\begin{aligned}
F(x|\mu,\sigma) &= P(X \le x|\mu,\sigma) \\
&= P(\ln X \le \ln x|\mu,\sigma) \\
&= P\left(Z \le \frac{\ln x - \mu}{\sigma}\right) \\
&= \Phi\left(\frac{\ln x - \mu}{\sigma}\right),
\end{aligned}
\quad (23.2)
$$

where $Z = (\ln X - \mu)/\sigma$ and Φ is the standard normal distribution function.

23.2 Moments

Mean:	$\exp[\mu + \sigma^2/2]$
Variance:	$\exp(\sigma^2)[\exp(\sigma^2) - 1]\exp(2\mu)$
Mode:	$\exp[\mu - \sigma^2]$
Median:	$\exp(\mu)$

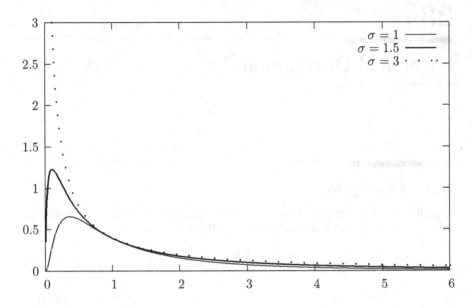

FIGURE 23.1: Lognormal probability density functions; $\mu = 0$

Coefficient of Variation:	$\sqrt{[\exp(\sigma^2) - 1]}$
Coefficient of Skewness:	$[\exp(\sigma^2) + 2]\sqrt{[\exp(\sigma^2) - 1]}$
Coefficient of Kurtosis:	$\exp(4\sigma^2) + 2\exp(3\sigma^2) + 3\exp(2\sigma^2) - 3$
Moments about the Origin:	$\exp\left[k\mu + k^2\sigma^2/2\right]$
Moments about the Mean:	$\exp\left[k(\mu + \sigma^2/2)\right] \sum_{i=0}^{k} (-1)^i \binom{k}{i} \exp\left[\frac{\sigma^2(k-i)(k-i-1)}{2}\right]$.
	[Johnson et al. (1994, p. 212)]

23.3   Probabilities, Percentiles, and Moments

The dialog box [StatCalc→Continuous →Lognormal] computes the cdf, percentiles, and moments of a lognormal(μ, σ^2) distribution.

To compute probabilities: Enter the values of the parameters μ, σ, and the observed value x; click $[P(X <= x)]$. For example, when $\mu = 1$, $\sigma = 2$, and $x = 2.3$, $P(X \leq 2.3) = 0.466709$ and $P(X > 2.3) = 0.533291$.

To compute percentiles: Enter the values of μ, σ, and the cumulative probability $P(X <= x)$; click on $[x]$. For example, when $\mu = 1$, $\sigma = 2$, and the cumulative probability $P(X <= x) = 0.95$, the 95th percentile is 72.9451. That is, $P(X \leq 72.9451) = 0.95$.

To compute μ: Enter the values of σ, x, and the cumulative probability $P(X <= x)$; click on $[U]$. For example, when $x = 2.3$, $\sigma = 2$, and the cumulative probability $P(X <= x) = 0.9$, the value of μ is -1.73019.

To compute σ: Enter the values of x, μ, and the cumulative probability $P(X <= x)$; click on $[S]$. For example, when $x = 3$, $\mu = 2$, and the cumulative probability $P(X <= x) = 0.1$, the value of σ is 0.703357.

To compute moments: Enter the values of μ and σ and click $[M]$.

23.4 Maximum Likelihood Estimators

Let $X_1, \ldots, X_n$ be a sample of independent observations from a lognormal(μ, σ) distribution. Let $Y_i = \ln(X_i)$, $i = 1, \ldots, n$. Then

$$\widehat{\mu} = \bar{Y} = \frac{1}{n} \sum_{i=1}^{n} Y_i \quad \text{and} \quad \widehat{\sigma} = \sqrt{\frac{1}{n} \sum_{i=1}^{n} (Y_i - \bar{Y})^2}$$

are the maximum likelihood estimates of μ and σ, respectively.

23.5 Confidence Interval and Test for the Mean

Let $X_1, \ldots, X_n$ be a sample from a lognormal(μ, σ). Let $Y_i = \ln(X_i)$, $i = 1, \ldots, n$. Let

$$\bar{Y} = \frac{1}{n} \sum_{i=1}^{n} Y_i \quad \text{and} \quad S^2 = \frac{1}{n-1} \sum_{i=1}^{n} (Y_i - \bar{Y})^2.$$

Recall that the mean of a lognormal(μ, σ) distribution is given by $\exp(\eta)$, where $\eta = \mu + \sigma^2/2$. Since the lognormal mean is a one-one function of η, it is enough to estimate or test about η. For example, if L is a 95% lower confidence limit for η, then $\exp(L)$ is a 95% lower limit for $\exp(\eta)$.

Generalized Confidence Intervals

The inferential procedures are based on the *generalized variable* approach given in Krishnamoorthy and Mathew (2003). Following Example 2.2, we obtain the generalized pivotal quantity (GPQ) for $\eta = \mu + \frac{1}{2}\sigma^2$ as

$$G_\eta = \bar{y} - \frac{Z}{\sqrt{\frac{\chi_{n-1}^2}{n-1}}} + \frac{1}{2} \frac{(n-1)s^2}{\chi_{n-1}^2},$$

where $Z \sim N(0,1)$ independently of χ^2_{n-1} random variable. The following algorithm can be used to compute the generalized confidence interval for η, and thereby for the lognormal mean.

Algorithm 23.1. (Calculation of generalized p-value)

For $j = 1, m$
Generate $Z \sim N(0,1)$ and $U^2 \sim \chi^2_{n-1}$
Set $T_j = \bar{y} - \sqrt{\frac{n-1}{n}} \frac{Z}{U} s + \frac{1}{2} \frac{(n-1)}{U^2} s^2$
(end loop)

The percentiles of the T_js generated above can be used to find confidence intervals for η. Let T_p denote the $100p$th percentile of the T_js. Then $(T_{.025}, T_{.975})$ is a 95% confidence interval for η. Furthermore, $(\exp(T_{.025}), \exp(T_{.975}))$ is a 95% confidence interval for the lognormal mean $\exp(\eta)$. A 95% lower limit for $\exp(\eta)$ is given by $\exp(T_{0.05})$.

Suppose we are interested in testing

$$H_0 : \exp(\eta) \leq c \quad \text{vs.} \quad H_a : \exp(\eta) > c,$$

where c is a specified number. For a given level of significance α, the H_0 is rejected if $\exp(T_\alpha) > c$ or $T_\alpha > \ln(c)$. Notice that

$$T_\alpha > \ln(c) \text{ if and only if } P\left(T < \ln(c)\right) < \alpha.$$

The above probability is the p-value of the test.

MOVER Approach

The MOVER confidence interval for $\eta = \mu + \frac{1}{2}\sigma^2$ is obtained by combining the confidence intervals for μ and σ^2. See Section 2.8 and Zou et al. (2009b). Let

$$(L_\mu, U_\mu) = \bar{Y} \mp z_{1-\alpha/2} \frac{S}{\sqrt{n}},$$

and let

$$(L_\sigma^2, U_\sigma^2) = \left(\frac{(n-1)S^2}{\chi^2_{n-1;1-\alpha/2}}, \frac{(n-1)S^2}{\chi^2_{n-1;\alpha/2}} \right).$$

The $1-\alpha$ MOVER confidence interval for η is given by (L_η, U_η), where

$$L_\eta = \bar{Y} + \frac{1}{2}S^2 - \sqrt{\left(\bar{Y} - L_\mu\right)^2 + \frac{1}{4}\left(S^2 - L_\sigma^2\right)^2} \qquad (23.3)$$

and

$$U_\eta = \bar{Y} + \frac{1}{2}S^2 + \sqrt{\left(\bar{Y} - U_\mu\right)^2 + \frac{1}{4}\left(S^2 - U_\sigma^2\right)^2}. \qquad (23.4)$$

Finally, an approximate $1 - \alpha$ confidence interval for the lognormal mean $\exp(\mu + \sigma^2/2)$ is $(\exp(L_\eta), \exp(U_\eta))$.

For a given sample size, mean, and standard deviation of the logged data, *StatCalc* computes confidence intervals and the p-values for testing about a lognormal mean using Algorithm 23.1 with a specified number of simulation runs and the MOVER confidence intervals.

Example 23.1. Suppose that a sample of 15 observations from a lognormal(μ, σ) distribution produced the mean of log-transformed sample as $\bar{y} = 1.2$ and the standard deviation of log-transformed sample $s = 1.5$. It is desired to find a 95% confidence interval for the lognormal mean $\exp(\mu + \sigma^2/2)$. To compute a 95% confidence interval using *StatCalc*, select [StatCalc→Continuous→Lognormal→CI and Test for Mean], enter 15 for the sample size, 1.2 for [Mean of ln(x)], 1.5 for [Std Dev of ln(x)], 0.95 for the confidence level, and click [GV CI] to get (4.37, 70.34), and click [MOVER CI] to get the MOVER confidence interval as (4.07, 64.22).
Suppose we want to test

$$H_0 : \exp(\eta) \leq 4.85 \quad \text{vs.} \quad H_a : \exp(\eta) > 4.85.$$

To find the p-value, enter 4.85 for [H0: M = M0] and click [p-values for] to get 0.045. Thus, at 5% level, we can reject the null hypothesis and conclude that the true mean is greater than 4.85.

23.6 Methods for Comparing Two Means

Suppose that we have a sample of n_i observations from a lognormal(μ, σ^2) population, $i = 1, 2$. Let $\eta_i = \mu_i + \sigma_i^2/2$, $i = 1, 2$. Let $\bar{Y}_i$ and S_i denote, respectively, the mean and standard deviation of the log-transformed measurements in the ith sample, $i = 1, 2$.

Generalized Variable Approach

Let $(\bar{y}_i, s_i)$ be an observed value of $(\bar{Y}_i, S_i)$, $i = 1, 2$. For a given $(n_1, \bar{y}_1, s_1, n_2, \bar{y}_2, s_2)$, generalized confidence intervals for the difference $\exp(\eta_1) - \exp(\eta_2)$ or for the ratio $\exp(\eta_1)/\exp(\eta_2)$ can be obtained using the following algorithm.

Algorithm 23.2. Calculation of p-values for comparing two means

For $j = 1, m$
Generate independent random numbers Z_1, Z_2, U_1^2, and U_2^2 such that
$Z_i \sim N(0, 1)$ and $U_i^2 \sim \chi_{n_i-1}^2$, $i = 1, 2$.
Set
$G_i = \bar{y}_i - \sqrt{\frac{n_i-1}{n_i}} \frac{Z_i s_i}{U_i} + \frac{1}{2} \frac{(n_i-1)s_i^2}{U_i^2}, i = 1, 2.$
$T_{2j} = \exp(G_1) - \exp(G_2)$
$R_j = \exp(G_1)/\exp(G_2)$
(end loop)

The percentiles of the T_{2j}'s generated above can be used to construct confidence intervals for $\exp(\eta_1) - \exp(\eta_2)$. Let $T_{2,p}$ denote the $100p$th percentile of the T_{2j}s. Then, $(T_{2,.025}, T_{2,.975})$ is a 95% confidence interval for $\exp(\eta_1) - \exp(\eta_2)$; $T_{2,.05}$ is a 95% lower confidence limit for $\exp(\eta_1) - \exp(\eta_2)$. Similarly, using appropriate percentiles of R_js, we can find confidence intervals for the ratio $\exp(\eta_1)/\exp(\eta_2)$ of the means.
    Suppose we are interested in testing

$$H_0 : \exp(\eta_1) - \exp(\eta_2) \leq 0 \quad \text{vs.} \quad H_a : \exp(\eta_1) - \exp(\eta_2) > 0.$$

Then, an estimate of the p-value based on the generalized variable approach is the proportion of the T_{2j}s that are less than zero.

MOVER Confidence Intervals for the Ratio of Means

A MOVER confidence interval for the ratio of means $\exp(\eta_1)/\exp(\eta_2) = \exp(\eta_1 - \eta_2)$ can be deduced from those of individual means. In particular, we can use the confidence intervals of the form (23.3) and (23.4) for η_1 and η_2, to find a MOVER confidence interval for $\eta_1 - \eta_2$. Let (L_{η_i}, U_{η_i}) be the $1-\alpha$ MOVER confidence interval for η_i based on $(\bar{Y}_i, S_i)$, $i = 1, 2$. Let

$$\widehat{\eta}_i = \bar{Y}_i + \frac{1}{2}S_i^2, \; i = 1, 2.$$

Then, the MOVER confidence interval for $\eta_1 - \eta_2$ is given by (L_D, U_D), where

$$L_D = \widehat{\eta}_1 - \widehat{\eta}_2 - \sqrt{(\widehat{\eta}_1 - L_{\eta_1})^2 + (\widehat{\eta}_2 - U_{\eta_2})^2}$$

and

$$U_D = \widehat{\eta}_1 - \widehat{\eta}_2 + \sqrt{(\widehat{\eta}_1 - U_{\eta_1})^2 + (\widehat{\eta}_2 - L_{\eta_2})^2}.$$

In terms of (L_D, U_D), the MOVER confidence interval for the ratio of means is given by $(\exp(L_D), \exp(U_D))$.

For given sample sizes, sample means, and standard deviations of the log-transformed data, *StatCalc* computes the confidence intervals and the p-values for testing about the difference between two lognormal means using Algorithm 23.2 and the MOVER confidence interval.

Example 23.2.   The data for this example are taken from the website http://lib.stat.cmu.edu/DASL/. An oil refinery conducted a series of 31 daily measurements of the carbon monoxide levels arising from one of their stacks. The measurements were submitted as evidence for establishing a baseline to the Bay Area Air Quality Management District (BAAQMD). BAAQMD personnel also made nine independent measurements of the carbon monoxide concentration from the same stack. The data are given below:

Carbon Monoxide Measurements by the Refinery (in ppm):
45, 30, 38, 42, 63, 43, 102, 86, 99, 63, 58, 34, 37, 55, 58, 153, 75 58, 36, 59, 43, 102, 52, 30, 21, 40, 141, 85, 161, 86, 161, 86, 71

Carbon Monoxide Measurements by the BAAQMD (in ppm):
12.5, 20, 4, 20, 25, 170, 15, 20, 15
The assumption of lognormality is tenable. The hypotheses to be tested are

$$H_0 : \exp(\eta_1) \leq \exp(\eta_2) \quad \text{vs.} \quad H_a : \exp(\eta_1) > \exp(\eta_2),$$

where $\exp(\eta_1) = \exp(\mu_1 + \sigma_1^2/2)$ and $\exp(\eta_2) = \exp(\mu_2 + \sigma_2^2/2)$ denote, respectively, the population mean of the refinery measurements and the mean of the BAAQMD measurements. For log-transformed measurements taken by the refinery, we have: $n_1 = 31$, sample mean $\bar{y}_1 = 4.0743$, and $s_1 = 0.5021$; for log-transformed measurements collected by the BAAQMD, $n_2 = 9$, $\bar{y}_2 = 2.963$, and $s_2 = 0.974$. To find the p-value for testing the above hypotheses using *StatCalc*, enter the sample sizes and the summary statistics, and click [p-values for] to get 0.112. Thus, we cannot conclude that the true mean of the oil refinery measurements is greater than that of BAAQMD measurements. To get a 95% confidence intervals for the difference between two means using *StatCalc*, enter the sample sizes, the summary statistics and 0.95 for confidence level; click [GV CI] to get $(-79.6, 57.3)$, and [MOVER CI] to get $(-62.3, 57.1)$

Ratio of Two Means

Suppose that we have a sample of n_i observations from a lognormal population with parameters μ_i and σ_i, $i = 1, 2$. Let $\bar{Y}_i$ and S_i denote, respectively, the mean and standard deviation of the logged measurements from the ith sample, $i = 1, 2$. For given $(n_1, \bar{y}_1, s_1, n_2, \bar{y}_2, s_2)$, *StatCalc* computes confidence intervals for the ratio $\exp(\eta_1)/\exp(\eta_2)$, where $\eta_i = \mu_i + \sigma_i^2/2$, $i = 1, 2$. *StatCalc* uses Algorithm 23.2 with $R_j = \exp(G_1)/\exp(G_2) = \exp(G_1 - G_2)$.

Example 23.3. Let us construct a 95% confidence interval for the ratio of the population means in Example 23.2. We have $n_1 = 31$ and $n_2 = 9$. For log-transformed measurements, $\bar{y}_1 = 4.0743$, $s_1 = 0.5021$, $\bar{y}_2 = 2.963$, and $s_2 = 0.974$. To get a 95% confidence interval for the ratio of two means using *StatCalc*, select [StatCalc→ Continuous→Lognormal→CI for Mean1/Mean2], enter the sample sizes, the summary statistics, and 0.95 for confidence level; click [GV CI] to get $(0.46, 4.16)$, and click [MOVER CI] to get $(.514, 4.39)$. Because both confidence intervals include 1, we cannot conclude that the means are significantly different.

23.7 Tolerance Limits, Prediction Limits, and Survival Probability

As the log-transformed samples from a lognormal distribution follow a normal distribution, we can use normal-based methods for log-transformed samples to find tolerance limits, prediction limits, and confidence interval for a survival probability. Specifically, we simply find normal tolerance intervals (or prediction intervals) based on log-transformed samples, and then taking exponentiation, we can find tolerance intervals for the sampled lognormal population.

    As an example, let $X_1, ..., X_n$ be a sample from a lognormal distribution, and let $\bar{Y}$ and S_y be the mean and standard deviation of Y_is, where $Y_i = \ln(X_i)$, $i = 1, ..., n$. The $(p, 1 - \alpha)$ normal tolerance interval is

$$\bar{Y} \pm kS_y,$$

where k is the factor as defined in (11.24). The interval

$$\left(\exp(\bar{Y} - kS_y), \exp(\bar{Y} + kS_y)\right)$$

is a $(p, 1 - \alpha)$ tolerance interval for the lognormal distribution.

    A $1 - \alpha$ prediction interval for a future observation from a lognormal population can be found similarly using the formula (11.9).

    To obtain a confidence interval for a survival probability $P(X > t) = P(\ln(X) > \ln(t))$, we simply find normal-based confidence interval based on log-transformed data for the probability evaluated at $\ln(t)$. See Section 11.4.3.

Normal-based methods for censored samples are also applicable for censored samples from a lognormal population to find tolerance limits and prediction intervals. See Example 23.5.

Example 23.4. To illustrate the normal-based methods for a lognormal distribution, let us use the following 20 simulated data from a lognormal distribution.

```
6.620    6.475   5.661   6.347   9.496   7.297   5.329    7.389   6.937   7.321
10.150   8.802   6.296   6.312   8.393   7.211   5.277   11.091   4.586   5.921
```

The mean of the log-transformed sample is $\bar{Y} = 1.941$ with the standard deviation $S_y = 0.2274$. Suppose it is desired to find a (.90, .95) tolerance intervals, then the required tolerance factor (using [StatCalc→Continuous →Normal→Tolernace...]) is 2.319. So the tolerance interval based on the log-transformed data is

$$1.941 \pm 2.319 \times .2274 = (1.414, 2.468).$$

Thus, the (.90, .95) tolerance interval for the lognormal distribution is

$$(\exp(1.414), \; \exp(2.468)) = (4.112, 11.799).$$

To find a 95% confidence interval for the probability $P(X > 4.5)$, select the dialog box [StatCalc→Continuous →Normal→Coefficient...], enter 20 for sample size, 1.941 for the sample mean, .2274 for the sample SD, .95 for the confidence level, $\ln(t) = \ln(4.5) = 1.504$ for [Value of t], and click [CI for P(X > t)] to get (.878, .996).

Example 23.5. The data in Table 22.1 represent failure r (in units of 1000 miles) of different locomotive controls in a life test involving 96 locomotive controls. Recall that for these data, the censoring value is 135,000 miles, and the number of censored observation is 37. It has been noted by Schmee and Nelson (1977) and Lawless (2003, Section 5.3) that a lognormal distribution gives a good fit to the data. In this type of situations, a lower tolerance limit is desired to assess the reliability of the controls and to estimate the lifetime at certain mileage.

Since the lognormal distribution is applicable here, we shall use normal-based methods to log-transformed data, and transforms the result back by taking exponentiation. To find tolerance limits, we first find the mean of the log-transformed data as 4.4226 with standard deviation 0.4087, and $\ln(135) = 4.905$. Using these numbers, we can find tolerance limits as follows. Select the dialog box [StatCalc→Continuous →Normal→Censored...], enter 2 for [right-censored], 96 for sample size, 58 for [No. censored], $\ln(135) = 4.905$ for [X1 or DL], 4.4226 for the the [Mean (uncens)], 0.4087 for [SD (uncens)], .9 for [Cont Level], and .9 for [Coverage Level], and click [Tol Interval] to get (3.66, 6.53). Thus, the (.90, .90) tolerance interval is $(\exp(3.66), \exp(6.53)) = (38.86, 685.4)$ and we are 90% confident that at least 90% of locomotive controls work 38,860 to 685,400 miles.

It should be noted that we used 58 for the number of censored observations (instead of 59), because we are using the method for the type II censored samples as an approximation for type I censored samples, and the accuracy of the approximation may be improved by using $k - 1 = 58$; see Remark 22.1.

23.8 Applications

The lognormal distribution can be postulated in physical problems when the random variable X assumes only positive values and its histogram is remarkably skewed to the right. In particular, lognormal model is appropriate for a physical problem if the natural logarithmic transformation of the data satisfy normality assumption. Although lognormal and gamma distributions are interchangeable in many practical situations, a situation where they could produce different results is studied by Wiens (1999).

Practical examples where lognormal model is applicable vary from modeling raindrop sizes (Mantra and Gibbins, 1999) to modeling the global position data (Kobayashi, 1999). The latter article shows that the position data of selected vehicles measured by global positioning system (GPS) follow a lognormal distribution. Application of lognormal distribution in wind speed study is given in Garcia et al. (1998) and Burlaga and Lazarus (2000). In exposure data analysis (data collected from employees who are exposed to workplace contaminants or chemicals) the applications of lognormal distributions are shown in Schulz and Griffin (1999), Borjanovic, et al. (1999), Saltzman (1997), Nieuwenhuijsen (1997), and Roig Navarro et al. (1997). In particular, the one-sided tolerance limits of a lognormal distribution is useful in assessing the workplace exposure to toxic chemicals (Tuggle, 1982). Wang and Wang (1998) showed that lognormal distributions fit very well to the fiber diameter data as well as the fiber strength data of merino wool. Lognormal distribution is also useful to describe the distribution of grain sizes (Jones et al., 1999). Nabe et al. (1998) analyzed data on inter-arrival time and the access frequency of World Wide Web traffic. They found that the document size and the request inter–arrival time follow lognormal distributions, and the access frequencies follow a Pareto distribution.

23.9 Properties and Results

The following results can be proved using the relation between the lognormal and normal distributions.

1. Let X_1 and X_2 be independent random variables with $X_i \sim$ lognormal(μ_i, σ_i^2), $i = 1, 2$. Then

$$X_1 X_2 \sim \text{lognormal}(\mu_1 + \mu_2, \sigma_1^2 + \sigma_2^2)$$

and

$$X_1 / X_2 \sim \text{lognormal}(\mu_1 - \mu_2, \sigma_1^2 + \sigma_2^2).$$

2. Let $X_1, \ldots, X_n$ be independent lognormal random variables with parameters (μ, σ). Then

$$\text{Geometric Mean} = \left(\prod_{i=1}^{n} X_i \right)^{1/n} \sim \text{lognormal} \left(\mu, \frac{\sigma^2}{n} \right).$$

3. Let $X_1, \ldots, X_n$ be independent lognormal random variables with $X_i \sim \text{lognormal}(\mu_i, \sigma_i^2)$, $i = 1, \ldots, n$. For any positive numbers $c_1, \ldots, c_n$,

$$\prod_{i=1}^{n} c_i X_i \sim \text{lognormal} \left(\sum_{i=1}^{n} (\ln c_i + \mu_i), \sum_{i=1}^{n} \sigma_i^2 \right).$$

For more results and properties, see Crow and Shimizu (1988).

23.10 Random Number Generation

Algorithm 23.3. Lognormal variate generator

For given μ and σ:
Generate z from N(0, 1)
Set $y = z * \sigma + \mu$
return $x = \exp(y)$
x is a pseudo random number from the lognormal(μ, σ^2) distribution.

23.11 Calculation of Probabilities and Percentiles

Using the relation that

$$P(X \le x) = P(\ln(X) \le \ln(x)) = P \left(Z \le \frac{\ln(x) - \mu}{\sigma} \right),$$

where Z is the standard normal random variable, the cumulative probabilities, and the percentiles of a lognormal distribution can be easily computed. Specifically, if z_p denotes the pth quantile of the standard normal distribution, then $\exp(\mu + z_p \sigma)$ is the pth quantile of the lognormal(μ, σ^2) distribution.

24

Pareto Distribution

24.1 Description

The probability density function (pdf) of a Pareto distribution with parameters σ and λ is given by

$$f(x|\sigma, \lambda) = \frac{\lambda \sigma^\lambda}{x^{\lambda+1}}, \ x \geq \sigma > 0, \ \lambda > 0. \tag{24.1}$$

The cumulative distribution function (cdf) is given by

$$F(x|\sigma, \lambda) = P(X \leq x|\sigma, \lambda) = 1 - \left(\frac{\sigma}{x}\right)^\lambda, \ x \geq \sigma.$$

For any given $0 < p < 1$, the inverse distribution function is

$$F^{-1}(p|\sigma, \lambda) = \frac{\sigma}{(1-p)^{1/\lambda}}.$$

Plots of the pdf are given in Figure 24.1 for $\lambda = 1, 2, 3$, and $\sigma = 1$. All the plots show long right tail; this distribution may be postulated if the data exhibit a long right tail.

If X has Pareto distribution, then

$$Y = \ln(X) \sim \exp\left(a = \ln(\sigma), b = \frac{1}{\lambda}\right),$$

where a is the location parameter and b is the scale parameter. This transformation is quite useful to find pivotal quantities, thereby confidence intervals and tolerance limits. In fact, various inferential results for exponential distribution can be extended to the Pareto distribution in a straightforward manner.

24.2 Moments

Mean:	$\frac{\sigma\lambda}{\lambda-1}$, $\lambda > 1$
Variance:	$\frac{\lambda\sigma^2}{(\lambda-1)^2(\lambda-2)}$, $\lambda > 2$

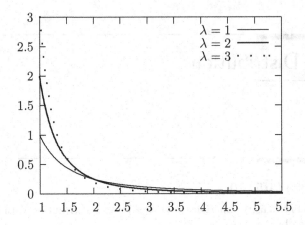

FIGURE 24.1: Pareto probability density functions; $\sigma = 1$

Mode:	σ
Median:	$\sigma 2^{1/\lambda}$
Mean Deviation:	$\frac{2\sigma\lambda^{\lambda-1}}{(\lambda-1)^{\lambda}}, \quad \lambda > 1$
Coefficient of Variation:	$\sqrt{\frac{1}{\lambda(\lambda-2)}}, \quad \lambda > 2$
Coefficient of Skewness:	$\frac{2(\lambda+1)}{(\lambda-3)}\sqrt{\frac{\lambda-2}{\lambda}}, \quad \lambda > 3$
Coefficient of Kurtosis:	$\frac{3(\lambda-2)(3\lambda^2+\lambda+2)}{\lambda(\lambda-3)(\lambda-4)}, \quad \lambda > 4$
Moments about the Origin:	$E(X^k) = \frac{\lambda\sigma^k}{(\lambda-k)}, \quad \lambda > k$
Moment Generating Function:	does not exist
Survival Function:	$(\sigma/x)^{\lambda}$
Hazard Function:	$\lambda \ln(x/\sigma)$

24.3   Probabilities, Percentiles, and Moments

The dialog box [StatCalc→Continuous →Pareto] computes the cdf, percentiles, and moments of a Pareto(σ, λ) distribution.

To compute probabilities: Enter the values of the parameters a, λ, and x; click [P(X <= x)]. For example, when $\sigma = 2$, $\lambda = 3$, and the value of $x = 3.4$, $P(X \le 3.4) = 0.796458$ and $P(X > 3.4) = 0.203542$.

To compute percentiles: Enter the values of σ, λ, and the cumulative probability; click [x]. For example, when $a = 2$, $\lambda = 3$, and the cumulative probability $= 0.15$, the 15th percentile is 2.11133. That is, $P(X \le 2.11133) = 0.15$.

To compute other parameters: Enter the values of one of the parameters, cumulative probability, and x. Click on the missing parameter. For example, when $\lambda = 4$, cumulative probability $= 0.15$, and $x = 2.4$, the value of σ is 2.30444.

To compute moments: Enter the values σ and λ and click [M].

24.4 Confidence Intervals

Let $X_1, \ldots, X_n$ be a sample of independent observations from a Pareto(σ, λ) distribution with pdf in (24.1). Let

$$X_{(1)} = \min\{X_1, \ldots, X_n\} \quad \text{and} \quad \tilde{G} = \left(\prod_{i=1}^{n} X_i\right)^{1/n}.$$

Consider the transformation $Y_i = \ln(X_i)$, $i = 1, ..., n$. Then $Y_1, ..., Y_n$ can be regarded as a sample from an exponential$(a = \ln(\sigma)$, $b = 1/\lambda)$ distribution. Furthermore, the maximum likelihood estimates (MLEs) based on exponential model are

$$\hat{a} = Y_{(1)} \quad \text{and} \quad \hat{b} = \bar{Y} - Y_{(1)}, \tag{24.2}$$

which give the MLEs

$$\hat{\sigma} = X_{(1)} \quad \text{and} \quad \hat{\lambda} = \frac{1}{\hat{b}} = \frac{1}{\ln\left(\tilde{G}/X_{(1)}\right)}. \tag{24.3}$$

Unbiased estimators, in terms of the MLEs, are

$$\hat{\lambda}_u = \left(1 - \frac{1}{2n}\right)\hat{\lambda} \quad \text{and} \quad \hat{\sigma}_u = \left(1 - \frac{1}{(n-1)\hat{\lambda}}\right)\hat{\sigma}.$$

It follows from the distributional results (15.6) that

$$2n\lambda \ln\left(\frac{\hat{\sigma}}{\sigma}\right) \sim \chi_2^2 \quad \text{and} \quad \frac{2n\lambda}{\hat{\lambda}} \sim \chi_{2n-2}^2, \tag{24.4}$$

where the chi-square random variables are independent.

Confidence Interval for λ

A $1 - \alpha$ confidence interval for λ based on the distributional result (24.4) is given by

$$\left(\frac{\hat{\lambda}}{2n}\chi_{2(n-1),\alpha/2}^2, \; \frac{\hat{\lambda}}{2n}\chi_{2(n-1),1-\alpha/2}^2\right).$$

Confidence Interval for σ

Let F_α^* denote $F_{2,2n-2;\alpha}/(n-1)$. A $1-\alpha$ confidence interval for σ based on the confidence interval (see Section 15.5) for the location parameter of the exponential distribution is

$$\left(\exp\left(\widehat{a}-F_{1-\alpha/2}^*\widehat{b}\right), \exp\left(\widehat{a}-F_{\alpha/2}^*\widehat{b}\right)\right),$$

where $\widehat{a}$ and $\widehat{b}$ are as defined in (24.2). After simplification, it can be written as

$$\left(X_{(1)}\left(\frac{X_{(1)}}{\widetilde{G}}\right)^{F_{1-\alpha/2}^*}, \quad X_{(1)}\left(\frac{X_{(1)}}{\widetilde{G}}\right)^{F_{\alpha/2}^*}\right). \tag{24.5}$$

24.5 Prediction Intervals and Tolerance Limits

A $1-\alpha$ prediction interval for a future observation X from a Pareto(λ,σ) distribution can be obtained from the one for an exponential distribution as follows. Let $X_1,...,X_n$ be a sample from a Pareto distribution, and let $Y_i=\ln(X_i)$, $i=1,...,n$. Construct a $1-\alpha$ exponential prediction interval (15.11)

$$(L,U)=(\widehat{a}+R_{n;\alpha}\widehat{b}, \ \widehat{a}+R_{n;1-\alpha}\widehat{b}),$$

where $R_{n;\alpha}$ is defined in (15.12), based on $Y_1,...,Y_n$. Then $(\exp(L),\exp(U))$ is a $1-\alpha$ prediction interval for X.

Tolerance Limits

One-sided tolerance limits can also obtained via log-transformation. Let L denote the $(p,1-\alpha)$ exponential lower tolerance limit based on a log-transformed sample from a Pareto distribution. Then $\exp(L)$ is a $(p,1-\alpha)$ lower tolerance limit for the sampled Pareto distribution. An upper tolerance limit can be found similarly.

24.6 Applications

The Pareto distribution is often used to model the data on personal incomes and city population sizes. Pareto distributions are useful to model a wide variety of socioeconomic data. It is often used to model the size distribution of income. In this context, Pareto (1897) introduced the concept in his well-known economics text. Even though there are criticisms on application of a Pareto model to income distributions, on the basis of empirical evidence it is generally accepted that most income distributions indeed fit Pareto distributions. In general, this distribution may be postulated if the histogram of the data from a physical problem has a long tail. Nabe et al. (1998) studied the traffic data of World Wide Web (www), and they found that the access frequencies of www follow a Pareto distribution. Atteia and Kozel (1997) showed that water particle sizes fit a Pareto distribution. The Pareto distribution is also used to describe the lifetimes of components. Aki and Hirano

(1996) mentioned a situation where the lifetimes of components in a conservative-k-out-of-n-F system follow a Pareto distribution.

24.7 Random Number Generation

For a given a and λ:
Generate u from uniform$(0, 1)$
Set $x = \sigma/(1 - u) ** (1/\lambda)$
x is a pseudo random number from the Pareto(a, λ) distribution.

25

Weibull Distribution

25.1 Description

Let Y be a standard exponential random variable with probability density function (pdf)

$$f(y) = e^{-y}, \quad y > 0.$$

Define

$$X = bY^{1/c} + m, \quad b > 0, \ c > 0.$$

The distribution of X is known as the Weibull distribution with shape parameter c, scale parameter b, and the location parameter m. Its probability density is given by

$$f(x|b,c,m) = \frac{c}{b}\left(\frac{x-m}{b}\right)^{c-1} \exp\left\{-\left[\frac{x-m}{b}\right]^c\right\}, \quad x > m, \ b > 0, \ c > 0. \quad (25.1)$$

The cumulative distribution function (cdf) is given by

$$F(x|b,c,m) = 1 - \exp\left\{-\left[\frac{x-m}{b}\right]^c\right\}, \quad x > m, \ b > 0, \ c > 0. \quad (25.2)$$

For $0 < p < 1$, the inverse distribution function is

$$F^{-1}(p|b,c,m) = m + b(-\ln(1-p))^{\frac{1}{c}}. \quad (25.3)$$

Let us denote the three-parameter distribution by $\text{Weibull}(b,c,m)$. A two-parameter Weibull distribution is denoted by $\text{Weibull}(a,b)$.

25.2 Moments

The following formulas are valid when $m = 0$.

Mean:	$b\Gamma(1 + 1/c)$
Variance:	$b^2\Gamma(1 + 2/c) - [\Gamma(1 + 1/c)]^2$
Mode:	$b\left(1 - \frac{1}{c}\right)^{1/c}, \ c \geq 1$
Median:	$b[\ln(2)]^{1/c}$

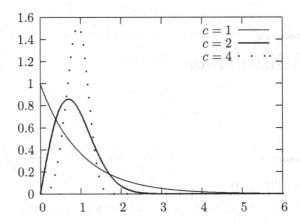

FIGURE 25.1: Weibull probability density functions; $m = 0$ and $b = 1$

Coefficient of Variation:	$\dfrac{\sqrt{\Gamma(1+2/c)-[\Gamma(1+1/c)]^2}}{\Gamma(1+1/c)}$
Coefficient of Skewness:	$\dfrac{\Gamma(1+3/c)-3\Gamma(1+1/c)\Gamma(1+2/c)+2[\Gamma(1+1/c)]^3}{\left[\Gamma(1+2/c)-\{\Gamma(1+1/c)\}^2\right]^{3/2}}$
Moments about the Origin:	$E(X^k) = b^k\Gamma(1 + k/c)$
Inverse Distribution Function(p):	$b\{-\ln(1 - p)\}^{1/c}$
Survival Function:	$P(X > x) = \exp\{-(x/b)^c\}$
Inverse Survival Function(p):	$b\{(1/c)\ln(-p)\}$
Hazard Rate:	cx^{c-1}/b^c
Hazard Function:	$(x/b)^c$

25.3   Probabilities, Percentiles, and Moments

The dialog box [StatCalc→Continuous→Weibull] computes the cdf, percentiles, and moments of a Weibull(b, c, m) distribution.

To compute probabilities: Enter the values of m, c, b, and the cumulative probability; click [P(X <= x)]. For example, when $m = 0$, $c = 2.3$, $b = 2$, and $x = 3.4$, $P(X \leq 3.4) = 0.966247$ and $P(X > 3.4) = 0.033753$.

To compute percentiles: Enter the values of m, c, b, and the cumulative probability; click [x]. For example, when $m = 0$, $c = 2.3$, $b = 2$, and the cumulative probability $= 0.95$, the 95th percentile is 3.22259. That is, $P(X \leq 3.22259) = 0.95$.

To compute other parameters: Enter the values of any two of m, c, b, cumulative probability, and x. Click on the missing parameter. For example, when $m = 1$, $c = 2.3$, $x = 3.4$, and the cumulative probability $= 0.9$, the value of b is 1.67004.

To compute moments: Enter the values of c and b and click [M]. The moments are computed assuming that $m = 0$.

25.4 Maximum Likelihood Estimators and Pivotal Quantities

Let $X_1, \ldots, X_n$ be a sample from a two-parameter Weibull(b, c). Let $Y_i = \ln(X_i)$. An asymptotically unbiased estimator of $\theta = (1/c)$ is given by

$$\widehat{\theta} = \frac{\sqrt{6}}{\pi} \sqrt{\frac{\sum\limits_{i=1}^{n} (Y_i - \bar{Y})^2}{n-1}}. \tag{25.4}$$

Further, the estimator is asymptotically distributed as normal with variance $= 1.1/(c^2 n)$ [Menon, 1963]. The maximum likelihood estimate (MLE) of c is the solution to the equation

$$\frac{1}{\widehat{c}} - \frac{\sum\limits_{i=1}^{n} X_i^{\widehat{c}} Y_i}{\sum\limits_{i=1}^{n} X_i^{\widehat{c}}} + \frac{1}{n} \sum\limits_{i=1}^{n} Y_i = 0, \tag{25.5}$$

and the MLE of b is given by

$$\widehat{b} = \left(\frac{1}{n} \sum\limits_{i=1}^{n} X_i^{\widehat{c}} \right)^{1/\widehat{c}}. \tag{25.6}$$

The R function 25.1 based on the Newton–Raphson iterative method with initial value (25.4) can be used to find the MLEs.

Pivotal Quantities

Pivotal quantities can be obtained using the one-one relation between the Weibull and extreme-value distributions, and then applying the distributional results of the MLEs for a location-scale family. Specifically, $Y_i = -\ln(X_i)$, $i = 1, \ldots, n$, can be regarded as a sample from an extreme-value(μ, σ) distribution, where $\mu = -\ln(b)$ is the location parameter, and $\sigma = 1/c$ is the scale parameter. The MLEs of μ and σ are

$$\widehat{\mu} = \ln(\widehat{b}) \quad \text{and} \quad \widehat{\sigma} = 1/\widehat{c},$$

respectively. Since the family of extreme-value distributions is a location-scale family,

$$\frac{\widehat{\mu} - \mu}{\widehat{\sigma}}, \quad \frac{\widehat{\sigma}}{\sigma} \quad \text{and} \quad \frac{\widehat{\mu} - \mu}{\sigma}$$

are pivotal quantities. Replacing $(\mu, \sigma, \widehat{\mu}, \widehat{\sigma})$ by $\left(-\ln(b), \frac{1}{c}, -\ln(\widehat{b}), \frac{1}{\widehat{c}}\right)$, we see that

$$\frac{\widehat{c}}{c} \quad \text{and} \quad \widehat{c}\ln\left(\frac{\widehat{b}}{b}\right) \tag{25.7}$$

are pivotal quantities. This means that the empirical distribution of these pivotal quantities can be evaluated using simulated samples from a Weibull$(1, 1)$ = exponential$(0, 1)$ distribution. In other words,

$$\frac{\widehat{c}}{c} \sim \widehat{c}^* \quad \text{and} \quad \widehat{c}\ln\left(\frac{\widehat{b}}{b}\right) \sim \widehat{c}^* \ln(\widehat{b}^*), \tag{25.8}$$

where $\widehat{c}^*$ and $\widehat{b}^*$ are determined by (25.5) and (25.6) with $(X_1, ..., X_n)$ being a sample from the standard exponential distribution.

R function 25.1. Calculation of the Weibull MLEs

```
mles.weibull.nr = function(x){
y = log(x); yb = mean(y)
cho = 1.28255/sd(y)
j = 1
repeat{
xc = x**cho
s1 = sum(xc); v = xc/s1; s2 = sum(y*v); s3 = sum(y*y*v)
fc = 1.0/cho+yb-s2; fpc= 1.0/cho**2 +(s3-s2**2)
ch  = cho + fc/fpc
if(abs(fc) <= 0.000001 | j >= 30){break}
cho = ch; j = j + 1}
bh = (mean(x**ch))**(1/ch)
return(c(ch,bh,j))
}
```

25.5   Confidence Intervals and Prediction Intervals

Let $(\widehat{c}, \widehat{b})$ be the MLEs of (c, b) based on a sample of size n from a Weibull(b, c) distribution.

Confidence Interval for c

On the basis of the pivotal quantities in (25.7), we find a $1 - \alpha$ confidence interval for c as

$$\left(\frac{\widehat{c}}{c_l}, \frac{\widehat{c}}{c_u}\right),$$

where c_l and c_u are determined so that

$$P\left(c_l \leq \widehat{c}^* \leq c_u\right) = 1 - \alpha,$$

and $\widehat{c}^*$ is as defined in (25.8).

Confidence Interval for b

Let b_l and b_u be determined by

$$P\left(b_l \leq \widehat{c}^* \ln(\widehat{b}^*) \leq b_u\right) = 1 - \alpha.$$

A $1 - \alpha$ confidence interval for b is given by

$$\left(\widehat{b}\exp\left(-\frac{b_u}{\widehat{c}}\right),\ \widehat{b}\exp\left(-\frac{b_l}{\widehat{c}}\right)\right). \tag{25.9}$$

The above confidence intervals for the parameters are exact, except for the simulation errors involved in estimating the percentiles.

Mean

Let $\widehat{b}_0$ and $\widehat{c}_0$ be the observed values of the MLEs based on a sample of n observations from a Weibull(b, c) distribution. A generalized pivotal quantity (GPQ) for the scale parameter b is given by

$$G_b = \left(\frac{b}{\widehat{b}}\right)^{\frac{\widehat{c}}{\widehat{c}_0}} \widehat{b}_0 = \left(\frac{1}{\widehat{b}^*}\right)^{\frac{\widehat{c}^*}{\widehat{c}_0}} \widehat{b}_0. \tag{25.10}$$

A GPQ for c can be obtained as

$$G_c = \frac{c}{\widehat{c}}\widehat{c}_0 = \frac{\widehat{c}_0}{\widehat{c}^*}. \tag{25.11}$$

It can be easily checked that the GPQs G_b and G_c satisfy the two conditions in Section 2.7.2.

By substituting the above GPQs in the expression for the mean, we find the GPQ for the mean as

$$G_\mu = G_b\, \Gamma\left(1 + 1/G_c\right). \tag{25.12}$$

Note that for a given $(\widehat{c}_0, \widehat{b}_0)$, the distribution of G_μ does not depend on any unknown parameters, and so its percentiles can be estimated using Monte Carlo simulated samples from a Weibull($1, 1$) distribution. Let $G_{\mu;p}$ denote the pth quantile of G_μ. Then

$$\left(G_{\mu;\frac{\alpha}{2}},\ G_{\mu;1-\frac{\alpha}{2}}\right)$$

is a $1 - \alpha$ confidence interval for the Weibull mean. Simulation studies by Krishnamoorthy, Lin and Xia (2009) indicate that the above generalized confidence intervals are quite accurate even for small samples.

Prediction Intervals

Let X be a future observation form a Weibull(a, b) distribution, and let $\widehat{a}$ and $\widehat{b}$ be the MLEs based on a sample of size n from the Weibull(a, b) distribution. The pivotal quantity for finding a prediction interval is

$$\frac{\ln(X) - \ln(\widehat{\mu})}{\widehat{\sigma}},$$

where $\widehat{\mu} = -\ln(\widehat{b})$ and $\widehat{\sigma}/\widehat{c}$. Let P_l and P_u be determined such that

$$P\left(P_l \leq \frac{\ln(X) - \ln(\widehat{\mu})}{\widehat{\sigma}} \leq P_u\right) = 1 - \alpha,$$

where $1 - \alpha$ is a specified value. Then

$$\left(\widehat{b}\exp\left(\frac{P_l}{\widehat{c}}\right), \widehat{b}\exp\left(\frac{P_u}{\widehat{c}}\right)\right) \tag{25.13}$$

is a $1 - \alpha$ prediction interval for X.

25.6   One-Sided Tolerance Limits

Recall that the $(p, 1 - \alpha)$ upper tolerance limit is a $1 - \alpha$ upper confidence limit for the pth quantile q_p of the Weibull(b, c) distribution,

$$q_p = b(-\ln(1 - p))^{\frac{1}{c}}. \tag{25.14}$$

Substituting the GPQs for b and c (see (25.10) and (25.11)) in the above expression, we obtain GPQ for q_p, denoted by G_{q_p}. Letting $\kappa_p = -\ln(1 - p)$, we can write

$$\ln(G_{q_p}) = (\widehat{c}_0)^{-1}\left[\widehat{c}^*\left(-\ln(\widehat{b}^*) + \ln(\kappa_p)\right)\right] + \ln(\widehat{b}_0),$$

where $(\widehat{b}_0, \widehat{c}_0)$ is an observed value of the MLE $(\widehat{b}, \widehat{c})$. Let

$$w_p = \widehat{c}^*\left(-\ln(\widehat{b}^*) + \ln(\kappa_p)\right),$$

and let $w_{p;q}$ denote the qth quantile w_p. Then a $(p, 1 - \alpha)$ upper tolerance limit can be expressed as

$$\widehat{b}_0 \exp\left(\frac{w_{p;1-\alpha}}{\widehat{c}_0}\right). \tag{25.15}$$

Note that the distribution of w_p does not depend on any unknown parameters, and so its percentiles can be estimated based on simulated samples from a Weibull$(1, 1)$ distribution. The one-sided lower tolerance limit can be obtained similarly as

$$\widehat{b}_0 \exp\left(\frac{w_{1-p;\alpha}}{\widehat{c}_0}\right). \tag{25.16}$$

25.7   Survival Probability

For a given t, the survival probability

$$S_t = P(X > t) = e^{-(t/b)^c}.$$

Let $S_t^* = \ln[-\ln(S(t))] = c\ln(t/b)$. A GPQ for S_t^* can be expressed as

$$G_{S_t^*} = G_c \ln(t/G_b) = \frac{\ln(-\ln(\widehat{S}_t))}{\widehat{c}^*} + \ln(\widehat{b}^*),$$

where $\widehat{S}_t = \exp\left[-\left(\frac{t}{\widehat{b}_0}\right)^{\widehat{c}_0}\right]$. For a given $\widehat{b}_0$ and $\widehat{c}_0$, let L and U be determined so that

$$P(L \leq G_{S_t^*} \leq U) = 1 - \alpha.$$

In terms of L and U, a $1 - \alpha$ confidence interval for S_t can be expressed as

$$(\exp[-\exp(U)],\ \exp[-\exp(L)]).$$

For a given $(\widehat{b}_0, \widehat{c}_0)$, the distribution of $G_{S_t^*}$ does not depend on any unknown parameters, and so the percentiles L and U can be estimated by simulation.

Example 25.1. The data in Table 25.1 represent the number of million revolutions before failure for each of 23 ball bearings. The data were analyzed by Thoman et al. (1969) and others using a Weibull distribution. The MLEs $\widehat{c}_0 = 2.103$ and $\widehat{b}_0 = 81.876$.

TABLE 25.1: Numbers of Millions of Revolutions of 23 Ball Bearings before Failure

17.88	28.92	33.00	41.52	42.12	45.60	48.48	51.84
51.96	54.12	55.56	67.80	68.64	68.64	68.88	84.12
93.12	98.64	105.12	105.84	127.92	128.04	173.40	

The MLE $\widehat{\eta}$ for the Weibull mean $\widehat{b}_0\Gamma(1 + 1/\widehat{c}_0) = 72.52$. After entering these sample size, MLEs, and .95 for [Conf Level] in the dialog box [StatCalc→Continuous→Weibull→CIs for para...], click on [CI for Mean] to get (58.58, 91.11). In the same dialog box, by clicking [CI for c] and [CI for b], we can estimate confidence intervals for c and b, respectively. For this example, 95% confidence interval for c is (1.413, 2.728), and 95% confidence interval for b is (65.35, 102.2).

Suppose it is desired to find a 95% confidence interval for the probability that a ball bearing lasts 50 million or more revolutions. To find the confidence interval, enter 50 [Mission time t], .95 for [Conf Level], and click [CI for P(X ¿ t)] to get [.56, .81]. To find a 95% lower confidence bound for $P(X > 50)$, enter .90 for [Conf Level] and click [CI for P(X > t)] to get 0.59. That is, at least 59% of the ball bearings last 50 million or more revolutions with confidence 95%.

25.8 One-sided Prediction Limits for at Least l of m Observations at Each of r Locations

We shall first describe a pivotal-based approach to find a prediction interval for a single future observation X from a Weibull(b, c) distribution. Let $X_1, ..., X_n$ be a sample from a Weibull(b, c) distribution. Using the one-one relation between the Weibull and extreme value distribution, we find the following pivotal quantity:

$$\frac{\ln X - \ln \widehat{b}}{1/\widehat{c}} = \widehat{c}\ln\left(\frac{X}{\widehat{b}}\right) \sim \widehat{c}^* \ln\left(\frac{X^*}{\widehat{b}^*}\right), \tag{25.17}$$

where $X^* \sim$ Weibull$(1, 1)$ independently of $\widehat{b}^*$ and $\widehat{c}^*$, which are the MLEs based on a sample of size n from a Weibull$(1, 1)$ distribution. Let P_l and P_u be determined so that

$$P\left(P_l \leq \widehat{c}\ln(X/\widehat{b}) \leq P_u\right) = 1 - \alpha.$$

A $1 - \alpha$ prediction interval for a future observation X is given by

$$\left(\widehat{b}\exp(P_l/\widehat{c}),\ \widehat{b}\exp(P_u/\widehat{c})\right). \tag{25.18}$$

We describe an empirical method of constructing a $1 - \alpha$ upper prediction limit for an extreme value distribution so that it will include at least l of m samples from each of r locations. We shall consider the upper prediction limit of the form $\widehat{\mu} + u_{n,r,m,l}\widehat{\sigma}$, where $u_{n,r,m,l}$ is the factor to be determined so that the coverage probability is $1 - \alpha$. Let $\widehat{\mu}^*$ and $\widehat{\sigma}^*$ be the MLEs based on a sample $y_1^*, ..., y_n^*$ from an extreme-value$(0, 1)$ distribution, and $y_{11}^*, ..., y_{1m}^*, y_{21}^*, ..., y_{2m}^*, \cdots , y_{r1}^*, ..., y_{rm}^*$ be independent samples from an extreme-value$(0, 1)$ distribution. Assume that y_i^*s and y_{ij}^*s are all mutually independent. Let $y_{i(l)}^*$ be the lth order statistic, $i = 1, ..., r$, $y_{r,m,l}^* = \max_{1 \leq i \leq r} y_{i(l)}^*$, and $u = (y_{r,m,l}^* - \widehat{\mu}^*)/\widehat{\sigma}^*$. The $1 - \alpha$ quantile of u is the desired factor for constructing upper prediction limit. A $1 - \alpha$ upper prediction limit that will include at least l of m observations from each of r locations is given by

$$\exp\left(\widehat{\mu} + u_{n,r,m,l}\widehat{\sigma}\right) = \widehat{b}\exp\left(\frac{u_{n,r,m,l}}{\widehat{c}}\right). \tag{25.19}$$

As the distribution of u does not depend on any parameter, its percentiles can be obtained using Monte Carlo simulation, as explained in the following algorithm.

Algorithm 25.1. Calculation of the factor $u_{n,r,m,l}$

For a given n, r, m, l and $1 - \alpha$:
For $k = 1, N$
    Generate $x_1^*, ..., x_n^*$ from a Weibull$(1, 1)$ distribution
    Compute the MLEs $\widehat{b}^*$ and $\widehat{c}^*$
    Generate $x_{i1}^*, ..., x_{im}^*$ from a Weibull$(1, 1)$ distribution, $j = 1, ..., m.$ $i = 1, ..., r$
    Find the lth order statistic $x_{i(l)}$, $i = 1, ..., r$
    Set $x_{r,m,l}^* = \max_{1 \leq i \leq r} x_{i(l)}^*$
    Set $u_k = \widehat{c}^* \ln(x_{r,m,l}^*/\widehat{b}^*)$
(end k loop)
The $100(1-\alpha)$th percentile of the u_js is a Monte Carlo estimate of the factor $u_{n,r,m,l}$.

Example 25.2. To illustrate the construction of an upper prediction limit, we use the vinyl chloride data given in Table 16.1. A Q–Q plot in Figure 25.2 indicates that the assumption of Weibull model is tenable for the data.

We shall calculate upper prediction limit of the form $\widehat{b}\exp(u_{n,r,m,l}/\widehat{c})$ so that at least l of m measurements from each of r locations are less than the upper prediction

TABLE 25.2: 95% Upper Prediction Limits for the Vinyl Chloride Data
$n = 34$, $\widehat{b} = 1.8842$ and $\widehat{c} = 1.1010$

r	m	l	$u_{n,r,m,l}$	Weibull UPL	Gamma UPL
1	2	1	0.450	2.942	2.893
10	2	1	1.079	5.483	5.203
10	3	1	0.650	3.586	3.479
10	3	2	1.295	6.791	6.369

limit. In the following Table 25.2, we provided 95% factors for some values of (r, m, l), and the corresponding Weibull upper prediction limits (UPLs) along with gamma UPLs (Example 16.8).

The values of the factor $u_{n,r,m,l}$ for constructing 95% upper prediction limits for a few combinations of r, m, and l are given in Table 25.2. To find, for example, the 95% factor $u_{34,1,2,1}$, select the dialog box [StatCalc→Continuous→Weibull→Prediction ...], enter $(34, 1, 2, 1)$ for (n, r, m, l), .95 for [Conf Level], and click [Factor k] to get 0.450. The corresponding upper prediction limit is

$$\widehat{b}\exp(.450/\widehat{c}) = 2.942.$$

The other upper prediction limits are computed similarly. The Weibull-based prediction limits, along with those based on a gamma distribution are given in Table 25.2. We observe from this table that the upper prediction limits based on a Weibull model are slightly larger than those based on a gamma model for all the cases considered. It should be noted that the upper prediction limits based on the Weibull distribution are exact except for simulation errors, whereas the ones based on a gamma distribution are approximate.

25.9   Properties and Results

1. Let X be a Weibull(b, c, m) random variable. Then,

$$\left(\frac{X - m}{b}\right)^c \sim \exp(0, 1),$$

that is, the exponential distribution with mean 1.

2. It follows from (1) and the probability integral transform that

$$1 - \exp\left[-\left(\frac{X - m}{b}\right)^c\right] \sim \text{uniform}(0, 1),$$

and hence

$$X = m + b[-\ln(1 - U)]^{1/c} \sim \text{Weibull}(b, c, m),$$

where U denotes the uniform$(0, 1)$ random variable.

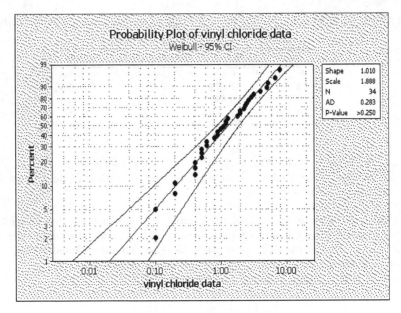

FIGURE 25.2: Weibull probability plot of the vinyl chloride data

25.10   Random Number Generation

For a given m, b, and c:
Generate u from uniform(0, 1)
return $x = m + b * (-\ln(1 - u)) * *(1/c)$

26

Extreme Value Distribution

26.1 Description

The probability density function (pdf) of the extreme value distribution with the location parameter μ and the scale parameter σ is given by

$$f(x|\mu,\sigma) = \frac{1}{\sigma}\exp\left(-z - \exp(-z)\right), \qquad (26.1)$$

where $z = \frac{x-\mu}{\sigma}$. The cumulative distribution function (cdf) is given by

$$F(x|\mu,\sigma) = \exp\{-\exp[-z]\}, \quad -\infty < x < \infty, \sigma > 0. \qquad (26.2)$$

The inverse distribution function is given by

$$F^{-1}(p|\mu,\sigma) = \mu - \sigma\ln(-\ln(p)), \quad 0 < p < 1. \qquad (26.3)$$

The plots of the pdf in Figure 26.1 show that this distribution asymmetric and right-skewed. The family of distributions with the pdf in (26.1) is referred to as the type I largest extreme value distribution (LEV), as this is the sampling distribution of the largest observation in a sample from a continuous distribution. We shall refer to this distribution as LEV(μ,σ). We also note that if

$$Y \sim \text{Weibull}(b,c) \text{ then, } -\ln(Y) \sim \text{LEV}(\mu = -\ln b, \sigma = 1/c). \qquad (26.4)$$

This one-to-one relation allows us to find prediction intervals, tolerance limits, and confidence intervals for survival probability from those for a Weibull distribution.

The family of distributions with the pdf of the form

$$f(x|\mu,\sigma) = \frac{1}{\sigma}\exp\left(z - \exp(z)\right), \qquad (26.5)$$

where $z = \frac{x-\mu}{\sigma}$, is referred to as the type I smallest extreme value, as this is the sampling distribution of the smallest observation of a sample from a continuous distribution. If Y has the Weibull(b,c) distribution, then $\ln(Y)$ has the smallest extreme value distribution with $\mu = \ln(b)$ and $\sigma = 1/c$.

In the sequel, we shall consider only the Type I LEV distribution, which is also called the **Gumbel** distribution.

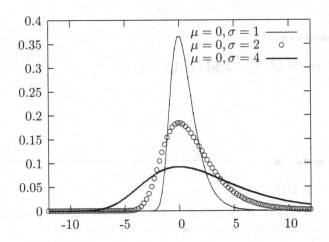

FIGURE 26.1: Extreme value probability density functions

26.2   Moments

Mean:	$\mu + \gamma\sigma$, where $\gamma = 0.5772\ 15664\ 9\ldots.$
Mode:	μ
Median:	$\mu - \sigma\ln(\ln 2)$
Variance:	$\sigma^2\pi^2/6$
Coefficient of Skewness:	1.139547
Coefficient of Kurtosis:	5.4
Moment Generating Function:	$\exp(\mu t)\,\Gamma(1 - \sigma t),\ \ t < 1/\sigma.$
Characteristic Function:	$\exp(i\mu t)\,\Gamma(1 - i\sigma t)$
Inverse Distribution Function:	$\mu - \sigma\ln(-\ln p)$
Inverse Survival Function:	$\mu - \sigma\ln(-\ln(1 - p))$
Hazard Function:	$\dfrac{\exp[-(x-\mu)/\sigma]}{\sigma\{\exp[\exp(-(x-\mu)/\sigma)]-1\}}$

26.3   Probabilities, Percentiles, and Moments

The dialog box [StatCalc→Continuous→Extreme→Probabilities ...] computes probabilities, percentiles, and moments of an extreme value distribution.

To compute probabilities: Enter the values of the parameters μ and σ, and of x; click [P(X <= x)]. For example, when $\mu = 2$, $\sigma = 3$, and $x = 2.3$, $P(X \le 2.3) = 0.404608$ and $P(X > 2.3) = 0.595392$.

To compute percentiles: Enter the values of μ, σ, and the cumulative probability; click [x]. For example, when $\mu = 1$, $\sigma = 2$, and the cumulative probability $= 0.15$, the 15th percentile is -0.280674. That is, $P(X \le -0.280674) = 0.15$.

For any given three of the four values μ, σ, cumulative probability, and x, *StatCalc* computes the missing one. For example, when $\sigma = 2$, $x = 1$, and P(X <=x) = 0.15, the value of μ is 2.28067.

To compute moments: Enter the values of μ and σ and click [M].

26.4   Maximum Likelihood Estimators

The maximum likelihood estimates (MLEs) of μ and σ can be readily obtained from those of Weibull MLEs based on e^{-X} transformation. Specifically, let $X_1, ..., X_n$ be a sample from an LEV(μ, σ) distribution. Transform the sample as $Y_i = \exp(-X_i)$, and calculate the Weibull MLEs $\widehat{b}$ and $\widehat{c}$. Then

$$\widehat{\mu} = -\ln(\widehat{b}) \quad \text{and} \quad \widehat{\sigma} = \frac{1}{\widehat{c}}$$

are the MLEs of μ and σ, respectively.

The MLEs can also be obtained by maximizing the log-likelihood based on LEV(μ, σ) distribution. Let $\theta = 1/\sigma$. The MLE $\widehat{\theta}$ of θ is the solution of the equation

$$\frac{1}{\widehat{\theta}} - \bar{X} + \frac{\sum_{i=1}^{n} X_i e^{-X_i \widehat{\theta}}}{\sum_{i=1}^{n} e^{-X_i \widehat{\theta}}} = 0, \tag{26.6}$$

and the MLE of μ is given by

$$\widehat{\mu} = -\frac{1}{\widehat{\theta}} \ln \left(\frac{1}{n} \sum_{i=1}^{n} e^{-X_i \widehat{\theta}} \right). \tag{26.7}$$

The root of (26.6) can be obtained iteratively using the scheme

$$\theta_{\text{new}} = \theta_0 - \frac{f(\theta_0)}{f'(\theta_0)},$$

where $f(\theta)$ is the function in (26.6) and

$$f'(\theta) = \frac{\left(\sum_{i=1}^{n} X_i e^{-X_i \theta} \right)^2}{\left(\sum_{i=1}^{n} e^{-X_i \theta} \right)^2} - \frac{\sum_{i=1}^{n} X_i^2 e^{-X_i \theta}}{\sum_{i=1}^{n} e^{-X_i \theta}} - \frac{1}{\theta^2},$$

and θ_0 is some starting value for the iterative scheme. The Menon (1963) estimate

$$\widehat{\theta} = \frac{\pi}{\sqrt{6}} \left(\frac{\sum_{i=1}^{n} (X_i - \bar{X})^2}{n-1} \right)^{-1/2} \tag{26.8}$$

can be used as a starting value.

The following R function can be used to calculate MLEs $(\widehat{\mu}, \widehat{\sigma})$ of the parameters (μ, σ).

R function 26.1. Calculation of the MLEs of μ and σ

```
mle.extreme = function(x){
theta0 = 1.0/(sqrt(6)*sd(x)/pi)
n = length(x); xb = mean(x);
j = 1
repeat{
y = exp(-x*theta0)
fc = 1.0/theta0 - xb + sum(x*y)/sum(y)
fpc = (sum(x*y))**2/(sum(y))**2-sum(x**2*y)/sum(y)-1.0/theta0**2
thetan = theta0-fc/fpc
if(abs(fc) <= 1.0e-5 | j > 30){break}
j = j + 1
theta0 = thetan
}
mu_hat = -log(mean(y))/thetan; sig_hat = 1.0/thetan
return(c(mu_hat,sig_hat))
}
```

26.5   Confidence Intervals

Let $\widehat{\mu}$ and $\widehat{\sigma}$ be the MLEs based on a sample $X_1, ..., X_n$ from an $\text{LEV}(\mu, \sigma)$ distribution. The pivotal quantities based on the MLEs are

$$\frac{\widehat{\mu} - \mu}{\widehat{\sigma}} \quad \text{and} \quad \frac{\widehat{\sigma}}{\sigma}.$$

This result implies that

$$\frac{\widehat{\mu} - \mu}{\widehat{\sigma}} \sim \frac{\widehat{\mu}^*}{\widehat{\sigma}^*} \quad \text{and} \quad \frac{\widehat{\sigma}}{\sigma} \sim \widehat{\sigma}^*, \tag{26.9}$$

where "$\sim$" means "distributed as," and $\widehat{\mu}^*$ and $\widehat{\sigma}^*$ are the MLEs based on a random sample of size n from an $\text{LEV}(0, 1)$ distribution.

Location Parameter

Let (c_l, c_u) be determined so that

$$P\left(c_l \leq \frac{\widehat{\mu}^*}{\widehat{\sigma}^*} \leq c_u\right) = 1 - \alpha.$$

Then

$$(\widehat{\mu} - c_u\widehat{\sigma}, \ \widehat{\mu} - c_l\widehat{\sigma}) \qquad\qquad (26.10)$$

is a $1 - \alpha$ confidence interval μ. The required percentiles μ_l and μ_u can be estimated based on simulated samples from an LEV$(0, 1)$ distribution.

Scale Parameter

Let (k_l, k_u) be determined so that

$$P\left(k_l \leq \widehat{\sigma}^* \leq k_u\right) = 1 - \alpha.$$

Then

$$\left(\frac{\widehat{\sigma}}{k_u}, \ \frac{\widehat{\sigma}}{k_l}\right)$$

is a $1 - \alpha$ confidence interval for σ.

Mean

Recall that the mean of an LEV(μ, σ) distribution is given by $\mu + \gamma\sigma$, where γ is the Euler constant. The pivotal for the mean is given by

$$Q = \frac{\widehat{\mu} - (\mu + \gamma\sigma)}{\widehat{\sigma}} \sim \frac{\widehat{\mu}^* - \gamma}{\widehat{\sigma}^*},$$

where $(\widehat{\mu}^*, \widehat{\sigma}^*)$ is as defined in (26.9). If Q_α denotes the α quantile of Q, then

$$(\widehat{\mu} - Q_{1-\alpha}\widehat{\sigma}, \widehat{\mu} - Q_\alpha\widehat{\sigma})$$

is an exact $1 - 2\alpha$ confidence interval for the mean. The percentiles of Q can be estimated using simulated samples from an LEV$(0, 1)$ distribution.

Example 26.1. We shall illustrate the interval estimation methods using the survival data on 20 rats given in Example 16.6. The Q–Q plot of the data with an LEV distribution is shown in Figure 26.2. This plot indicates that the assumption of an extreme value distribution for the data is tenable. To compute the MLEs, we used the R function 26.1:

```
x = c(152, 152, 115, 109, 137, 88, 94, 77, 160, 165, 125, 40, 128,
      123, 136, 101, 62, 153, 83, 69)
mle.extreme(x)
[1] 95.55  34.82
```

Thus, $\widehat{\mu} = 95.55$ and $\widehat{\sigma} = 34.82$. As noted earlier, we could also use Weibull MLEs (R function 25.1) as follows:

```
> y = exp(-x)
> mle.weibull.nr(y)
[1] 2.872219e-02 3.200613e-42
> -log(3.200613e-42)
[1] 95.54523 # mle of the location parameter
> 1/2.872219e-02
[1] 34.81629 # mle of the scale parameter
```

To compute the 95% parametric bootstrap (PB) confidence interval for the mean survival time, select the dialog box [StatCalc→Continuous→ Extreme→Conf ...], enter 20 for the sample size, 95.55 for the [MLE of mu], 34.82 for [MLE of sig], .95 for [Conf Level], and click [CI] to get (97.7, 139.8). Note that the confidence interval for the mean based on a gamma model (see Example 16.6) is (97, 134), not appreciably different from the one based on an extreme value distribution.

To find a 95% confidence interval for the location parameter, click on [CI for mu] (under CI for Parameters) to get (78.12, 113.1); click [CI for sig] to get 95% confidence interval for the scale parameter σ as (25.96, 53.41).

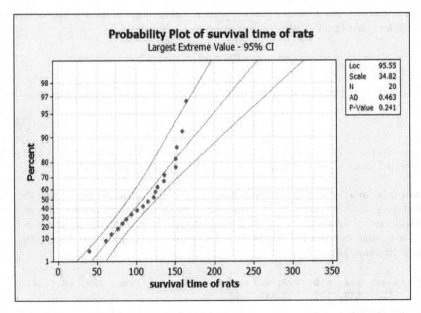

FIGURE 26.2: Extreme value Q–Q plot for survival times of 20 rats

26.6 Prediction Interval, Tolerance Limits, and Survival Probability

Using the one-to-one relation between the Weibull and extreme value distribution (see Section 26.4), prediction intervals, tolerance limits, and confidence intervals for survival probability can be readily obtained from those based on the Weibull distribution. Specifically, we transform the sample using the transformation $Y = \exp(-X)$, simply apply the Weibull-based methods to the transformed samples, and then transform the results back by applying negative log transformation.

Prediction Interval

Suppose it is desired to find a 95% prediction interval for a future observation X based on a sample $X_1, ..., X_n$ from an $\text{LEV}(\mu, \sigma)$ distribution. The following steps can be used to find the prediction interval.

1. Transform the sample $Y_i = \exp(-X_i)$, $i = 1, ..., n$

2. Based on the sample $Y_1, ..., Y_n$, find the 95% Weibull-based prediction interval, say, (L, U).

3. Noting that the transformation is monotone decreasing, we obtain

$$(-\ln(U), -\ln(L))$$

   as a 95% prediction interval for a future observation X from the sampled population.

The above steps lead to the prediction interval

$$(\hat{\mu} - U_w \hat{\sigma}, \hat{\mu} - L_w \hat{\sigma}),$$

where L_w and U_w are, respectively, lower and upper prediction factors for Weibull prediction intervals. See Section 25.8.

Tolerance Limits

One-sided tolerance limits can be found similarly. For example, to find a (.90, .95) upper tolerance limit, use the following steps.

1. Transform the sample $Y_i = \exp(-X_i)$, $i = 1, ..., n$.

2. Based on the sample $Y_1, ..., Y_n$, find the (.90, .95) Weibull-based lower tolerance limit, say, T_L.

3. Noting that the transformation is monotone decreasing, $-\ln(T_L)$ is the (.90, .95) upper tolerance limit for the extreme value distribution.

For a given sample size, MLEs $\hat{\mu}$ and $\hat{\sigma}$, the dialog box [StatCalc→Continuous→Extreme →Tolerance ...] calculates the prediction intervals and one-sided tolerance limits using the above methods. These methods are exact except for simulation errors.

Survival Probability

Let t denote the mission time, and consider the problem of estimating $P(X > t)$, where X is an $\text{LEV}(\mu, \sigma)$ random variable. Note that

$$\begin{aligned} P(X > t) &= P\left(e^{-X} < e^{-t}\right) \\ &= 1 - P\left(Y > e^{-t}\right), \end{aligned}$$

where Y is the $\text{Weibull}(\exp(-\mu), 1/\sigma)$ random variable. Thus, the problem of estimating $P(X > t)$ is equivalent to estimating the Weibull survival probability at $\exp(-t)$. If (L, U) is a $1 - \alpha$ confidence interval for $P\left(Y > e^{-t}\right)$, then $(1 - U, 1 - L)$ is a $1 - \alpha$ confidence interval for $P(X > t)$.

Example 26.2. We shall use again the survival data on 20 rats given in Example 16.6. From Example 26.1, we found

$$n = 20, \quad \hat{\mu} = 95.55, \quad \text{and} \quad \hat{\sigma} = 34.42.$$

To find a $(.90, .95)$ lower tolerance limit, select the dialog box [StatCalc→Continuous→ Extreme→Tolerance ...], enter 20 for [Sample Size], .9 for [Content Level], .95 for [Coverage Level], 95.55 for [MLE of mu], 34.42 for [MLE of sig], and click on [1-sided TLs] to get 47.88.

To find a 95% prediction interval for survival time of a rat, enter .95 for [Conf Level] (under Prediction Interval) and click [PI] to get (53, 208).

To find a 95% confidence interval for the survival probability $P(X > 70)$, select the dialog box [StatCalc→Continuous→Extreme→Confidence ...], enter the values of $(n, \hat{\mu}, \hat{\sigma}, t)$, and confidence level, click [CI for P(X>t)] to get $(.74, .95)$.

26.7   Applications

Extreme value distributions are often used to describe the limiting distribution of the maximum or minimum of n observations selected from an exponential family of distributions, such as normal, gamma, and exponential. They are also used to model the distributions of breaking strength of metals, capacitor breakdown voltage, and gust velocities encountered by airplanes. Parsons and Lal (1991) studied thirteen sets of flexural strength data on different kinds of ice and found that between the three-parameter Weibull and the extreme value distributions, the latter fits the data better. Belzer and Kellog (1993) used the extreme value distribution to analyze the sources of uncertainty in forecasting peak power loads. Onoz and Bayazit (1995) showed that the extreme value distribution fits the flood flow data (collected from 1819 site-years from all over the world) best among seven distributions considered. Cannarozzo et al. (1995), Karim and Chowdhury (1995), and Sivapalan and Bloschl (1998) also used extreme value distributions to model the rainfall and flood flow data. Xu (1995) used the extreme value distribution to study the stochastic characteristics of wind pressures on the Texas Tech University Experimental Building.

Extreme value distributions are also used in stress-strength models. Herrington

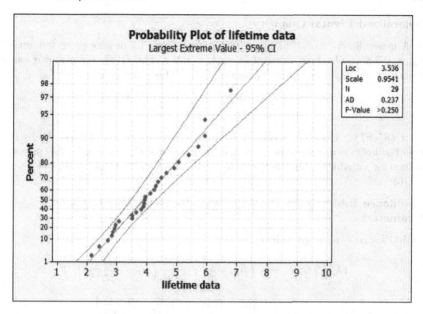

FIGURE 26.3: Extreme value Q–Q plot for the lifetime data

(1995) pointed out that if failure of a structural component is caused by the maximum of a sequence of applied loads, then the applied load distribution is an extreme value distribution. When strength of individual fibers is determined by the largest defect, an extreme value distribution describes the distribution of the size of the maximum defect of fibers. Lawson and Chen (1999) used an extreme value distribution to model the distribution of the longest possible microcracks in specimens of a fatigues aluminum-matrix silicon carbide whisker composite.

Kuchenhoff and Thamerus (1996) modeled extreme values of daily air pollution data by an extreme value distribution. Sharma et al. (1999) used an extreme value distribution for making predictions of the expected number of violations of the National Ambient Air Quality Standards, as prescribed by the Central Pollution Control Board of India for hourly and eight-hourly average carbon monoxide concentration in an urban road intersection region. Application of an extreme value distribution for setting the margin level in future markets is given in Longin (1999).

26.8   Two-Sample Inference

Let $(\widehat{\mu}_i, \widehat{\sigma}_i)$ denote the MLE of (μ_i, σ_i) based on a sample of size n_i from an $\mathrm{LEV}(\mu_i, \sigma_i)$ distribution, $i = 1, 2..$ We shall see some generalized inferential methods for comparing the location parameters, scale parameters, and the means.

Generalized Pivotal Quantity

A generalized pivotal quantity (GPQ) for μ_i can be developed as follows. Let $(\widehat{\mu}_{i0}, \widehat{\sigma}_{i0})$ be an observed value of $(\widehat{\mu}_i, \widehat{\sigma}_i)$, $i = 1, 2$. The GPQs for μ and σ can be obtained from (2.9) and (2.10) of Section 2.7.2. These GPQs are

$$G_{\mu_i} = \widehat{\mu}_{i0} - \frac{\widehat{\mu}_i^*}{\widehat{\sigma}_i^*}\widehat{\sigma}_{i0} \quad \text{and} \quad G_{\sigma_i} = \frac{\widehat{\sigma}_{0i}}{\widehat{\sigma}_i^*}, \tag{26.11}$$

where $(\widehat{\mu}_i^*, \widehat{\sigma}_i^*)$ is the MLE based on a sample of size n_i from an LEV$(1, 1)$ distribution. The lower α quantile and the upper α quantile of G_{μ_i} form a $1 - 2\alpha$ confidence interval for μ_i, which is an exact confidence interval for μ_i; see the confidence interval (26.10).

Confidence Intervals for the Difference Between Two Location Parameters

A GPQ for $\mu_1 - \mu_2$ is given by

$$\begin{aligned}
G_{\mu_1} - G_{\mu_2} &= \left(\widehat{\mu}_{10} - \frac{\widehat{\mu}_1^*}{\widehat{\sigma}_1^*}\widehat{\sigma}_{10}\right) - \left(\widehat{\mu}_{20} - \frac{\widehat{\mu}_2^*}{\widehat{\sigma}_2^*}\widehat{\sigma}_{20}\right) \\
&= \widehat{\mu}_{10} - \widehat{\mu}_{20} - \left(\widehat{\sigma}_{10}\frac{\widehat{\mu}_1^*}{\widehat{\sigma}_1^*} - \widehat{\sigma}_{20}\frac{\widehat{\mu}_2^*}{\widehat{\sigma}_2^*}\right) \\
&= \widehat{\mu}_{10} - \widehat{\mu}_{20} - Q,
\end{aligned} \tag{26.12}$$

where $Q = \left(\widehat{\sigma}_{10}\frac{\widehat{\mu}_1^*}{\widehat{\sigma}_1^*} - \widehat{\sigma}_{20}\frac{\widehat{\mu}_2^*}{\widehat{\sigma}_2^*}\right)$. Let Q_α denote the α quantile of Q. Then

$$(\widehat{\mu}_{10} - \widehat{\mu}_{20} - Q_{1-\alpha}, \ \widehat{\mu}_{10} - \widehat{\mu}_{20} - Q_\alpha)$$

is a $1 - 2\alpha$ confidence interval for $\mu_1 - \mu_2$. Notice that for a given $(n_i, \widehat{\mu}_{i0}, \widehat{\sigma}_{i0})$, $i = 1, 2$, the percentiles of Q can be estimated using simulated samples from an LEV$(0, 1)$ distribution.

Confidence Intervals for the Ratio of Two Scale Parameters

A GPQ for σ_1/σ_2 is given by

$$\frac{G_{\sigma_1}}{G_{\sigma_2}} = \frac{\widehat{\sigma}_{10}}{\widehat{\sigma}_{20}}\frac{\widehat{\sigma}_2^*}{\widehat{\sigma}_1^*} = \frac{\widehat{\sigma}_{10}}{\widehat{\sigma}_{20}}R, \tag{26.13}$$

where $R = \frac{\widehat{\sigma}_2^*}{\widehat{\sigma}_1^*}$. Let R_α denote the α quantile of R. Then

$$\left(\frac{\widehat{\sigma}_{10}}{\widehat{\sigma}_{20}}R_\alpha, \ \frac{\widehat{\sigma}_{10}}{\widehat{\sigma}_{20}}R_{1-\alpha}\right)$$

is a $1 - 2\alpha$ confidence interval for the ratio σ_1/σ_2.

Confidence Intervals for the Difference Between Two Means

Recall that the mean of an LEV(μ, σ) is given by $\eta = \mu + \gamma\sigma$. A GPQ for $\eta_1 - \eta_2$, where $\eta_i = \mu_i + \gamma\sigma_i$ and γ is the Euler constant, is given by

$$\begin{aligned}
G_{\eta_1} - G_{\eta_2} &= (G_{\mu_1} + \gamma G_{\sigma_1}) - (G_{\mu_2} + \gamma G_{\sigma_2}) \\
&= \widehat{\mu}_{10} - \widehat{\mu}_{20} - \left(\widehat{\sigma}_{10}\frac{\widehat{\mu}_1^*}{\widehat{\sigma}_1^*} - \widehat{\sigma}_{20}\frac{\widehat{\mu}_2^*}{\widehat{\sigma}_2^*}\right) + \gamma\left(\frac{\widehat{\sigma}_{10}}{\widehat{\sigma}_1^*} - \frac{\widehat{\sigma}_{20}}{\widehat{\sigma}_2^*}\right). \tag{26.14}
\end{aligned}$$

For a given $(\widehat{\mu}_{10}, \widehat{\sigma}_{10}, \widehat{\mu}_{20}, \widehat{\sigma}_{20})$, appropriate percentiles of $G_{\eta_1} - G_{\eta_2}$ form a confidence interval for $\eta_1 - \eta_2$. These percentiles can be estimated by Monte Carlo simulation.

The dialog box [StatCalc→Continuous→Extreme→CIs for ...] computes the preceding generalized confidence intervals using 100,000 simulation runs.

Example 26.3. We shall use the following simulated samples to illustrate the preceding two-sample confidence intervals. Sample 1 is generated from LEV$(3, 2)$ distribution, and Sample 2 is generated from LEV$(1, 1)$ distribution.

	Sample 1									
	11.39	3.18	3.82	9.32	3.98	9.86	0.290	1.97	2.79	4.62
	6.38	3.12	3.09	1.66	3.76	3.44	6.26	4.47	7.49	5.15

	Sample 2								
	2.25	1.24	0.32	3.99	1.21	3.93	2.68	1.01	0.97	0.48
	0.74	-0.21	0.67	0.93	1.04				

For sample 1, $n_1 = 20$, $\widehat{\mu}_1 = 3.5259$, and $\widehat{\sigma}_1 = 2.1876$. For sample 2, $n_2 = 15$, $\widehat{\mu}_2 = 0.8887$, and $\widehat{\sigma}_2 = 0.8478$. To find a 95% confidence interval for the difference $\mu_1 - \mu_2$, select the dialog box [StatCalc→Continuous→Extreme→CIs for ...], enter the sample sizes, and the MLEs, enter .95 for [Conf Level] under (CI for mu1 - mu2), and click [CI] to get (1.42, 3.88). To find a 95% confidence interval for the difference between two means, enter .95 for [Conf Level] under (CI for M1 - M2), and click [CI] to get [2.08, 5.03]. By clicking [CI] under (CI for sig1/sig2), we obtain the 95% confidence interval for σ_1/σ_2 as (1.76, 4.40).

26.9 Properties and Results

1. If X is an exponential$(0, \sigma)$, then $\mu - \sigma \ln(X) \sim$ extreme(μ, σ).
2. If X and Y are independently distributed as extreme(μ, σ) random variable, then
$$X - Y \sim \text{logistic}(0, \sigma).$$
3. If X is an LEV$(0, 1)$ variable, then
$$b \exp(-X/c) \sim \text{Weibull}(b, c)$$
and
$$\exp[-\exp(-X/b)] \sim \text{Pareto}(a, b).$$

26.10 Random Number Generation

For a given μ and σ:
Generate u from uniform$(0, 1)$
Set $x = \mu - \sigma * \ln(-\ln(u))$
x is a pseudo random number from the extreme(μ, σ) distribution.

27

Cauchy Distribution

27.1 Description

The probability density function (pdf) of a Cauchy distribution with the location parameter a and the scale parameter b is given by

$$f(x|a,b) = \frac{1}{\pi\, b[1 + ((x-a)/b)^2]}, \quad -\infty < a < \infty, \ b > 0.$$

The cumulative distribution function (cdf) can be expressed as

$$F(x|a,b) = \frac{1}{2} + \frac{1}{\pi}\tan^{-1}\left(\frac{x-a}{b}\right), \quad b > 0. \tag{27.1}$$

We refer to this distribution as Cauchy(a, b). The standard forms of the pdf and the cumulative distribution function can be obtained by replacing a with 0 and b with 1.

The inverse distribution function can be expressed as

$$F^{-1}(p|a,b) = a + b\tan(\pi(p - 0.5)), \quad 0 < p < 1. \tag{27.2}$$

Using the above inverse distribution function, random variates from a Cauchy(a,b) distribution can be generated in a straightforward manner. Specifically, if u is uniform$(0, 1)$ variate, then $x = a + b\tan(\pi(u - 0.5))$ is a Cauchy(a,b) random number.

27.2 Moments

Mean:	does not exist
Median:	a
Mode:	a
First Quartile:	$a - b$
Third Quartile:	$a + b$
Moments:	do not exist
Characteristic Function:	$\exp(ita - |t|b)$

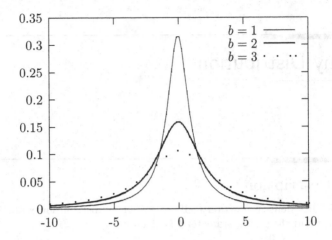

FIGURE 27.1: Cauchy probability density functions; $a = 0$

27.3 Probabilities and Percentiles

The dialog box [StatCalc→Continuous→Cauchy] computes the cumulative probabilities and percentiles of a Cauchy distribution.

To compute probabilities: Enter the values of the parameters a and b, and of x; click
[P(X <= x)]. For example, when $a = 1$, $b = 2$, and $x = 1.2$, $P(X \leq 1.2) = 0.531726$
and $P(X > 1.2) = 0.468274$.

To compute percentiles: Enter the values of a, b, and cumulative probability; click
[x]. For example, when $a = 1$, $b = 2$, and the cumulative probability $= 0.95$, the
95th percentile is 13.6275. That is, $P(X \leq 13.6275) = 0.95$.

To compute parameters: Enter the value of one of the parameters, cumulative probability, and x; click on the missing parameter. For example, when $b = 3$, cumulative
probability $= 0.5$, and $x = 1.25$, the value of a is 1.25.

27.4 Inference

Let $X_1, ..., X_n$ be a sample from a Cauchy(a, b) distribution. For $0.5 < p < 1$, let X_p
and X_{1-p} denote the sample quantiles.

27.4.1 Estimation Based on Sample Quantiles

The point estimators of a and b based on the sample quantiles X_p and X_{1-p} and their variances are as follows.

$$\widehat{a} = \frac{X_p + X_{1-p}}{2}$$

with

$$\mathrm{Var}(\widehat{a}) \simeq \frac{\widehat{b}^2}{n}\left[\frac{\pi^2}{2}(1-p)\right]\cosec^4(\pi p), \tag{27.3}$$

and

$$\widehat{b} = 0.5(x_p - x_{1-p})\tan[\pi(1-p)]$$

with

$$\mathrm{Var}(\widehat{b}) \simeq \frac{\widehat{b}^2}{n}\left[2\pi^2(1-p)(2p-1)\right]\cosec^2(2\pi p).$$

27.4.2 Maximum Likelihood Estimators

Maximum likelihood estimators of a and b are the solutions of the equations

$$\frac{1}{n}\sum_{i=1}^{n}\frac{2}{1+[(x_i-a)/b]^2} = 1$$

and

$$\frac{1}{n}\sum_{i=1}^{n}\frac{2x_i}{1+[(x_i-a)/b]^2} = a.$$

27.5 Applications

The Cauchy distribution represents an extreme case and serves as counter examples for some well-accepted results and concepts in statistics. For example, the central limit theorem does not hold for the limiting distribution of the mean of a random sample from a Cauchy distribution (see Section 26.6, Property 4). Because of this special nature, some authors consider the Cauchy distribution as a pathological case. However, it can be postulated as a model for describing data that arise as n realizations of the ratio of two normal random variables. Other applications given in the recent literature: Min et al. (1996) found that Cauchy distribution describes the distribution of velocity differences induced by different vortex elements. An application of the Cauchy distribution to study the polar and nonpolar liquids in porous glasses is given in Stapf et al. (1996). Kagan (1992) pointed out that the Cauchy distribution describes the distribution of hypocenters on focal spheres of earthquakes. It is shown in the paper by Winterton et al. (1992) that the source of fluctuations in contact window dimensions is variation in contact resistivity, and the contact resistivity is distributed as a Cauchy random variable.

27.6   Properties and Results

1. If X and Y are independent standard normal random variables, then $X/Y \sim$ Cauchy$(0,1)$.

2. If $X \sim$ Cauchy$(0,1)$, then $2X/(1-X^2)$ also follows a Cauchy$(0,1)$ distribution.

3. Student's t distribution with df $= 1$ specializes to the Cauchy$(0,1)$ distribution.

4. If $X_1, \ldots, X_k$ are independent random variables with $X_j \sim$ Cauchy(a_j, b_j), $j = 1, \ldots, k$. Then

$$\sum_{j=1}^{k} c_j X_j \sim \text{Cauchy} \left(\sum_{j=1}^{k} c_j a_j, \ \sum_{j=1}^{k} |c_j| \, b_j \right).$$

5. It follows from (4) that the mean of a random sample of n independent observations from a Cauchy distribution follows the same distribution.

28

Inverse Gaussian Distribution

28.1 Description

The probability density function (pdf) of X is given by

$$f(x|\mu,\sigma) = \left(\frac{\lambda}{2\pi x^3}\right)^{\frac{1}{2}} \exp\left(\frac{-\lambda(x-\mu)^2}{2\mu^2 x}\right), \quad x > 0, \ \lambda > 0, \ \mu > 0. \qquad (28.1)$$

This distribution is usually denoted by $\mathrm{IG}(\mu, \lambda)$. Using the standard normal distribution function Φ, the cumulative distribution function (cdf) of an $\mathrm{IG}(\mu, \lambda)$ can be expressed as

$$F(x|\mu,\lambda) = \Phi\left(\sqrt{\frac{\lambda}{x}}\left(\frac{x}{\mu}-1\right)\right) + e^{2\lambda/\mu}\Phi\left(-\sqrt{\frac{\lambda}{x}}\left(\frac{x}{\mu}+1\right)\right), \quad x > 0, \qquad (28.2)$$

where $\Phi(x)$ is the standard normal distribution function.

Inverse Gaussian (IG) distributions offer a convenient modeling for positive right skewed data. The IG family is often used as an alternative to the normal family because of the similarities between the inference methods for these two families.

28.2 Moments

Mean:	μ
Variance:	$\frac{\mu^3}{\lambda}$
Mode:	$\mu\left[\left(1+\frac{9\mu^2}{4\lambda^2}\right)^{1/2} - \frac{3\mu}{2\lambda}\right]$
Coefficient of Variation:	$\sqrt{\frac{\mu}{\lambda}}$
Coefficient of Skewness:	$3\sqrt{\frac{\mu}{\lambda}}$
Coefficient of Kurtosis:	$3 + 15\mu/\lambda$

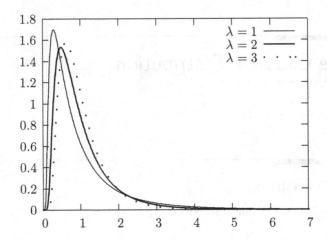

FIGURE 28.1: Inverse Gaussian probability density functions; $\mu = 1$

Moments about the Origin: $\mu^k \sum_{i=0}^{k-1} \frac{(k-1+i)!}{(k-1-i)!} \left(\frac{\mu}{2\lambda}\right)^i, \quad k \geq 2$

Moment Generating Function: $\exp\left[\frac{\lambda}{\mu}\left(1 - \left(1 - \frac{2\mu^2 t}{\lambda}\right)^{1/2}\right)\right]$

Mean Deviation: $4\sqrt{\frac{\lambda}{\mu}} \exp\left(2\sqrt{\frac{\lambda}{\mu}}\right) \Phi\left(-2\sqrt{\frac{\lambda}{\mu}}\right) \sqrt{\frac{\mu^3}{\lambda}},$
where Φ is the standard normal cdf

28.3 Probabilities, Percentiles, and Moments

The dialog box [StatCalc→Continuous→Inv Gau→Probabilities, Percentiles and Moments]

To compute probabilities: Enter the values of the parameters μ and λ and the observed value x; click [P]. For examples, when $\mu = 2$, $\lambda = 1$, and $x = 3$,

$$P(X \leq 3) = 0.815981 \text{ and } P(X > 3) = 0.184019.$$

To compute percentiles: Enter the values of μ and λ, and the cumulative probability P(X <= x); click on [x]. For examples, when $\mu = 1$, $\lambda = 2$, and the cumulative probability $P(X \leq x) = 0.95$, the 95th percentile is 2.37739. That is, $P(X \leq 2.37739) = 0.95$.

To compute moments: Enter the values of μ and λ; click [M].

28.4 One-Sample Inference

Let $X_1, \ldots, X_n$ be a sample from a IG(μ, λ) distribution. Define

$$\bar{X} = \frac{1}{n} \sum_{i=1}^{n} X_i \quad \text{and} \quad V = \frac{1}{n} \sum_{i=1}^{n} (1/X_i - 1/\bar{X}). \qquad (28.3)$$

The sample mean $\bar{X}$ is the maximum likelihood estimate (MLE) as well as unbiased estimate of μ and V^{-1} is the MLE of λ. The minimum variance unbiased estimator of $1/\lambda$ is given by $nV/(n-1)$. The mean $\bar{X}$ and V are independent with $\bar{X} \sim$ IG(μ, $n\lambda$), and $\lambda nV \sim \chi^2_{n-1}$. Furthermore,

$$\left| \frac{\sqrt{(n-1)}(\bar{X} - \mu)}{\mu\sqrt{\bar{X}V}} \right| \sim |t_{n-1}|,$$

where t_m is a Student's t variable with df $= m$.

28.4.1 A Test for the Mean

Let $\bar{X}$ and V be as defined in (28.3). Define

$$S_1 = \sum_{i=1}^{n} (X_i + \mu_0)^2/X_i \quad \text{and} \quad S_2 = \sum_{i=1}^{n} (X_i - \mu_0)^2/X_i.$$

The p-value for testing

$$H_0 : \mu \leq \mu_0 \quad \text{vs.} \quad H_a : \mu > \mu_0$$

is given by

$$F_{n-1}(-w_0) + \left(\frac{S_1}{S_2} \right)^{(n-2)/2} F_{n-1}\left(-\sqrt{4n + w_0^2 \mu_0 S_1} \right), \qquad (28.4)$$

where F_m denotes the cdf of Student's t variable with df $= m$, and w_0 is an observed value of

$$W = \frac{\sqrt{(n-1)}(\bar{X} - \mu_0)}{\mu_0\sqrt{\bar{X}V}}.$$

See Chhikara and Folks (1989). The p-value for testing

$$H_0 : \mu = \mu_0 \quad \text{vs.} \quad H_a : \mu \neq \mu_0$$

is given by

$$P(|t_{n-1}| > |w_0|),$$

where t_m denotes the t variable with df $= m$.

28.4.2   Confidence Interval for the Mean

A $1 - \alpha$ confidence interval for μ is given by

$$\left(\frac{\bar{X}}{1 + t_{n-1,1-\alpha/2}\sqrt{\frac{V\bar{X}}{n-1}}}, \frac{\bar{X}}{\max\left\{0, 1 + t_{n-1,\alpha/2}\sqrt{\frac{V\bar{X}}{n-1}}\right\}} \right).$$

Example 28.1. Suppose that a sample of 18 observations from an IG(μ, λ) distribution yielded $\bar{X} = 2.5$ and $V = 0.65$. To find a 95% confidence interval for the mean μ, select [Continuous→Inv Gau→CI and Test for Mean], enter 0.95 for the confidence level, 18 for the sample size, 2.5 for the mean, and 0.65 for V; click [2-sided] to get (1.51304, 7.19009).

Suppose we want to test $H_0 : \mu = 1.4$ vs. $H_a : \mu \neq 1.4$. Enter 1.4 for [H0: M=M0], click [p-values for] to get 0.0210821.

28.5   Two-Sample Inference

The following two-sample inferential procedures are based on the generalized variable approach given in Krishnamoorthy and Tian (2008). This approach is valid only for two-sided hypothesis testing about $\mu_1 - \mu_2$, and constructing confidence intervals for $\mu_1 - \mu_2$ (not one-sided limits). More details can be found in the above mentioned paper.

Let $X_{i1}, ..., X_{in_i}$ be a sample from an IG(μ_i, λ_i) distribution, $i = 1, 2$. Let

$$\bar{X}_i = \frac{1}{n_i} \sum_{j=1}^{n_i} X_{ij} \text{ and } V_i = \frac{1}{n_i} \sum_{j=1}^{n_i} (1/X_{ij} - 1/\bar{X}_i), i = 1, 2. \quad (28.5)$$

The generalized variable of μ_i is given by

$$G_i = \frac{\bar{x}_i}{\max\left\{0, 1 + t_{n_i-1}\sqrt{\frac{\bar{x}_i v_i}{n_i-1}}\right\}}, \quad i = 1, 2, \quad (28.6)$$

where $(\bar{x}_i, v_i)$ is an observed value of $(\bar{X}_i, V_i)$, $i = 1, 2$, and t_{n_1-1} and t_{n_2-1} are independent Student's t variables.

28.5.1   Inferences for the Difference between Two Means

Notice that for a given $(n_i, \bar{x}_i, v_i)$, the distribution of the generalized variable G_i in (28.6) does not depend on any unknown parameters. Therefore, the Monte Carlo method can be used to estimate the p-value for testing about $\mu_1 - \mu_2$ or to find confidence intervals for $\mu_1 - \mu_2$. The procedure is given in the following algorithm.

Algorithm 28.1. Generalized CIs for the difference between two IG means

---

For a given $(n_1, \bar{x}_1, v_1, n_2, \bar{x}_2, v_2)$:

For $j = 1, m$
Generate t_{n_1-1} and t_{n_2-1}
Compute G_1 and G_2 using (28.6)
Set $T_j = G_1 - G_2$
(end loop)

Suppose we are interested in testing

$$H_0 : \mu_1 = \mu_2 \quad \text{vs.} \quad H_a : \mu_1 \neq \mu_2.$$

Then, the generalized p-value for the above hypotheses is given by

$$2 \min \{ P(G_1 - G_2 < 0), P(G_1 - G_2 > 0) \}.$$

The null hypothesis will be rejected when the above p-value is less than a specified nominal level α. Notice that $P(G_1 - G_2 < 0)$ can be estimated by the proportion of the T_js in Algorithm 27.5.1 that are less than zero; similarly, $P(G_1 - G_2 > 0)$ can be estimated by the proportion of the T_js that are greater than zero.

For a given $0 < \alpha < 1$, let T_α denote the αth quantile of the T_js in Algorithm 27.5.1. Then, $(T_{\alpha/2}, T_{1-\alpha/2})$ is a $1 - \alpha$ confidence interval for the mean difference.

StatCalc uses Algorithm 28.1 with $m = 1,000,000$ to compute the generalized p-value and generalized confidence interval. The results are almost exact. (see Krishnamoorthy and Tian, 2008).

Example 28.2. Suppose that a sample of 18 observations from an $IG(\mu_1, \lambda_1)$ distribution yielded $\bar{x}_1 = 2.5$ and $v_1 = 0.65$. Another sample of 11 observations from an $IG(\mu_2, \lambda_2)$ distribution yielded $\bar{x}_2 = 0.5$ and $v_2 = 1.15$. To find a 95% confidence interval for the mean difference $\mu_1 - \mu_2$, enter these statistics in the dialog box [StatCalc→Continuous→Inv Gau→CI and Test for Mean1-Mean2], 0.95 for the confidence level, and click [2-sided] to get (0.85, 6.65).

The p-value for testing $H_0 : \mu_1 = \mu_2$ vs. $H_a : \mu_1 \neq \mu_2$ is given by 0.008.

28.5.2   Inferences for the Ratio of Two Means

Let G_1 and G_2 be as defined in (28.6), and let $R = G_1/G_2$. The generalized p-value for testing

$$H_0 : \frac{\mu_1}{\mu_2} = 1 \quad \text{vs.} \quad H_a : \frac{\mu_1}{\mu_2} \neq 1$$

is given by

$$2 \min \{ P(R < 1), P(R > 1) \}.$$

Let R_p denote the $100p$th percentile of R. Then, $(R_{\alpha/2}, R_{1-\alpha/2})$ is a $1-\alpha$ confidence interval for the ratio of the IG means.

The generalized p-value and confidence limits for μ_1/μ_2 can be estimated using Monte Carlo method similar to the one given in Algorithm 27.5.1. The results are very accurate for practical purposes (see Krishnamoorthy and Tian, 2008).

Example 28.3. Suppose that a sample of 18 observations from an $IG(\mu_1, \lambda_1)$ distribution yielded $\bar{x}_1 = 2.5$ and $v_1 = 0.65$. Another sample of 11 observations from an $IG(\mu_2, \lambda_2)$ distribution yielded $\bar{x}_2 = 0.5$ and $v_2 = 1.15$. To find a 95% confidence interval for the ratio μ_1/μ_2, enter these statistics and 0.95 for the confidence level in [StatCalc→ Continuous→Inv Gau→CI and Test for Mean1/Mean2], and click [2-sided] to get (2.05, 15.35). The p-value for testing $H_0 : \mu_1 = \mu_2$ vs. $H_a : \mu_1 \neq \mu_2$ is given by 0.008.

28.6   Random Number Generation

The following algorithm is due to Taraldsen and Lindqvist (2005).

Algorithm 28.2. Inverse Gaussian variate generator

For a given μ and λ:

Generate $w \sim \text{uniform}(0, 1)$ and $z \sim N(0, 1)$

$\qquad$ set $v = z^2$; $d = \lambda/\mu$

$\qquad\quad y = 1 - 0.5(\sqrt{v^2 + 4dv} - v)/d$

$\qquad\quad x = y\mu$

$\qquad\quad$ if $(1 + y)w > 1$, set $x = \mu/y$

x is a random number from the $\text{IG}(\mu, \lambda)$ distribution.

29

Rayleigh Distribution

29.1 Description

The Rayleigh distribution with the scale parameter b has the probability density function (pdf)

$$f(x|b) = \frac{x}{b^2} \exp\left(-\frac{1}{2}\frac{x^2}{b^2}\right), \quad x > 0, \ b > 0.$$

The cumulative distribution function (cdf) is given by

$$F(x|b) = 1 - \exp\left(-\frac{1}{2}\frac{x^2}{b^2}\right), \quad x > 0, \ b > 0. \tag{29.1}$$

Letting $F(x|b) = p$, and solving (29.1) for x, we get the inverse distribution function as

$$F^{-1}(p|b) = b\sqrt{-2\ln(1-p)}, \quad 0 < p < 1, \ b > 0. \tag{29.2}$$

We observe from the plots of pdfs in Figure 28.1 that the Rayleigh distribution is always right skewed.

If X has a Rayleigh distribution, then X^2 follows an exponential distribution with mean $2b^2$. This transformation is useful to find confidence interval, prediction interval, and tolerance interval based on the results for an exponential distribution.

29.2 Moments

Mean:	$b\sqrt{\frac{\pi}{2}}$	Variance:	$\left(2 - \frac{\pi}{2}\right)b^2$
Mode:	b	Median:	$b\sqrt{\ln(4)}$
Coefft. of Variation:	$\sqrt{(4/\pi - 1)}$	Coefft. of Skewness:	$\frac{2(\pi-3)\sqrt{\pi}}{(4-\pi)^{3/2}}$
Coefft. of Kurtosis:	$\frac{(32-3\pi^2)}{(4-\pi)^2}$	Moments about the Origin:	$2^{k/2}b^k\Gamma(k/2+1)$

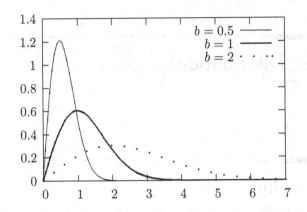

FIGURE 29.1: Rayleigh probability density functions

29.3 Probabilities, Percentiles, and Moments

The dialog box [StatCalc→Continuous→Rayleigh] computes the tail probabilities, percentiles, and moments of a Rayleigh distribution.

To compute probabilities: Enter the values of the parameter b and x; click [P(X <= x)]. For example, when $b = 2$ and $x = 2.3$, $P(X \leq 2.3) = 0.733532$ and $P(X > 2.3) = 0.266468$.

To compute percentiles: Enter the values of b and the cumulative probability $P(X <= x)$; click on [x]. For example, when $b = 1.2$, and the cumulative probability $P(X <= x) = 0.95$, the 95th percentile is 2.07698. That is, $P(X \leq 2.07698) = 0.95$.

To compute the Value of b: Enter the values of x and the cumulative probability P(X <= x); click on [b]. For example, when $x = 3$, and the cumulative probability P(X <= x) = 0.9, the value of b is 1.97703. That is, $P(X \leq 3 | b = 1.97703) = 0.9$.

To compute moments: Enter the value of b and click [M].

29.4 Confidence Interval

Let $X_1, ..., X_n$ be a sample from Rayleigh distribution with parameter b. The squared sample $X_1^2, ..., X_n^2$ can be regarded as a sample from an exponential distribution with mean $2b^2$. Thus, maximum likelihood estimate (MLE) of b, follows from the MLE of the scale parameter of an exponential distribution (see Section 15.4), and is given by

$$\widehat{b} = \sqrt{\frac{1}{2n}\sum_{i=1}^{n}X_i^2}.$$

A confidence interval for b can be obtained from the one for $2b^2$. The $1-\alpha$ confidence interval for b can be obtained from the one for $2b^2$ based on $X_1^2, ..., X_n^2$ and the result for exponential distribution (see Section 15.4). It is given by

$$\left(\sqrt{\frac{\sum_{i=1}^{n} X_i^2}{\chi_{2n;1-\alpha/2}^2}}, \sqrt{\frac{\sum_{i=1}^{n} X_i^2}{\chi_{2n;\alpha/2}^2}} \right).$$

29.5 Prediction Intervals and One-Sided Tolerance Limits

Recall that if X has a Rayleigh distribution with parameter b, then X^2 has an exponential distribution with mean $2b^2$, or equivalently, X^2/b^2 follows a chi-square distribution with df $= 2$. Let $X_1, ..., X_n$ be a sample from a Rayleigh distribution, and X follow the same distribution independently of the sample. Then

$$\frac{X^2}{\sum_{i=1}^{n} X_i^2} \sim \frac{1}{n} F_{2,2n}.$$

On the basis of the above distributional result, we find the $1 - \alpha$ prediction interval for X as

$$\left(\sqrt{\frac{1}{n} \sum_{i=1}^{n} X_i^2 F_{2,2n;\frac{\alpha}{2}}}, \sqrt{\frac{1}{n} \sum_{i=1}^{n} X_i^2 F_{2,2n;1-\frac{\alpha}{2}}} \right).$$

To find one-sided tolerance limits, we first note that the pth quantile of the Rayleigh distribution with parameter b is given by $q_p = b\sqrt{-2\ln(1-p)}$, $0 < p < 1$. For $0.5 < p < 1$, a $1-\alpha$ upper confidence limit for q_p is the $(p, 1-\alpha)$ upper tolerance limit. Noticing that q_p is an increasing function of b, the $(p, 1 - \alpha)$ upper tolerance limit is given by

$$\sqrt{-2\ln(1-p)} \sqrt{\frac{\sum_{i=1}^{n} X_i^2}{\chi_{2n;\alpha}^2}}.$$

For example, the $(.90, .95)$ upper tolerance limit is given by

$$\sqrt{-2\ln(0.1)} \sqrt{\frac{\sum_{i=1}^{n} X_i^2}{\chi_{2n;.05}^2}}.$$

For $0 < p < .5$, the $(p, 1 - \alpha)$ lower tolerance limit can be expressed as

$$\sqrt{-2\ln(1-p)} \sqrt{\frac{\sum_{i=1}^{n} X_i^2}{\chi_{2n;1-\alpha}^2}}.$$

For example, the $(.90, .95)$ upper tolerance limit is given by

$$\sqrt{-2\ln(0.9)} \sqrt{\frac{\sum_{i=1}^{n} X_i^2}{\chi_{2n;.95}^2}}.$$

29.6   Relation to Other Distributions

1. Let X_1 and X_2 be independent $N(0, \ b^2)$ random variables. Then, $Y = \sqrt{X_1^2 + X_2^2}$ follows a Rayleigh(b) distribution.

2. The Rayleigh(b) distribution is a special case of the Weibull distribution (see Chapter 24) with $b = \sqrt{2}b$, $c = 2$ and $m = 0$.

3. Let X be a Rayleigh(b) random variable. Then, $Y = X^2$ follows an exponential distribution with mean $2b^2$. That is, Y has the pdf

$$\frac{1}{2b^2} \exp\left(-\frac{y}{2b^2}\right), \ \ y > 0.$$

The above result also implies that X^2/b^2 has a chi-square distribution with degrees of freedom $= 2$.

29.7   Random Number Generation

Since the cdf has explicit from (see Section 28.1), random numbers can be generated using inverse transformation:

For a given b :

Generate u $\sim$ uniform(0,1)

Set $x = b * \sqrt{-2\ln(u)}$

x is a random number from the Rayleigh(b)

30

Bivariate Normal Distribution

30.1 Description

Let (Z_1, Z_2) be a bivariate normal random vector with

$$E(Z_1) = 0, \ \text{Var}(Z_1) = 1.0, \ E(Z_2) = 0, \ \text{Var}(Z_2) = 1.0$$

and

$$\text{Correlation}(Z_1, Z_2) = \rho.$$

The probability density function (pdf) of (Z_1, Z_2) is given by

$$f(z_1, z_2 | \rho) = \frac{1}{2\pi\sqrt{1-\rho^2}} \exp\left\{-\frac{1}{2(1-\rho^2)}\left(z_1^2 - 2\rho\,z_1 z_2 + z_2^2\right)\right\}, \quad (30.1)$$

$$-\infty < z_1 < \infty, \ -\infty < z_2 < \infty, \ -1 < \rho < 1.$$

Suppose that (X_1, X_2) is a bivariate normal random vector with

$$E(X_1) = \mu_1, \ \text{Var}(X_1) = \sigma_{11}, \ E(X_2) = \mu_2, \ \text{Var}(X_2) = \sigma_{22}$$

and the covariance $\text{Cov}(X_1, X_2) = \sigma_{12}$. Then

$$\left(\frac{X_1 - \mu_1}{\sqrt{\sigma_{11}}}, \ \frac{X_2 - \mu_2}{\sqrt{\sigma_{22}}}\right)$$

is distributed as (Z_1, Z_2) with correlation coefficient $\rho = \frac{\sigma_{12}}{\sqrt{\sigma_{11}\sigma_{22}}}$. That is,

$$P(X_1 \leq a, \ X_2 \leq b) = P\left(Z_1 \leq \frac{a - \mu_1}{\sqrt{\sigma_{11}}}, \ Z_2 \leq \frac{b - \mu_2}{\sqrt{\sigma_{22}}}\right).$$

The following relations are useful for computing probabilities over different regions.

1. $P(Z_1 \leq a, \ Z_2 > b) = \Phi(a) - P(Z_1 \leq a, \ Z_2 \leq b)$,

2. $P(Z_1 > a, \ Z_2 \leq b) = \Phi(b) - P(Z_1 \leq a, \ Z_2 \leq b)$,

3. $P(Z_1 > a, \ Z_2 > b) = 1 - \Phi(a) - \Phi(b) + P(Z_1 \leq a, \ Z_2 \leq b)$,

where Φ is the standard normal distribution function.

30.2   Computing Probabilities

Let (X, Y) be a bivariate normal random vector with mean $(0, 0)$ and the correlation coefficient ρ. For given x, y, and ρ, the dialog box [StatCalc→Continuous→ Biv Normal→All Tail Probabilities] computes the following probabilities:

a. $P(X \leq x, Y > y)$

b. $P(X > x, Y > y)$

c. $P(X > x, Y \leq y)$

d. $P(X \leq x, Y \leq y)$

e. $P(|X| < x, |Y| < y)$.

For example, when $x = 1.1$, $y = 0.8$, and $\rho = 0.6$,

a. $P(X \leq 1.1, Y > 0.8) = 0.133878$

b. $P(X > 1.1, Y > 0.8) = 0.077977$

c. $P(X > 1.1, Y \leq 0.8) = 0.057689$

d. $P(X \leq 1.1, Y \leq 0.8) = 0.730456$

e. $P(|X| < 1.1, |Y| < 0.8) = 0.465559$.

If (X, Y) is a normal random vector with mean $= (\mu_1, \mu_2)$ and covariance matrix

$$\Sigma = \begin{pmatrix} \sigma_{11} & \sigma_{12} \\ \sigma_{21} & \sigma_{22} \end{pmatrix},$$

then to compute the probabilities at (x, y), enter the standardized values $\frac{x - \mu_1}{\sqrt{\sigma_{11}}}$ for the x value, $\frac{y - \mu_2}{\sqrt{\sigma_{22}}}$ for the y value, and $\frac{\sigma_{12}}{\sqrt{\sigma_{11}\sigma_{22}}}$ for the correlation coefficient, and click on [P].

Example 30.1. The Fuel Economy Guide published by the Department of Energy reports that the average city mileage for the 1998 compact car is 22.8 with standard deviation 4.5, the average highway mileage is 31.1 with standard deviation 5.5. In addition, the correlation coefficient between the city and highway mileage is 0.95.

a. Find the percentage of 1998 compact cars that give city mileage greater than 20 and highway mileage greater than 28.

b. What is the average city mileage of a car that gives highway mileage of 25?

Solution: Let (X_1, X_2) denote the (city, highway) mileage of a randomly selected compact car. Assume that (X_1, X_2) follows a bivariate normal distribution with the means, standard deviation, and correlation coefficient given in the problem.

a.

$$
\begin{aligned}
P(X_1 > 20, X_2 > 28) &= P\left(Z_1 > \frac{20 - 22.8}{4.5}, Z_2 > \frac{28 - 31.1}{5.5}\right) \\
&= P(Z_1 > -0.62, Z_2 > -0.56) \\
&= 0.679158.
\end{aligned}
$$

That is, about 68% of the 1998 compact cars give at least 20 city mileage and at least 28 highway mileage. To find the above probability, select the dialog box [StatCalc→Continuous→Biv Normal→All Tail Probabilities] from *StatCalc*, enter -0.62 for the [x value], -0.56 for the [y value], and 0.95 for the correlation coefficient; click [P].

b. From Section 30.6, Property 4, we have

$$\mu_1 + \sqrt{\frac{\sigma_{11}}{\sigma_{22}}}\rho(x_2 - \mu_2) = 22.8 + \frac{4.5}{5.5} \times 0.95 \times (25 - 31.1)$$

$$= 18.06 \text{ miles.}$$

For other applications and more examples, see "Tables of the Bivariate Normal Distribution Function and Related Functions," National Bureau of Standards, Applied Mathematics Series 50, 1959.

30.3 Inferences on Correlation Coefficients

Let $(X_{11}, X_{21}), \ldots, (X_{1n}, X_{2n})$ be a sample of independent observations from a bivariate normal population with

$$\text{covariance matrix } \boldsymbol{\Sigma} = \begin{pmatrix} \sigma_{11} & \sigma_{12} \\ \sigma_{21} & \sigma_{22} \end{pmatrix}.$$

The population correlation coefficient is defined by

$$\rho = \frac{\sigma_{12}}{\sqrt{\sigma_{11}\sigma_{22}}}, \quad -1 \le \rho \le 1. \tag{30.2}$$

Define

$$\begin{pmatrix} \overline{X}_1 \\ \overline{X}_2 \end{pmatrix} = \begin{pmatrix} \frac{1}{n}\sum_{i=1}^{n} X_{1i} \\ \frac{1}{n}\sum_{i=1}^{n} X_{2i} \end{pmatrix} \quad \text{and} \quad \boldsymbol{S} = \begin{pmatrix} s_{11} & s_{12} \\ s_{21} & s_{22} \end{pmatrix}, \tag{30.3}$$

where s_{ii} denotes the sample variance of X_i, given by

$$s_{ii} = \frac{1}{n-1}\sum_{j=1}^{n}(X_{ij} - \overline{X}_i)^2, \quad i = 1, 2,$$

and s_{12} denotes the sample covariance between X_1 and X_2 and is computed as

$$s_{12} = \frac{1}{n-1}\sum_{j=1}^{n}(X_{1j} - \overline{X}_1)(X_{2j} - \overline{X}_2).$$

The sample correlation coefficient is defined by

$$r = \frac{s_{12}}{\sqrt{s_{11}s_{22}}}, \quad -1 \le r \le 1.$$

The pdf of r is given by

$$f(r|\rho) = \frac{2^{n-3}(1-\rho^2)^{(n-1)/2}(1-r^2)^{n/2-2}}{(n-3)!\pi} \sum_{i=0}^{\infty} \frac{(2r\rho)^i}{i!} \{\Gamma[(n+i-1)/2]\}^2. \quad (30.4)$$

Another form of the density function is given by Hotelling (1953):

$$\frac{(n-2)}{\sqrt{2\pi}} \frac{\Gamma(n-1)}{\Gamma(n-1/2)}(1-\rho^2)^{(n-1)/2}(1-r^2)^{n/2-2}(1-r\rho)^{-n+3/2}$$

$$\times \quad F\left(\frac{1}{2};\frac{1}{2};n-\frac{1}{2};\frac{1+r\rho}{2}\right), \quad (30.5)$$

where

$$F(a;b;c;x) = \sum_{j=0}^{\infty} \frac{\Gamma(a+j)}{\Gamma(a)} \frac{\Gamma(b+j)}{\Gamma(b)} \frac{\Gamma(c)}{\Gamma(c+j)} \frac{x^j}{j!}.$$

This series converges faster than the series in (30.4).

The sample correlation coefficient r is a biased estimate of the population correlation coefficient ρ. Specifically,

$$E(r) = \rho - \frac{\rho(1-\rho^2)}{2(n-1)} + O\left(1/n^2\right).$$

The estimator

$$U(r) = r + \frac{r(1-r^2)}{(n-2)}, \quad (30.6)$$

is an asymptotically unbiased estimator of ρ; the bias is of $O(1/n^2)$. [Olkin and Pratt, 1958]

An Exact Test

Consider the hypotheses

$$H_0 : \rho \le \rho_0 \quad \text{vs.} \quad H_a : \rho > \rho_0. \quad (30.7)$$

For a given n and an observed value r_0 of r, the test that rejects the null hypothesis whenever $P(r > r_0|n, \rho_0) < \alpha$ has exact size α. Furthermore, when

$$H_0 : \rho \ge \rho_0 \quad \text{vs.} \quad H_a : \rho < \rho_0, \quad (30.8)$$

the null hypothesis will be rejected if $P(r < r_0|n, \rho_0) < \alpha$. The null hypothesis of

$$H_0 : \rho = \rho_0 \quad \text{vs.} \quad H_a : \rho \ne \rho_0, \quad (30.9)$$

will be rejected whenever

$$P(r < r_0|n, \rho_0) < \alpha/2 \quad \text{or} \quad P(r > r_0|n, \rho_0) < \alpha/2.$$

The above tests are uniformly most powerful (UMP) among the scale and location invariant tests. (see Anderson 1984, p. 114.) The above p-values can be computed by numerically integrating the pdf in (30.5).

A simpler test is available for the special case of $\rho_0 = 0$. In this case, the statistic

$$r\sqrt{\frac{n-2}{1-r^2}} \sim t_{n-2}.$$

For example, for testing hypotheses in (30.7), the null hypothesis is rejected if

$$r_0 \sqrt{\frac{n-2}{1-r_0^2}} > t_{n-2;1-\alpha},$$

where r_0 is an observed value of r.

Fisher's z-Transformation

Fisher's z-transformation of a sample correlation coefficient r is given by

$$z = \frac{1}{2} \ln\left(\frac{1+r}{1-r}\right).$$

If the r is a correlation coefficient for a sample of size n from a bivariate normal distribution, then

$$z \sim N\left(\Lambda = \frac{1}{2}\ln\left(\frac{1+\rho}{1-\rho}\right), \frac{1}{n-3}\right), \text{ approximately.} \quad (30.10)$$

Let r_0 be an observed value of r. For testing $H_0 : \rho \le \rho_0$ vs. $H_a : \rho > \rho_0$, the p-value is given by

$$1 - \Phi\left(\sqrt{n-3}(z_0 - \Lambda_0)\right),$$

where

$$z_0 = \frac{1}{2}\ln\left(\frac{1+r_0}{1-r_0}\right) \text{ and } \Lambda_0 = \frac{1}{2}\ln\left(\frac{1+\rho_0}{1-\rho_0}\right),$$

and Φ is the standard normal cdf.

A Generalized Variable Test

The generalized test given in Krishnamoorthy and Xia (2007) involves Monte Carlo simulation, and is equivalent to the exact test described above. The following algorithm can be used to compute the p-value and confidence interval for ρ.

Algorithm 30.1. Generalized confidence interval for ρ

For a given n and r:
    Set $r^* = r/\sqrt{1-r^2}$
For $i = 1$ to m
    Generate $Z \sim N(0,1)$, $U_1 \sim \chi^2_{n-1}$ and $U_2 \sim \chi^2_{n-2}$
    Set $G_i = \dfrac{r^*\sqrt{U_2}-Z}{\sqrt{(r^*\sqrt{U_2}-Z)^2+U_1}}$
[end loop]

The generalized p-value for (30.7) is estimated by the proportion of G_is that are less than ρ_0. The H_0 in (30.7) will be rejected if this generalized p-value is less than α. Similarly, the generalized p-value for (30.8) is estimated by the proportion of G_is that are greater than ρ_0, and the generalized p-value for (30.9) is given by two times the minimum of one-sided p-values.

    The dialog box [StatCalc→Continuous→Biv Normal→Test...] uses Algorithm 30.1 with $m = 100,000$ for computing the *generalized p-values* for hypothesis test about ρ. Krishnamoorthy (2013) showed that this generalized variable test is exact.

Example 30.2. Suppose that a sample of 20 observations from a bivariate normal population produced the correlation coefficient $r = 0.75$, and we would like to test

$$H_0 : \rho \leq 0.5 \quad \text{vs.} \quad H_a : \rho > 0.5.$$

To compute the p-value using *StatCalc*, enter 20 for n, 0.75 for [Sam Corrl r] and 0.5 for [rho_0]. Click [GV p-val] to get 0.045, and click [Fisher's z] to get .040. Note that both methods produced p-values that are less than 0.05, and so we can conclude that the population correlation coefficient is larger than 0.5 at the 5% level.

Approximate Confidence Interval for ρ

An approximate confidence interval for ρ is based on the well-known Fisher's Z transformation of r. Let

$$Z = \frac{1}{2} \ln \left(\frac{1+r}{1-r} \right) \quad \text{and} \quad \mu_\rho = \frac{1}{2} \ln \left(\frac{1+\rho}{1-\rho} \right). \tag{30.11}$$

Then

$$Z \sim N(\mu_\rho, (n-3)^{-1}) \quad \text{asymptotically.} \tag{30.12}$$

The confidence interval for ρ is given by

$$\left(\tanh[Z - z_{1-\alpha/2}/\sqrt{n-3}], \ \tanh[Z + z_{1-\alpha/2}/\sqrt{n-3}] \right), \tag{30.13}$$

where $\tanh(x) = \frac{e^x - e^{-x}}{e^x + e^{-x}}$, and z_p is the pth quantile of the standard normal distribution.

Exact Confidence Interval for ρ

Let r_0 be an observed value of r based on a sample of n bivariate normal observations. For a given confidence level $1 - \alpha$, the upper limit ρ_U for ρ is the solution of the equation

$$P(r \leq r_0 | n, \rho_U) = \alpha/2, \tag{30.14}$$

and the lower limit ρ_L is the solution of the equation

$$P(r \geq r_0 | n, \rho_L) = \alpha/2. \tag{30.15}$$

See Anderson (1984, Section 4.2.2). One-sided limits can be obtained by replacing $\alpha/2$ by α in the above equations. Although (30.14) and (30.15) are difficult to solve for ρ_U and ρ_L, they can be used to assess the accuracy of the approximate confidence intervals (see Krishnamoorthy and Xia, 2007).

Generalized Confidence Interval for ρ

The generalized confidence interval due to Krishnamoorthy and Xia (2007) can be constructed using the percentiles of the G_is given in Algorithm 30.1. Specifically,

$$(G_{\alpha/2}, \ G_{1-\alpha/2}),$$

where G_p, $0 < p < 1$, denotes the pth quantile of G_is, is a $1 - \alpha$ confidence interval. Furthermore, these confidence intervals are exact because the endpoints satisfy the equations (30.15) and (30.14). For example, when $n = 20$, $r = 0.7$, the 95%

generalized confidence interval for ρ (using *StatCalc*) is given by $(0.365, 0.865)$. The probabilities in (30.15) and (30.14) are

$$P(r \geq 0.7 | 20, 0.365) = 0.025 \quad \text{and} \quad P(r \leq 0.7 | 20, 0.865) = 0.025.$$

In this sense, the generalized confidence limits are exact (see Krishnamoorthy and Xia, 2007, and Krishnamoorthy, 2013).

The dialog box [StatCalc→Continuous→Biv Normal→Test and CI for Correlation Coefficient] uses Algorithm 30.1 with $m = 100,000$ for computing the *generalized confidence intervals* for ρ. The number of simulation runs can be increased as you wish.

Example 30.3. Suppose that a sample of 20 observations from a bivariate normal population produced the correlation coefficient $r = 0.75$. To compute a 95% confidence interval for ρ using *StatCalc*, enter 20 for n, 0.75 for [Sam Corrl r], and 0.95 for [Conf Lev]. Click [1] to get one-sided confidence limits 0.509 and 0.872; click [2] to get confidence interval as $(0.450, 0.889)$.

Example 30.4. The marketing manager of a company wants to determine whether there is a positive association between the number of TV ads per week and the amount of sales (in \$1,000). He collected a sample of data from the records as shown in the following table. The sample correlation coefficient is given by

TABLE 30.1: TV Sales Data

No. of ads, x_1	5	7	4	4	6	7	9	6	6
Sales, x_2	24	30	18	20	21	29	31	22	25

$$r = \frac{\sum\limits_{i=1}^{n} (x_{1i} - \bar{x}_1)(x_{2i} - \bar{x}_2)}{\sqrt{\sum\limits_{i=1}^{n} (x_{1i} - \bar{x}_1)^2 \sum\limits_{i=1}^{n} (x_{2i} - \bar{x}_2)^2}} = 0.88.$$

An unbiased estimator using (30.6) is $0.88 + 0.88(1 - 0.88^2)/7 = 0.91$.

To compute a 95% confidence interval for the population correlation coefficient ρ, select [Continuous→Biv Normal→Test and CI for Correlation Coefficient] from *StatCalc*, enter 9 for n, 0.88 for sample correlation coefficient, 0.95 for the confidence level, and click [2] to get $(0.491, 0.969)$.

Suppose we want to test

$$H_0 : \rho \leq 0.70 \quad \text{vs.} \quad H_a : \rho > 0.70.$$

To compute the p-value, enter 9 for n, 0.88 for the sample correlation, and 0.70 for [rho_0]; click [p-values for] to get 0.122. Since this is not less than 0.05, at the 5% level, there is not sufficient evidence to indicate that the population correlation coefficient is greater than 0.70.

30.4 Inferences on the Difference Between Two Correlation Coefficients

Let r_i denote the correlation coefficient based on a sample of n_i observations from a bivariate normal population with covariance matrix Σ_i, $i = 1, 2$. Let ρ_i denote the population correlation coefficient based on Σ_i, $i = 1, 2$.

An Asymptotic Approach

The asymptotic approach is based on Fisher's Z transformation given for the one-sample case, and is mentioned in Anderson (1984, p. 114). Let

$$Z_k = \frac{1}{2} \ln \left(\frac{1 + r_k}{1 - r_k} \right) \quad \text{and} \quad \mu_{\rho_k} = \frac{1}{2} \ln \left(\frac{1 + \rho_k}{1 - \rho_k} \right), \quad k = 1, 2. \qquad (30.16)$$

Then it follows from the asymptotic result in (30.12) that,

$$\frac{(Z_1 - Z_2) - (\mu_{\rho_1} - \mu_{\rho_2})}{\sqrt{\frac{1}{n_1 - 3} + \frac{1}{n_2 - 3}}} \sim N(0, 1) \quad \text{asymptotically.} \qquad (30.17)$$

Using the above asymptotic result, one can easily develop test procedures for $\rho_1 - \rho_2$. Specifically, for an observed value (z_1, z_2) of (Z_1, Z_2), the p-value for testing

$$H_0 : \rho_1 \leq \rho_2 \quad \text{vs.} \quad H_a : \rho_1 > \rho_2 \qquad (30.18)$$

is given by $1.0 - \Phi \left((z_1 - z_2) / \sqrt{1/(n_1 - 3) + 1/(n_2 - 3)} \right)$, where Φ is the standard normal distribution function. Notice that, using the distributional result in (30.17), one can easily obtain confidence interval for $\mu_{\rho_1} - \mu_{\rho_2}$ but not for $\rho_1 - \rho_2$.

Generalized Test and Confidence Limits for $\rho_1 - \rho_2$

The generalized p-values and confidence intervals can be obtained using the following algorithm.

Algorithm 30.2. Generalized confidence interval for $\rho_1 - \rho_2$

For a given (r_1, n_1) and (r_2, n_2):
$\qquad$ Set $r_1^* = r_1 / \sqrt{1 - r_1^2}$ and $r_2^* = r_2 / \sqrt{1 - r_2^2}$
For $i = 1$ to m
$\qquad$ Generate $Z_{10} \sim N(0, 1)$, $U_{11} \sim \chi_{n_1 - 1}^2$ and $U_{12} \sim \chi_{n_1 - 2}^2$
$\qquad\qquad$ $Z_{20} \sim N(0, 1)$, $U_{21} \sim \chi_{n_2 - 1}^2$ and $U_{22} \sim \chi_{n_2 - 2}^2$
$\qquad$ Set $T_i = \dfrac{r_1^* \sqrt{U_{12}} - Z_{01}}{\sqrt{\left(r_1^* \sqrt{U_{12}} - Z_{01} \right)^2 + U_{11}}} - \dfrac{r_2^* \sqrt{U_{22}} - Z_{02}}{\sqrt{\left(r_2^* \sqrt{U_{22}} - Z_{02} \right)^2 + U_{21}}}$
(end loop)

Suppose we want to test

$$H_0 : \rho_1 - \rho_2 \leq c \quad \text{vs.} \quad H_a : \rho_1 - \rho_2 > c, \qquad (30.19)$$

where c is a specified number. The generalized p-value for (30.19) is estimated by the proportion of the T_is that are less than c. The H_0 in (30.19) will be rejected if this generalized p-value is less than α. Similarly, the generalized p-value for a left-tail test is estimated by the proportion of T_is that are greater than c, and the generalized p-value for a two-tail test is given by two times the minimum of the one-sided p-values.

Generalized confidence limits for $\rho_1 - \rho_2$ can be constructed using the percentiles of the T_is in Algorithm 29.4.2. In particular, $(T_{\alpha/2}, T_{1-\alpha/2})$, where T_p, $0 < p < 1$, is a $1 - \alpha$ confidence interval for $\rho_1 - \rho_2$.

The dialog box [StatCalc→Continuous→Biv Normal→Test and CI for rho1 - rho2] uses Algorithm 29.4.2 with $m = 500,000$ for computing the generalized p-values and generalized confidence intervals for $\rho_1 - \rho_2$.

Example 30.5. Suppose that a sample of 15 observations from a bivariate normal population produced the correlation coefficient $r_1 = 0.8$, and a sample from another bivariate normal population yielded $r_2 = 0.4$. It is desired to test

$$H_0 : \rho_1 - \rho_2 \leq 0.1 \quad \text{vs.} \quad H_a : \rho_1 - \rho_2 > 0.1.$$

To compute the p-value using *StatCalc*, enter the sample sizes and correlation coefficients in the appropriate edit boxes, and 0.1 for [H0: rho1 - rho2]; click on [p-values for] to get 0.091. Because this p-value is not less than 0.05, the H_0 can not be rejected at the level of significance 0.05.

To get confidence intervals for $\rho_1 - \rho_2$, click on [1] to get one-sided limits, and click on [2] to get confidence interval. For this example, 95% one-sided limits for $\rho_1 - \rho_2$ are 0.031 and 0.778, and 95% confidence interval is $(-0.037, 0.858)$.

30.5 Test and Confidence Interval for Variances

Let

$$\begin{pmatrix} x_{11} \\ x_{21} \end{pmatrix}, \ldots, \begin{pmatrix} x_{1n} \\ x_{2n} \end{pmatrix}$$

be a sample from a bivariate normal distribution with covariance matrix $\Sigma = \begin{pmatrix} \sigma_{11} & \sigma_{12} \\ \sigma_{21} & \sigma_{22} \end{pmatrix}$. The sample variance-covariance matrix is defined as

$$\begin{aligned} \mathbf{S} &= \begin{pmatrix} s_{11} & s_{12} \\ s_{12} & s_{22} \end{pmatrix} \\ &= \frac{1}{n-1} \begin{pmatrix} \sum_{i=1}^{n}(x_{1i} - \bar{x}_1)^2 & \sum_{i=1}^{n}(x_{1i} - \bar{x}_1)(x_{2i} - \bar{x}_2) \\ \sum_{i=1}^{n}(x_{1i} - \bar{x}_1)(x_{2i} - \bar{x}_2) & \sum_{i=1}^{n}(x_{2i} - \bar{x}_2)^2 \end{pmatrix}. \end{aligned}$$

Let $\omega = \sigma_{11}/\sigma_{22}$, $w = s_{11}/s_{22}$, and $r = s_{12}/\sqrt{s_{11}s_{22}}$. Consider testing

$$H_0 : \omega = \omega_0 \quad \text{vs.} \quad H_a : \omega \neq \omega_0, \tag{30.20}$$

where ω_0 is a specified positive value. Pitman (1939) showed that the statistic

$$t = \frac{(w - \omega_0)\sqrt{n-2}}{\sqrt{4(1 - r^2)w\omega_0}} \sim t_{n-2},$$

where t_m denotes the t-distribution with df $= m$ (Pitman, 1939). The H_0 in (30.20) is rejected at the level α if $|t| > t_{n-2;1-\alpha/2}$.

An exact $1 - \alpha$ confidence interval for $\omega = \sigma_{11}/\sigma_{22}$ is given by

$$\left(w\left[K_r - \sqrt{K_r^2 - 1}\right],\ w\left[K_r + \sqrt{K_r^2 - 1}\right]\right), \tag{30.21}$$

where $w = s_{11}/s_{22}$ and $K_r = 1 + 2(1 - r^2)t_{n-2;1-\alpha/2}^2/(n - 2)$.

For a given $(n, s_{11}, s_{22}, s_{12})$, the dialog box [StatCalc→Continuous→Biv Normal→Test and confidence interval for sig11/sig22] calculates the above confidence interval and p-values for testing the parameter σ_{11}/σ_{22}.

Example 30.6. The following data represent measurements on a variable before and after administration of a treatment on 10 subjects:

```
Before (x1):  7.4,  6.4, 14.9, 10.2, 32.5, 15.0, 17.1, 19.2, 22.0, 15.9
After  (x2): 11.3, 10.3, 16.2, 12.4, 45.0, 17.4, 16.8, 22.1, 23.9, 17.9
```

For the above data, $s_{11} = 57.96$, $s_{22} = 100.39$ and $s_{12} = 73.17$. To find a 95% confidence interval for the population variance ratio σ_{11}/σ_{22}, select the dialog box [StatCalc→Continuous→Biv Normal→Test and CI for sig11/sig22], enter these quantities, and $n = 10$. Click on [CI] to get (0.366, 0.912). This confidence interval indicates that σ_{22} is significantly larger than σ_{11}.

To find the p-value for testing $H_0 : \sigma_{11}/\sigma_{22} \geq .95$ vs. $H_a : \sigma_{11}/\sigma_{22} < .95$, enter 0.95 for [value of c] in the dialog box noted in the preceding paragraph, and click on [p-value] to get 0.0179637. Thus, the data provide sufficient evidence to indicate that $\sigma_{11}/\sigma_{22} < 0.95$.

30.6 Some Properties

Suppose that (X_1, X_2) is a bivariate normal random vector with

$$E(X_1) = \mu_1,\ \mathrm{Var}(X_1) = \sigma_{11},\ E(X_2) = \mu_2,\ \mathrm{Var}(X_2) = \sigma_{22}$$

and the covariance, $\mathrm{Cov}(X_1, X_2) = \sigma_{12}$.

1. The marginal distribution of X_1 is normal with mean μ_1 and variance σ_{11}.

2. The marginal distribution of X_2 is normal with mean μ_2 and variance σ_{22}.

3. The distribution of $aX_1 + bX_2$ is normal with

    $$\mathrm{mean} = a\mu_1 + b\mu_2 \text{ and variance} = a^2\sigma_{11} + b^2\sigma_{22} + 2ab\sigma_{12}.$$

4. The conditional distribution of X_1 given X_2 is normal with

$$\mathrm{mean} = \mu_1 + \frac{\sigma_{12}}{\sigma_{22}}(x_2 - \mu_2)$$

$$= \mu_1 + \rho\sqrt{\frac{\sigma_{11}}{\sigma_{22}}}(x_2 - \mu_2)$$

and

$$\mathrm{variance} = \sigma_{11} - \sigma_{12}^2/\sigma_{22}.$$

30.7   Random Number Generation

The following algorithm generates bivariate normal random vectors with mean (μ_1, μ_2) and variances σ_{11} and σ_{22}, and the correlation coefficient ρ.

Algorithm 30.3. Bivariate normal random numbers generator

Generate independent N(0, 1) variates u and v
Set
$$x_1 = \mu_1 + \sqrt{\sigma_{11}}\, u$$
$$x_2 = \sqrt{\sigma_{22}}(\rho u + v\sqrt{1 - \rho^2}) + \mu_2$$

For a given n and ρ, correlation coefficients r can be generated using the following results. Let $A = (n-1)S$, where S is the sample covariance matrix defined in (30.3). Write

$$A = VV', \quad \text{where} \quad V = \begin{pmatrix} v_{11} & 0 \\ v_{21} & v_{22} \end{pmatrix}, \quad v_{ii} > 0, \ i = 1, 2.$$

Similarly, let us write

$$\Sigma = \begin{pmatrix} \sigma_{11} & \sigma_{12} \\ \sigma_{21} & \sigma_{22} \end{pmatrix} = \theta\theta', \quad \text{where} \quad \theta = \begin{pmatrix} \theta_{11} & 0 \\ \theta_{21} & \theta_{22} \end{pmatrix}, \quad \theta_{ii} > 0, \ i = 1, 2.$$

Then, V is distributed as θT, where T is a lower triangular matrix whose elements t_{ij} are independent with $t_{11}^2 \sim \chi_{n-1}^2$, $t_{22}^2 \sim \chi_{n-2}^2$ and $t_{21} \sim N(0,1)$. Furthermore, note that the population correlation coefficient can be expressed as

$$\rho = \frac{\theta_{21}}{\sqrt{\theta_{21}^2 + \theta_{22}^2}} = \frac{\theta_{21}/\theta_{22}}{\sqrt{\theta_{21}^2/\theta_{22}^2 + 1}}.$$

The above equation implies that

$$\frac{\theta_{21}}{\theta_{22}} = \rho \bigg/ \sqrt{1 - \rho^2} = \rho^*, \text{ say.}$$

Similarly, the sample correlation coefficient can be expressed in terms of the elements v_{ij} of V as

$$r = \frac{v_{21}}{\sqrt{v_{21}^2 + v_{22}^2}} \sim \frac{\theta_{21}t_{11} + \theta_{22}t_{21}}{\sqrt{(\theta_{21}t_{11} + \theta_{22}t_{21})^2 + \theta_{22}^2 t_{22}^2}} = \frac{\rho^* t_{11} + t_{21}}{\sqrt{(\rho^* t_{11} + t_{21})^2 + t_{22}^2}}.$$

Using these results, we get the following algorithm for generating r.

Algorithm 30.4. Sample correlation coefficient generator

For a given sample size n and ρ:
Set $\rho^* = \rho \bigg/ \sqrt{1 - \rho^2}$
Generate $t_{11}^2 \sim \chi_{n-1}^2$, $t_{22}^2 \sim \chi_{n-2}^2$, and $t_{21} \sim N(0, 1)$
Set $r = \dfrac{\rho^* t_{11} + t_{21}}{\sqrt{(\rho^* t_{11} + t_{21})^2 + t_{22}^2}}$

30.8   A Computational Algorithm for Probabilities

In the following, $\Phi(x)$ denotes the standard normal distribution function and
Prob $= P(Z_1 > x, Z_2 > y)$. Define

$$f(x,y) = \frac{1}{\sqrt{2\pi}} \int_0^x \Phi\left(\frac{ty}{x}\right) \exp(-t^2/2)dt.$$

If $\rho = 0$, return Prob $= (1 - \Phi(x)) * (1 - \Phi(y))$
   If $x = 0$ and $y = 0$, return Prob $= 0.25 + $ arc $\sin(\rho)/(2\pi)$
   If $\rho = 1$ and $y \leq x$, return Prob $= 1 - \Phi(x)$
   If $\rho = 1$ and $y > x$, return Prob $= 1 - \Phi(y)$
   If $\rho = -1$ and $x + y \geq 0$, return Prob $= 0$
   If $\rho = -1$ and $x + y \leq 0$, return Prob $= 1 - \Phi(x) - \Phi(y)$
   $F = 0.25 - 0.5 * (\Phi(x) + \Phi(y) - 1) + \arcsin(\rho)/(2 * \pi)$
   Prob $= f\left(x, \frac{y - \rho x}{\sqrt{1-\rho^2}}\right) + f\left(y, \frac{x - \rho y}{\sqrt{1-\rho^2}}\right) + F$

[Tables of the Bivariate Normal Distribution Function and Related Functions; National Bureau of Standards, Applied Mathematics Series 50, 1959]

31

Some Nonparametric Methods

In all earlier chapters we dealt with parametric models, and described inferential methods for testing and estimating model parameters, prediction intervals and tolerance intervals. In this chapter, we shall see some basic inferential methods for continuous distributions without assuming any parametric models. In other words, the results in the following sections are applicable to any continuous population. Furthermore, if an inferential method is not available for a parametric continuous distribution, then one could safely use an appropriate nonparametric method. However, it should be noted that the nonparametric methods are less efficient than their parametric counterparts. If a sample data from a population fits a parametric model, such as a normal distribution, it is better to use the inferential methods applicable to that model.

31.1 Distribution of Runs

Consider a random arrangement of $m+n$ elements, m of them are one type, and n of them are of another type. A run is a sequence of symbols of the same type bounded by symbols of another type except for the first and last position. Let R denote the total number of runs in the sequence. The probability mass function of R is given by

$$P(R = r|m, n) = \frac{2\binom{m-1}{r/2-1}\binom{n-1}{r/2-1}}{\binom{m+n}{n}} \quad \text{for even } r,$$

and

$$P(R = r|m, n) = \frac{\binom{m-1}{(r-1)/2}\binom{n-1}{(r-3)/2} + \binom{m-1}{(r-3)/2}\binom{n-1}{(r-1)/2}}{\binom{m+n}{n}} \quad \text{for odd } r.$$

The distribution of runs is useful to test the hypothesis of randomness of an arrangement of elements. The hypotheses of interest are

H_0: arrangement is random   vs. H_a: arrangement is nonrandom.

Too many runs or too few runs provide evidence against the null hypothesis. Specifically, for a given m, n, the observed number of runs r of R, and the level of significance α, the H_0 will be rejected if the p-value

$$2\min\{P(R \leq r), P(R \geq r)\} \leq \alpha.$$

The plots of probability mass functions of runs in Figure 31.1 show that the run distribution is asymmetric when $m \neq n$ and symmetric when $m = n$. The mean and

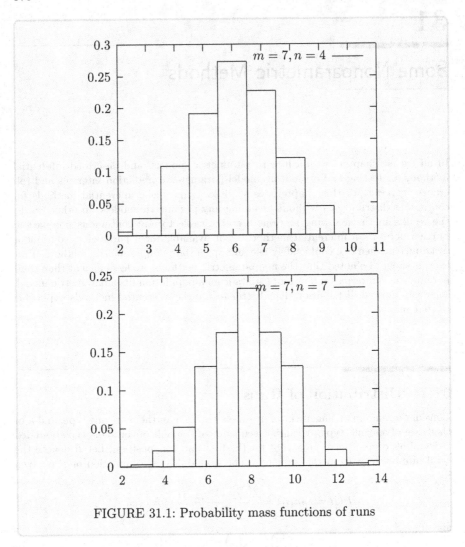

FIGURE 31.1: Probability mass functions of runs

variance of R are given by

$$E(R) = 1 + \frac{2mn}{m+n} \quad \text{and} \quad \text{Var}(R) = \frac{2mn(2mn - m - n)}{(m+n)^2(m+n-1)}.$$

For given m and n, the dialog box [StatCalc→Nonparametric→Distribution of Runs] evaluates the distribution function of R and critical points.

Example 31.1. Consider the following sequence:

$$\underline{a\,a\,a}\,\overline{b\,b}\,\underline{a}\,\overline{b\,b\,b}\,\underline{a\,a\,a\,a}\,\overline{b\,b}\,\underline{a\,a\,a\,a}\,\overline{b}.$$

Here $m = 12$, $n = 8$, and the observed number of runs $r = 8$. To compute the probabilities using *StatCalc*, enter 12 for the number of [1st type symbols], 8 for the number of [2nd type symbols], and 8 for the [Observed runs r]; click on [P(R <= r)] to get

$$P(R \leq 8) = 0.159085 \quad \text{and} \quad P(R \geq 8) = 0.932603.$$

Since the probability of observing 8 or fewer runs is not less than 0.025, the null hypothesis that the arrangement is random will be retained at the level 0.05. To find the critical value, enter 0.025 for the tail probability, and click on [Left Crt] to get

$$P(R \leq 6) = 0.024609 \quad \text{and} \quad P(R \geq 16) = 0.00655.$$

This means that the null hypothesis of randomness will be rejected (at the 5% level) if the observed number of runs is 6 or less, or, 16 or more.

Example 31.2. Consider the simple linear regression model

$$Y_i = \alpha + \beta X_i + \epsilon_i, \quad i = 1, ..., N.$$

It is usually assumed that the error terms ϵ_is are random. This assumption can be tested using the estimated errors

$$e_i = Y_i - \widehat{Y}_i, \quad i = 1, ..., N,$$

where Y_i and $\widehat{Y}_i$ denote, respectively, the observed value and the predicted value of the ith individual. Suppose that the estimated errors when $N = 18$ are:

-3.85	-0.11	6.63	-2.42	3.63	5.37	0.15	-2.76	-3.54
5.85	2.42	-6.37	-2.15	-0.37	3.24	3.48	-7.63	3.37

The arrangement of the signs of the errors is given by

$$- - + - + + + - - + + - - - + + - +$$

In this example, $m = 9$, $n = 9$, and the observed number of runs $r = 10$. We want to test the null hypothesis that the errors are randomly distributed at the level of significance 0.05. To compute the p-value, select the dialog box [StatCalc→Nonparametric→Distribution of Runs], enter 9 for the number of first type of symbols, 9 for the number of second type symbols, and 10 for the observed number of runs; click [P(R <= r)] to get

$$P(R \leq 10) = 0.600782 \quad \text{and} \quad P(R \geq 10) = 0.600782.$$

Since these probabilities are greater than $0.05/2 = 0.025$, the null hypothesis of randomness will be retained at the 5% level.

To get the critical values, enter 0.025 for the tail probability, and click [Left Crt] to get 5 and 15. That is, the left-tail critical value is 5 and the right-tail critical value is 15. Thus, the null hypothesis will be rejected at 0.05 level, if the total number of runs is 5 or less or 15 or more.

Example 31.3. Suppose that a sample of 20 students from a school is selected for some purpose. The genders of the students are recorded as they were selected:

M F M F F M F F M F M F F M F F M M F

We would like to test if this really is a random sample. In this example, $m = 8$ (number of male students), $n = 12$ (number of female students), and the total number of runs $= 14$. Since the mean number of runs is given by

$$1 + \frac{2mn}{m+n} = 10.6,$$

it appears that there are too many runs, and so we want to compute the probability of observing 14 or more runs under the hypothesis of randomness. To compute the p-value using *StatCalc*, select the dialog box [StatCalc→Nonparametric→Distribution of Runs], enter 8 for the number of first type of symbols, 12 for the number of second type symbols, and 14 for the observed number of runs; click [P(R <= r)] to get $P(R \geq 14) = 0.0799$. Since the probability of observing 14 or more runs is not less than 0.025, we can accept the sample as a random sample at the 5% level.

31.2 Sign Test and Confidence Interval for the Median

Let $X_1, \ldots, X_n$ be a sample of independent observations from a continuous population. Let M denote the median of the population. We want to test

$$H_0 : M = M_0 \quad \text{vs.} \quad H_a : M \neq M_0.$$

Let K denote the number of plus signs of the differences $X_1 - M_0, \ldots, X_n - M_0$. That is, K is the number of observations greater than M_0. Then, under the null hypothesis, K follows a binomial distribution with number of trials n and success probability

$$P(X_i > M_0) = 0.5.$$

If K is too large, then we conclude that the true median $M > M_0$; if K is too small, then we conclude that $M < M_0$. Let k be an observed value of K. For a given level of significance α, the null hypothesis will be rejected in favor of the alternative hypothesis $M > M_0$ if $P(K \geq k|n, 0.5) \leq \alpha$, and in favor of the alternative hypothesis $M < M_0$ if $P(K \leq k|n, 0.5) \leq \alpha$. If the alternative hypothesis is $M \neq M_0$, then the null hypothesis will be rejected if

$$2\min\{P(K \geq k|n, 0.5), P(K \leq k|n, 0.5)\} \leq \alpha.$$

Confidence Interval for the Median

Let $X_{(1)}, \ldots, X_{(n)}$ be the ordered statistics based on a sample $X_1, \ldots, X_n$. Let r be the largest integer such that

$$P(K \leq r) = \sum_{i=0}^{r} \binom{n}{i} (0.5)^n \leq \alpha/2 \tag{31.1}$$

and s is the smallest integer such that

$$P(K \geq s) = \sum_{i=s}^{n} \binom{n}{i} (0.5)^n \leq \alpha/2. \tag{31.2}$$

Then, the interval $(X_{(r+1)}, X_{(s)})$ is a $1 - \alpha$ confidence interval for the median M with coverage probability at least $1 - \alpha$.

For a given sample size n and confidence level $1 - \alpha$, the dialog box [StatCalc→Nonparametric→Sign Test and Confidence Interval for the Median] computes the integers r and s that satisfy (31.1) and (31.2), respectively.

Remark 31.1. If $X_i - M_0 = 0$ for some i, then simply discard those observations and reduce the sample size n accordingly. Zero differences can also be handled by assigning signs randomly (e.g. flip a coin; if the outcome is head, assign $+$, otherwise assign $-$).

The dialog box [StatCalc→Nonparametric→Signa Test and Confidence Interval for the Median] computes confidence intervals and p-values for testing the median.

Example 31.4. (Confidence Interval) To compute a 95% confidence interval for the median of a continuous population based on a sample of 40 observations, enter 40 for n, 0.95 for confidence level, click [CI] to get 14 and 27. That is, the required confidence interval is formed by the 14th and 27th order statistics from the sample.

Example 31.5. (p-value) Suppose that a sample of 40 observations yielded $k = 13$, the number of observations greater than the specified median. To obtain the p-value of the test when $H_a : M < M_0$, enter 40 for n, 13 for k, and click [P(X <= k)] to get $P(K \leq 13) = 0.0192387$. Since this p-value is less than 0.05, the null hypothesis that $H_0 : M = M_0$ will be rejected at the 5% level.

The nonparametric inferential procedures are applicable for any continuous distribution. However, in order to understand the efficiency of the procedures, we apply them to a normal distribution.

Example 31.6. A sample of 15 observations is generated from a normal population with mean 2 and standard deviation 1, and is given below.

1.01	3.00	1.12	1.68	0.82	4.01	2.85	2.49
1.58	2.30	2.84	2.32	3.01	1.77	2.10	

Recall that for a normal population, mean and median are the same. Let us test the hypotheses that

$$H_0 : M \leq 1.5 \quad \text{vs.} \quad H_a : M > 1.5.$$

Note that there are 12 data points that are greater than 1.5. So, the p-value, $P(K \geq 12) = 0.01758$, which is less than 0.05. Therefore, the null hypothesis is rejected at the 5% level.

To find a 95% confidence interval for the median, enter 15 for the sample size n, and 0.95 for the confidence level. Click [CI] to get $(X_{(4)}, X_{(12)})$. That is, the 4th smallest observation and the 12th smallest (or the 4th largest) observation form the required confidence interval; for this example, the 95% confidence interval is (1.58, 2.84). Note that this interval indeed contains the actual median 2. Furthermore, if the hypotheses are

$$H_0 : M = 1.5 \quad \text{vs.} \quad H_a : M \neq 1.5,$$

then the null hypothesis will be rejected, because the 95% confidence interval does not contain 1.5.

31.3   Wilcoxon Signed-Rank Test and Mann–Whitney U Statistic

Wilcoxon signed-rank statistic is useful to test the median of a continuous symmetric distribution. Let M denote the median of the population. Consider the null hypothesis $H_0 : M = M_0$, where M_0 is a specified value of the median. Let $X_1, \ldots, X_n$ be a sample from the population, and let

$$D_i = X_i - M_0, \quad i = 1, 2, \ldots, n.$$

Rank the absolute values $|D_1|, |D_2|, \ldots, |D_n|$. Let T^+ be the sum of the ranks of the positive D_is. The distribution function of T^+ under H_0 is symmetric about its mean $n(n+1)/4$ (see Figure 31.2). The null hypothesis will be rejected if T^+ is too large or too small.

Let t be an observed value of T^+. For a given level α, the null hypothesis will be rejected in favor of the alternative

$$H_a : M > M_0 \ \text{ if } \ P(T^+ \geq t) \leq \alpha,$$

and in favor of the alternative

$$H_a : M < M_0 \ \text{ if } \ P(T^+ \leq t) \leq \alpha.$$

Furthermore, for a two-sided test, the null hypothesis will be rejected in favor of the alternative

$$H_a : M \neq M_0 \ \text{ if } \ 2 \min\{P(T^+ \leq t), P(T^+ \geq t)\} \leq \alpha.$$

For a given n and an observed value t of T^+, the dialog box [StatCalc→Nonparametric→Wilcoxon Signed-Rank Test] computes the above tail probabilities. *StatCalc* also computes the critical point for a given n and nominal level α.

For $\alpha < 0.5$, the left tail critical point k is the largest number such that

$$P(T^+ \leq k) \leq \alpha,$$

and the right tail critical point k is the smallest integer such that

$$P(T^+ \geq k) \leq \alpha.$$

Mean:	$\dfrac{n(n+1)}{4}$
Variance:	$\dfrac{n(n+1)(2n+1)}{24}$

[Gibbons and Chakraborti, 1992, p. 156]

The variable

$$Z = \frac{(T^+) - \text{mean}}{\sqrt{\text{var}}}$$

is approximately distributed as the standard normal random variable. This approximation is satisfactory for n greater than or equal to 15.

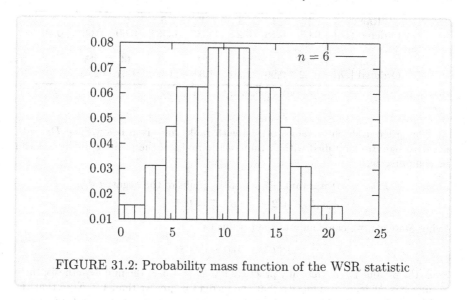

FIGURE 31.2: Probability mass function of the WSR statistic

Computing the Percentiles of Wilcoxon-Signed Rank Statistic

The dialog box [StatCalc→ Nonparametric→Wilcoxon Signed-Rank Test] computes the tail probabilities and critical points of the distribution of the signed-rank statistic T^+. As an example, suppose that a sample of 20 observations yielded $T^+ = 130$. To compute tail probabilities using *StatCalc*, enter 20 for n, 130 for T^+, and click on [P(X <= k)] to get $P(T^+ \leq 130) = 0.825595$ and $P(T^+ \geq 130) = 0.184138$.

To get the upper 5% critical point, enter 20 for n, 0.95 for cumulative probability and click [Critical Pt] to get 150. That is, $P(T^+ \geq 150) = 0.048654$. To get the lower 5% critical point, enter 0.05 for cumulative probability and click [Critical Pt] to get 60. Notice that, because of the discreteness of the distribution, we get $P(T^+ \leq 60) = 0.048654$, not exactly 0.05.

Example 31.7. We will illustrate the Wilcoxon Signed-Rank test for the data given in Example 31.4.1. The data set consists of 15 observations generated from a normal population with mean 2 and standard deviation 1, and they are

1.01	3.00	1.12	1.68	0.82	4.01	2.85	2.49
1.58	2.30	2.84	2.32	3.01	1.77	2.10	

Note that for a normal population mean and median are the same. For illustration purpose, let us test the hypotheses that

$$H_0 : M \leq 1.5 \quad \text{vs.} \quad H_a : M > 1.5.$$

The differences $D_i = X_i - 1.5$ are

−0.49	1.5	−0.38	0.18	−0.68	2.51	1.30	
0.99	0.08	0.80	1.34	0.82	1.51	0.27	0.60

Rank	1	2	3	4	5	6	7	8
Ordered $|D_i|$	0.08	0.18	0.27	0.38*	0.49*	0.60	0.68*	0.80

Rank	9	10	11	12	13	14	15
Ordered $|D_i|$	0.82	0.99	1.30	1.34	1.50	1.51	2.51

The ordered absolute values of D_is with ranks are given below. The D_is with negative sign are identified with *. Sum of the ranks of the positive differences can be computed as

$$T^+ = \text{Total rank} - \text{Sum of the ranks of the negative } D_i \text{s}$$
$$= 15(15+1)/2 - 4 - 5 - 7 = 104.$$

Using *StatCalc*, we can compute the p-value as

$$P(T^+ \geq 104) = 0.0051.$$

Since the p-value is less than any practical levels, we reject the null hypothesis; there is sufficient evidence to indicate that the median is greater than 1.5. Note that the Wilcoxon Signed-Rank test provides stronger evidence against the null hypothesis than the sign test (see Example 31.4.1).

To get the right-tail critical point using *StatCalc*, enter 15 for n, 0.95 for cumulative probability, and click [Critical Pt] to get

$$P(T^+ \geq 90) = 0.0473022.$$

Thus, any observed value of T^+ greater than or equal to 90 would lead to the rejection of H_0 at the 5% level of significance.

31.4   Wilcoxon Rank-Sum Test

The Wilcoxon rank-sum test is useful for comparing two continuous distributions. Let $X_1, \ldots, X_m$ be independent observations from a continuous distribution F_X, and $Y_1, \ldots, Y_n$ be independent observations from a continuous distribution F_Y. Let W denote the sum of the ranks of the X observations in the combined ordered arrangement of the two samples. The range of W is given by

$$\frac{m(m+1)}{2} \leq W \leq \frac{m(2(m+n) - m + 1)}{2}.$$

The W can be used as a test statistic for testing

$$H_0 : F_X(x) = F_Y(x) \text{ for all } x \quad \text{vs.} \quad H_a : F_X(x) = F_Y(x - c) \text{ for some } c \neq 0.$$

The alternative hypothesis means that the distributions of X and Y are the same except for location. If W is too large or too small, then the null hypothesis will be rejected in favor of the alternative. If the hypotheses are

$$H_0 : F_X(x) = F_Y(x) \text{ for all } x \quad \text{vs.} \quad H_a : F_X(x) = F_Y(x - c) \text{ for some } c > 0,$$

then the alternative hypothesis implies that the distribution of Y is shifted to the left from the distribution of X by c. That is, X values are more likely to be larger than Y values. In this case, the null hypothesis will be rejected for larger values of W. If the hypotheses are

$$H_0 : F_X(x) = F_Y(x) \text{ for all } x \quad \text{vs.} \quad H_a : F_X(x) = F_Y(x+c) \quad \text{for some } c > 0,$$

then the alternative hypothesis implies that the distribution of Y is shifted to the right from the distribution of X by c. That is, X values are more likely to be smaller than Y values. In this case, the null hypothesis will be rejected for smaller values of W.

Let W_0 be an observed value of W and α be a specified level of significance. If $P(W \geq W_0) \leq \alpha$, then we conclude that $F_X(x) = F_Y(x-c)$, $c > 0$; if $P(W \leq W_0) \leq \alpha$, then we conclude that $H_a : F_X(x) = F_Y(x+c)$, $c > 0$. If $P(W \leq W_0) \leq \alpha/2$ or $P(W \geq W_0) \leq \alpha/2$, then we have evidence to conclude that $F_X(x) = F_Y(x-c)$, $c \neq 0$.

Mean:	$\frac{m(m+n+1)}{2}$
Variance:	$\frac{mn(m+n+1)}{12}$
	[Gibbons and Chakraborti, 1992, p. 241]

The variable

$$Z = \frac{W - \text{mean}}{\sqrt{\text{var}}} \sim N(0, 1) \quad \text{approximately.}$$

Mann–Whitney U Statistic

Wilcoxon rank-sum statistic and Mann–Whitney U statistic differ only by a constant, and, therefore, the test results based on these two statistics are the same. The Mann–Whitney U statistic is defined as follows: Let

$$D_{ij} = \begin{cases} 1 & \text{if } Y_j < X_i, \\ 0 & \text{otherwise}, \end{cases}$$

for $i = 1, \ldots, m$ and $j = 1, \ldots, n$. The statistic U is defined as

$$U = \sum_{i=1}^{m} \sum_{j=1}^{n} D_{ij},$$

or equivalently,

$$U = \sum_{i=1}^{n} (\text{number of } X's \text{ greater than the } Y_i).$$

It can be shown that

$$U = W - \frac{m(m+1)}{2},$$

and hence

$$P(U \leq u) = P\left(W \leq u + \frac{m(m+1)}{2}\right).$$

Calculation of Percentiles of the Wilcoxon Rank-Sum Statistic

The dialog box [StatCalc→Nonparametric →Wilcoxon Rank-Sum Test] computes probabilities and percentiles of the rank-sum statistic W.

To compute probabilities: Enter the values of m, n, and the observed value w; click on [P(W <= w)]. For example, when $m = 13$, $n = 12$, and the observed value w is 180, $P(W \leq 180) = 0.730877$ and $P(W \geq 180) = 0.287146$.

To compute the critical values: Enter the values of m, n, and the cumulative probability; click on [Critical Pt]. For example, when $m = 13$, $n = 12$, and the cumulative probability is 0.05, the left-tail critical value is 138. Because of the discreteness of the distribution, the actual probability is 0.048821; that is $P(W \leq 138) = 0.048821$.

Example 31.8. It is desired to compare two types of treatments, A and B, based on recovery times of patients. Given below are recovery times of a sample of 9 patients who received treatment A, and recovery times of a sample of 9 patients who received treatment B. We want to see whether the data provide sufficient evidence to indicate

Recovery times (in days)

A:	17	19	20	24	13	18	21	22	25
B:	14	18	19	23	16	15	13	22	16

that the recovery times of A are more likely longer than the recovery times of B. In notation,

$$H_0 : F_A(x) = F_B(x) \text{ for all } x, \quad \text{vs.} \quad H_a : F_A(x) = F_B(x - c) \quad \text{for some } c > 0.$$

Note that the alternative hypothesis implies that the values of the random variable associated with A are more likely to be larger than those associated with B. The pooled sample data and their ranks (the average of the ranks is assigned to tied observations) are as follow. The sum of the ranks of A in the pooled

Pooled sample and ranks

Treatment	B	A	B	B	B	B	A	A	B
data	13	13	14	15	16	16	17	18	18
ranks	1.5	1.5	3	4	5.5	5.5	7	8.5	8.5
Treatment	A	B	A	A	B	A	B	A	A
data	19	19	20	21	22	22	23	24	25
ranks	10.5	10.5	12	13	14.5	14.5	16	17	18

sample is 102. The null hypothesis will be rejected if this sum is too large. Using [StatCalc→Nonparametric →Wilcoxon Rank-Sum Test], we get

$$P(W \geq 102) = 0.0807,$$

which is less than 0.1. Hence, the null hypothesis will be rejected at the 10% level.

Remark 31.2. *StatCalc* also computes critical values; for the above example, to compute right-tail 5% critical point, enter 9 for m, 9 for n, and 0.95 for the cumulative probability; click [Critical Pt] to get 105. That is,

$$P(W \geq 105) = 0.04696.$$

Thus, for the above example, had the observed value of W been 105 or more, then we would have rejected the null hypothesis at the 5% level of significance.

31.5 Quantile Estimation and Nonparametric Tolerance Interval

Let $X_1, \ldots, X_n$ be a sample from a continuous distribution F_X. Let $X_{(k)}$ denote the kth order statistic. Let κ_p denote the pth quantile of F_X, and let $X_{(1)}, \ldots, X_{(n)}$ be the order statistics for the sample $X_1, \ldots, X_n$ from F_X. The sample pth quantile is the rth order statistic, where $r = np$ if np is an integer, and is $\lfloor np \rfloor$ otherwise, where $\lfloor x \rfloor$ is the largest integer not exceeding x (floor function). For example, $[5.9] = 5$ and $[5.1] = 5$. The order statistic $X_{(r)}$ is the sample pth quantile, which is a point estimate of κ_p. Point estimate is of limited use, and for practical applications one needs a confidence interval for κ_p.

To find a $1 - \alpha$ confidence interval for κ_p based on order statistics, we need to find the values of r and s, $r < s$, so that

$$
\begin{aligned}
P(X_{(r)} < \kappa_p < X_{(s)}) &= P(X_{(r)} < \kappa_p) - P(X_{(s)} < \kappa_p) \\
&= \sum_{i=r}^{s-1} \binom{n}{i} p^i [1-p]^{n-i} \quad \text{(because } F(\kappa_p) = p) \\
&= 1 - \alpha.
\end{aligned}
\tag{31.3}
$$

For a given confidence level $1-\alpha$, choose r and s so that $s-r$ small. Then $(X_{(r)}, X_{(s)})$ is a $1 - \alpha$ confidence interval for κ_p. For a given n, there may not exist r and s so that $P(X_{(r)} < \kappa_p < X_{(s)}) \geq 1 - \alpha$. However, the above expression is useful to find the percentile of the population contained in $(X_{(r)}, X_{(s)})$ when r, s, and $1 - \alpha$ are given.

Example 31.9. The following data represent lifetimes (in days) of a sample of 40 electronic components.

```
35.11, 38.45, 39.05, 39.83, 39.97, 40.98, 41.55, 41.72, 42.46, 44.41
45.49, 46.91, 47.90, 48.29, 48.48, 48.97, 49.24, 50.11, 50.17, 50.41
50.51, 51.18, 51.28, 51.46, 52.80, 53.25, 53.61, 53.69, 55.61, 55.92
56.08, 56.27, 57.82, 59.04, 60.24, 60.45, 60.46, 61.41, 65.98, 68.10
```

What percentile does the interval (39, 60) include with 95% confidence?

Solution: For this example, $r \simeq 3$ and $s \simeq 35$ so that $(X_{(3)}, X_{(35)}) \simeq (39, 60)$. We need to find the value of p so that

$$\sum_{i=3}^{34} \binom{40}{i} p^i [1-p]^{n-i} = .95.$$

The following R can be used to find the value of p.

R function 31.1. (Content estimate)

```
cont.in.rs = function(n, r, s, p){
u = pbinom(s-1,n,p)-pbinom(r-1,n,p)
return(u)
}
```

```
> cont.in.rs(40,3,35,.7)
[1] 0.9913819
> cont.in.rs(40,3,35,.8)
[1] 0.8386712
> cont.in.rs(40,3,35,.75)
[1] 0.956726
> cont.in.rs(40,3,35,.74)
[1] 0.9679278
> cont.in.rs(40,3,35,.755)
[1] 0.9499561
```

So, the 75.5th percentile of the lifetime distribution is somewhere between 39 and 60 days with confidence 95%.

Nonparametric Tolerance Intervals

A p-content $-(1-\alpha)$ coverage nonparametric tolerance interval is formed by a pair of order statistics $X_{(r)}$ and $X_{(s)}$, where the integers $1 \leq r < s \leq n$ should be determined so that

$$P_{X_{(r)}, X_{(s)}} \left\{ F(X_{(s)}) - F(X_{(r)}) \geq p \right\} = 1 - \alpha, \tag{31.4}$$

or equivalently,

$$P_{U_{(r)}, U_{(s)}} \left\{ U_{(s)} - U_{(r)} \geq p \right\} = P(X \leq s - r - 1 | n, p), \tag{31.5}$$

where $U_{(m)}$, the mth order statistic for a sample of size n from a uniform$(0, 1)$ distribution, and $X \sim$ binomial(n, p). Notice that the above probability depends only on the difference $s - r$ not the actual values of s and r. Let $k = s - r$ is the least value for which

$$P(X \leq s - r - 1) \geq 1 - \alpha. \tag{31.6}$$

Then any interval $(X_{(l)}, X_{(m)})$ is a $(p, 1-\alpha)$ tolerance interval provided $1 \leq l < m \leq n$ and $m - l = k$. It is customary to take $s = n - r + 1$, so that $(X_{(r)}, X_{(n-r+1)})$ is a $(p, 1-\alpha)$ tolerance interval. However, we note that we can find intervals of the form $(X_{(r)}, X_{(n-r)})$, which is shorter than $(X_{(r)}, X_{(n-r+1)})$, in many cases. Specifically, for a given m, let n be the least sample size that satisfies

$$P(X \leq n - m - 1) \geq 1 - \alpha. \tag{31.7}$$

Then $(X_{(m)}, X_{(n)})$, $\left(X_{\left(\frac{m}{2}\right)}, X_{\left(n-\frac{m}{2}\right)} \right)$ if m is even, and $\left(X_{\left(\frac{m+1}{2}\right)}, X_{\left(n-\frac{m+1}{2}+1\right)} \right)$ if m is odd, are $(p, 1-\alpha)$ tolerance intervals.

Sample Size for Computing Tolerance Intervals

The required sample size n so that the smallest order statistic and the largest order statistic form a $(p, 1 - \alpha)$ tolerance interval is the smallest integer that satisfies

$$(n - 1)p^n - np^{n-1} + 1 \geq 1 - \alpha. \tag{31.8}$$

For a one-sided limit, the value of the n is the smallest integer such that

$$1 - p^n \geq 1 - \alpha. \tag{31.9}$$

For a one-sided limit, the sample size is determined so that at least proportion p of the population data are greater than or equal to $X_{(1)}$; furthermore, at least proportion p of the population data are less than or equal to $X_{(n)}$.

Because of the discreteness of the sample size, the true coverage probability will be slightly more than the specified probability g. For example, when $p = 0.90$ and $1 - \alpha = 0.90$, the required sample size for the two-sided tolerance limits is 38. Substituting 38 for n, 0.90 for p in (31.8), we get 0.9047, which is the actual coverage probability.

For a given value of $(p, 1-\alpha = g, m)$, the dialog box [StatCalc→Nonparametric→ Sample Size for NP Tolerance Interval] computes the required sample size so that

$$(p, 1 - \alpha) \text{ tolerance interval} = \begin{cases} \left(X_{(m)}, X_{(n)}\right), & \\ \left(X_{\left(\frac{m}{2}\right)}, X_{\left(n - \frac{m}{2}\right)}\right), & \text{if } m \text{ is even,} \\ \left(X_{\left(\frac{m+1}{2}\right)}, X_{\left(n - \frac{m+1}{2} + 1\right)}\right), & \text{if } m \text{ is odd.} \end{cases}$$

Example 31.10. When $p = 0.90$ and $g = 0.95$, the value of n is 46; that is, the interval $(X_{(1)}, X_{(46)})$ would contain at least 90% of the population with confidence 95%. Furthermore, the sample size required for a one-sided tolerance limit is 29; that is, 90% of the population data are less than or equal to $X_{(29)}$ — the largest observation in a sample of 29 observations, with confidence 95%. Similarly, 90% of the population data are greater than or equal to $X_{(1)}$ — the smallest observation in a sample of 29 observations, with confidence 95%.

Example 31.11. Suppose that one desires to find a tolerance interval so that it would contain 99% of household incomes in a large city with coverage probability 0.95. How large should the sample be so that the interval (smallest order statistic, largest order statistic) would contain at least 99% of the household incomes with confidence 95%?

To compute the required sample size, enter 0.99 for p, 0.95 for g, and click [Required Sample Size] to get 473. That is, if we take a random sample of 473 households from the city and record their incomes, then at least 99% of the household incomes in the city fall between the lowest and the highest household incomes in the sample with 95% confidence. For one-sided limits, the required sample size is 299. That is, if we take a sample of 299 households from the city and record the incomes, then at least 99% of the household incomes are greater than the smallest income in the sample; further, at least 99% of the household incomes are less than the highest income in the sample.

Example 31.12. Suppose it is desired to find a sample size so that the smallest third order statistic and the largest order statistic form a $(.90, .95)$ tolerance interval.

To find the desired sample size enter (.90, .95, 3) for (p, g, m) and click on [sample size TI] to get 76. That is, the interval $(X_{(3)}, X_{(76)})$ from a sample of size 76 form a (.90, .95) tolerance interval. Furthermore, because $m = 3$ is odd, $(X_{(2)}, X_{(75)})$ is also a (.90, .95) tolerance interval.

References

Abramowitz, M. and Stegun, I. A. (1965). *Handbook of Mathematical Functions.* Dover Publications, New York.

Agresti A. and Coull, B. A. (1998). Approximate is better than "exact" for interval estimation of binomial proportion. *The American Statistician,* 52, 119–125.

Aki, S. and Hirano, K. (1996). Lifetime distribution and estimation problems of consecutive–k–out–of–n: F systems. *Annals of the Institute of Statistical Mathematics,* 48, 185–199.

Anderson T. W. (1984). *An Introduction to Multivariate Statistical Analysis.* Hoboken, NJ: John Wiley & Sons.

Andrés, A. M. and Hernández, M. A. (2011). Inferences about a linear combination of proportions. *Statistical Methods in Medical Research,* 20, 369–387.

Atteia, O. and Kozel, R. (1997). Particle size distributions in waters from a karstic aquifer: From particles to colloids. *Journal of Hydrology,* 201, 102–119.

Bain, L. J. (1969). Moments of noncentral t and noncentral F distributions. *American Statistician,* 23, 33–34.

Bain, L. J. and Engelhardt, M. (1973). Interval estimation for the two-parameter double exponential distribution. *Technometrics,* 15, 875–887.

Bain, L. J. and Patel, J. K. (1993). Prediction intervals based on partial observations for some discrete distributions. *IEEE Transactions on Reliability,* 42, 459–463.

Balakrishnan, N. (ed.) (1991). *Handbook of the Logistic Distribution.* New York: Marcel Dekker.

Bedrick, E.J. (1987). A family of confidence intervals for the ratio of two binomial proportions. *Biometrics,* 43, 993–998.

Bhaumik, D. K. and Gibbons, R. D. (2006). One-sided approximate prediction intervals for at least p of m observations from a gamma population at each of r locations. *Technometrics,* 48, 112–119.

Bhaumik, D. K., Kapur, K., and Gibbons, R. D. (2009). Testing parameters of a gamma distribution for small samples. *Technometrics,* 51, 326–334.

Belzer, D. B. and Kellogg, M. A. (1993). Incorporating sources of uncertainty in forecasting peak power loads—a Monte Carlo analysis using the extreme value distribution. *IEEE Transactions on Power Systems,* 8, 730–737.

Benton, D. and Krishnamoorthy, K. (2003). Computing discrete mixtures of continuous distributions: Noncentral chi-square, noncentral t and the distribution of the square of the sample multiple correlation coefficient. *Computational Statistics and Data Analysis,* 43, 249–267.

Benton, D., Krishnamoorthy, K., and Mathew, T. (2003). Inferences in multivariate–univariate calibration problems. *The Statistician* (JRSS-D), 52, 15–39.

Blaesild, P. and Granfeldt, J. (2003). *Statistics with Applications in Biology and Geology*. Boca Raton, FL: Chapman & Hall/CRC.

Borjanovic, S. S., Djordjevic, S. V., and Vukovic-Pal, M. D. (1999). A method for evaluating exposure to nitrous oxides by application of lognormal distribution. *Journal of Occupational Health*, 41, 27–32.

Bortkiewicz, L. von (1898). *Das Gesetz der Kleinen Zahlen*. Leipzig: Teubner.

Braselton, J., Rafter, J., Humphrey, P. and Abell, M. (1999). Randomly walking through Wall Street comparing lump-sum versus dollar-cost average investment strategies. *Mathematics and Computers in Simulation*, 49, 297–318.

Brown L. D., Cai T., and Das Gupta A. (2001). Interval estimation for a binomial proportion (with discussion). *Statistical Science*, 16, 101-133.

Burlaga, L. F. and Lazarus A. J. (2000). Lognormal distributions and spectra of solar wind plasma fluctuations: Wind 1995–1998. *Journal of Geophysical Research-Space Physics*, 105, 2357–2364.

Burr, I. W. (1973). Some approximate relations between the terms of the hypergeometric, binomial and Poisson distributions. *Communications in Statistics–Theory and Methods*, 1, 297–301.

Burstein, H. (1975). Finite population correction for binomial confidence limits. *Journal of the American Statistical Association,* 70, 67–69.

Cacoullos, T. (1965). A relation between t and F distributions. *Journal of the American Statistical Association*, 60, 528–531.

Cannarozzo, M., Dasaro, F., and Ferro, V. (1995). Regional rainfall and flood frequency-analysis for Sicily using the 2–component extreme-value distribution. *Hydrological Sciences Journal-Journal des Sciences Hydrologiques*, 40, 19–42.

Chapman, D. G. (1952). On tests and estimates of the ratio of Poisson means. *Annals of the Institute of Statistical Mathematics*, 4, 45–49.

Chatfield, C., Ehrenberg, A. S. C. and Goodhardt, G. J. (1966). Progress on a simplified model of stationary purchasing behaviour. *Journal of the Royal Statistical Society, Series A*, 129, 317–367.

Chattamvelli, R. and Shanmugam, R. (1997). Computing the noncentral beta distribution function. *Applied Statistics*, 46, 146–156.

Cheng, R. C. H. (1978). Generating beta variates with nonintegral shape parameters. *Communications ACM*, 21, 317–322.

Chhikara, R. S. and Folks, J. L. (1989). *The Inverse Gaussian Distribution*. New York: Marcel Dekker.

Chia, E. and Hutchinson, M. F. (1991). The beta distribution as a probability model for daily cloud duration. *Agricultural and Forest Meteorology*, 56, 195–208.

Childs, A. and Balakrishnan, N. (1997). Maximum likelihood estimation of laplace parameters based on general type-II censored samples. *Statistical Papers*, 38, 343–349.

Clopper, C. J. and Pearsons E. S. (1934). The use of confidence or fiducial limits illustrated in the case of the binomial. *Biometrika*, 26, 404–413.

Cohen, A. C. (1961). Tables for maximum likelihood estimates: Singly truncated and singly censored samples. *Technometrics*, 3, 535–541.

Cohen, L. A., Kendall, M. E., Zang, E., Meschter, C., and Rose, D. P. (1991). Modulation of N-Nitrosomethylurea-Induced mammary tumor promotion by dietary fibre and fat. *Journal of National Cancer Institution*, 83, 496–501.

Cornfield, J. A. (1956). Statistical problem arising from retrospective studies. In Neyman, J. (ed.), *Proceedings of the Third Berkeley Symposium on Mathematical Statistics and Probability*, 4, 135–148.

Craig, C. C. (1941). Note on the distribution of noncentral t with an application. *Annals of Mathematical Statistics*, 17, 193–194.

Crow, E. L. and Shimizu, K. (eds.) (1988). *Lognormal Distribution: Theory and Applications*. New York: Marcel Dekker.

Daniel, W. W. (1990). *Applied Nonparametric Statistics*. PWS-KENT Publishing Company, Boston.

Davis, C. B. and McNichols, R. T. (1987). One-sided intervals for at least p of m observations from a normal population on each of r future occasions. *Technometrics*, 29, 359–370.

Das, S. C. (1955). Fitting truncated type III curves to rainfall data. *Australian Journal of Physics*, 8, 298–304.

de Visser, C. L. M. and van den Berg, W. (1998). A method to calculate the size distribution of onions and its use in an onion growth model. *Sciatica Horticulturae*, 77, 129–143.

Dobson, A. J., Kuulasmmaa, K., Ebrerle, E., and Scherer, J. (1991). Confidence intervals for weighted sums of Poisson parameters. *Statistics in Medicine*, 10, 457–462.

Duncan, A. J. (1974). *Quality Control and Industrial Statistics*, Homewood, IL: Irwin.

Forkman, J. (2009). Estimator and tests for common coefficients of variation in normal distributions. *Communications in Statistics—Theory and Methods*, 38, 233–251.

Fraser, D. A. S., Reid, N., and Wong A. (1997). Simple and accurate inference for the mean of the gamma model. *The Canadian Journal of Statistics* 25, 91–99.

Garcia, A., Torres, J. L., Prieto, E., and De Francisco, A. (1998). Fitting wind speed distributions: A case study. *Solar Energy*, 62, 139–144.

Gibbons, R. D. (1994). *Statistical Methods for Groundwater Monitoring*, Hoboken, NJ: John Wiley & Sons.

Gibbons, J. D. and Chakraborti, S. (1992). *Nonparametric Statistical Inference*. New York: Marcel Dekker.

Graybill, F. A. and Wang, C.M. (1980). Confidence intervals on nonnegative linear combinations of variances. *Journal of the American Statistical Association*, 75, 869–873.

Graham, P. L., Mengersen, K., and Mortan, A. P. (2003). Confidence limits for the ratio of two rates based on likelihood scores: Non-iterative method. *Statistics in Medicine*, 22, 2071–2083.

Grice, J. V. and Bain, L. J. (1980). Inferences concerning the mean of the gamma distribution. *Journal of the American Statistical Association*, 75, 929–933.

Gross, A. J. and Clark, V. A. (1975). *Survival Distributions: Reliability Applications in the Biomedical Services*. Hoboken, NJ: John Wiley & Sons.

Grubbs, F. E. (1971). Approximate fiducial bounds on reliability for the two parameter negative exponential distribution. *Technometrics*, 13, 873–876

Guenther, W. C. (1969). Shortest confidence intervals. *American Statistician*, 23, 22–25.

Guenther, W. C. (1971). Unbiased confidence intervals. *American Statistician*, 25, 18–20.

Guenther, W. C. (1978). Evaluation of probabilities for noncentral distributions and the difference of two t-variables with a desk calculator. *Journal of Statistical Computation and Simulation*, 6, 199–206.

Guenther, W. C., Patil, S. A., and Uppuluri, V. R. R. (1976). One-sided β-content tolerance factors for the two parameter exponential distribution. *Technometrics*, 18, 333–340.

Gumbel, E. J. and Mustafi, C. K. (1967). Some analytical properties of bivariate extremal distributions. *Journal of the American Statistical Association*, 62, 569–588.

Haff, L. R. (1979). An identity for the Wishart distribution with applications. *Journal of Multivariate Analysis*, 9, 531–544.

Hahn, G. J. and Chandra, R. (1981). Tolerance intervals for Poisson and binomial variables. *Journal of Quality Technology*, 13, 100–110.

Hannig, J., Iyer, H. K. and Patterson, P. (2006). Fiducial generalized confidence intervals. *Journal of the American Statistical Association*, 101, 254–269.

Harley, B. I. (1957). Relation between the distributions of noncentral t and a transformed correlation coefficient. *Biometrika*, 44, 219–224.

Hart, J.F., Cheney, E. W., Lawson, C. L., Maehly, H. J., Mesztenyi, H. J., Rice, J. R., Thacher, Jr., H. G., and Witzgall, C. (1968). *Computer Approximations*. Hoboken, NJ: John Wiley & Sons.

Harter, H. L. and Moore, A. H. (1967). Maximum-likelihood estimation, from censored samples, of the parameters of logistic distribution. *Journal of the American Statistical Association* 62, 675–684.

Herrington, P. D. (1995). Stress-strength interference theory for a pin-loaded composite joint. *Composites Engineering*, 5, 975–982.

Hotelling, H (1953). New light on the correlation coefficient and its transforms. *Journal of the Royal Statistical Society B*, 15, 193–232.

Hwang, T. J. (1982). Improving on standard estimators in discrete exponential families with application to Poisson and negative binomial case. *The Annals of Statistics*, 10, 868–881.

Jaech, J. L. (1970). Comparing two methods of obtaining a confidence interval for the ratio of Poisson parameters. *Technometrics*, 12, 383–387.

Johnson, D. (1997). The triangular distribution as a proxy for the beta distribution in risk analysis. *Statistician*, 46, 387–398.

Jöhnk, M. D. (1964). Erzeugung von Betaverteilter und Gammaverteilter Zufallszahlen. *Metrika*, 8, 5–15.

Johnson, N. L. and Kotz, S. (1970). *Continuous univariate distributions—2*. New York: Houghton Mifflin Company.

Johnson, N. L., Kotz, S., and Kemp, A. W. (1992). *Univariate Discrete Distributions*. Hoboken, NJ: John Wiley & Sons.

Johnson, N. L., Kotz, S., and Balakrishnan, N. (1994). *Continuous Univariate Distributions*. Hoboken, NJ: John Wiley & Sons.

Johnson, N. L. and Welch, B. L. (1940). Application of the noncentral t-distribution. *Biometrika*, 31, 362–389.

Jones, G. R. and Jackson M., and O'Grady, K. (1999). Determination of grain size distributions in thin films. *Journal of Magnetism and Magnetic Materials*, 193, 75–78.

Jonhk, M. D. (1964). Erzeugung von Betaverteilter und Gammaverteilter Zufallszahlen. *Metrika*, 8, 5–15.

Kachitvichyanukul, V. and Schmeiser, B. (1985). Computer generation of hypergeometric random variates. *Journal of Statistical Computation and Simulation*, 22, 127–145.

Kachitvichyanukul, V. and Schmeiser, B. (1988). Binomial random variate generation. *Communications of the ACM*, 31, 216–222.

Kagan, Y. Y. (1992). Correlations of earthquake focal mechanisms. *Geophysical Journal International*, 110, 305–320.

Kamat, A. R. (1965). Incomplete and absolute moments of some discrete distributions, classical and contagious discrete distributions, 45–64. Oxford: Pergamon Press.

Kappenman, R. F. (1977). Tolerance Intervals for the double-exponential distribution. *Journal of the American Statistical Association*, 72, 908–909.

Karim, M. A. and Chowdhury, J. U. (1995). A comparison of four distributions used in flood frequency-analysis in Bangladesh. *Hydrological Sciences Journal-Journal des Siences Hydrologiques*, 40, 55–66.

Keating, J. P., Glaser, R. E., and Ketchum, N. S. (1990). Testing hypotheses about the shape parameter of a gamma distribution. *Technometrics*, 32, 67–82.

Kendall, M. G. (1943). *The Advance Theory of Statistics*, Vol. 1. London: Griffin.

Kendall, M. G. and Stuart, A. (1958). *The Advanced Theory of Statistics*, Vol. 1. New York: Hafner Publishing Company.

Kendall, M. G. and Stuart, A. (1973). *The Advanced Theory of Statistics*, Vol. 2. New York: Hafner Publishing Company.

Kennedy Jr. W. J. and Gentle, J. E. (1980). *Statistical Computing*. New York: Marcel Dekker.

Kinderman, A. J. and Ramage, J. G. (1976). Computer generation of normal random variates. *Journal of the American Statistical Association*, 71, 893–896.

Kobayashi, T., Shinagawa, N., and Watanabe, Y. (1999). Vehicle mobility characterization based on measurements and its application to cellular communication systems. *IEICE transactions on communications*, E82B, 2055–2060.

Korteoja, M., Salminen, L. I., Niskanen, K. J., and Alava, M. J. (1998). Strength distribution in paper. *Materials Science and Engineering a Structural Materials Properties Microstructure and Processing*, 248, 173–180.

Kramer, K. H. (1963). Tables for constructing confidence limits on the multiple correlation coefficient. *Journal of the American Statistical Association*, 58, 1082–1085.

Krishnamoorthy, K. and Mathew, T. (1999). Comparison of approximate methods for computing tolerance factors for a multivariate normal population. *Technometrics*, 41, 234–249.

Krishnamoorthy, K., Kulkarni, P., and Mathew, T. (2001). Hypothesis testing in calibration. *Journal of Statistical Planning and Inference*, 93, 211–223.

Krishnamoorthy, K., Mathew, T., and Mukherjee, S. (2008). Normal based methods for a gamma distribution: Prediction and tolerance interval and stress-strength reliability. *Technometrics*, 50, 69–78

Krishnamoorthy, K., Mallick, A., and Mathew, T. (2009). Model based imputation approach for data analysis in the presence of non-detectable values: Normal and related distributions. *Annals of Occupational Hygiene*, 59, 249–68.

Krishnamoorthy, K., and Lin, Y. (2010). Confidence limits for stressstrength reliability involving Weibull models. *Journal of Statistical Planning and Inference*, 140, 1754-1764.

Krishnamoorthy, K., Lin, Y., and Xia, Y. (2009). Confidence limits and prediction limits for a Weibull distribution based on the generalized variable approach. *Journal of Statistical Planning and Inference*, 139, 2675–2684.

Krishnamoorthy, K. and Lee, M. (2010). Inference for functions of parameters in discrete distributions based on fiducial approach: Binomial and Poisson cases. *Journal of Statistical Planning and Inference*, 140, 1182–1192.

Krishnamoorthy, K. and Xu, Z. (2011). Confidence limits for lognormal percentiles and for lognormal mean based on samples with multiple detection limits. *Annals of Occupational Hygiene*, 55, 495–509.

Krishnamoorthy, K. and Xie, F. (2011). Tolerance intervals for symmetric location-scale distributions based on censored or uncensored data. *Journal of Statistical Planning Inference*, 141, 1170–1182.

Krishnamoorthy, K. and Lee, M. (2013). New approximate confidence intervals for the difference between two Poisson means and comparison. *Journal of Statistical Computation and Simulation*, 83, 2232–2243.

Krishnamoorthy, K. and Lee, M. (2013). Improved tests for the equality of normal coefficients of variation. *Computational Statistics*, 29, 215–232.

Krishnamoorthy, K., Lee, M., and Zhang, D. (2014). Closed-form fiducial confidence intervals for some functions of independent binomial parameters with comparisons. *Statistical Methods in Medical Research*. DOI: 10.1177/0962280214537809

Krishnamoorthy, K., Lee, M., and Wang, X. (2015). Likelihood ratio tests for comparing several gamma distributions. To appear in *Environmetrics*.

Krishnamoorthy, K. and Zhang, D. (2015). Approximate and fiducial confidence intervals for the difference between two binomial proportions. *Communications in Statistics — Theory and Methods*, 44, 1745–1759.

Krishnamoorthy, K. and Thomson, J. (2002). Hypothesis testing about proportions in two finite populations. *The American Statistician*, 56, 215–222.

Krishnamoorthy, K. and Mathew, T. (2003). Inferences on the means of lognormal distributions using generalized p-values and generalized confidence intervals. *Journal of Statistical Planning and Inference*, 115, 103–121.

Krishnamoorthy, K. and Mathew, T. (2004). One-sided tolerance limits in balanced and unbalanced one-way random models based on generalized confidence limits. *Technometrics*, (2004), 46, 44–52.

Krishnamoorthy, K. and Peng, J. (2007). Some properties of the exact and score methods for a binomial proportion and sample size calculation. *Communications in Statistics—Simulation and Computation*, 36, 1171-1186.

Krishnamoorthy, K. and Peng, J. (2008). Exact properties of a new test and other tests for differences between several binomial proportions. *Journal of Applied Statistical Science*, 16, 23-35.

Krishnamoorthy, K. and Mathew, T. (2009). *Statistical Tolerance Regions: Theory, Applications and Computation.* Hoboken, NJ: John Wiley & Sons.

Krishnamoorthy, K. and Peng, J. (2011). Improved closed-form prediction intervals for a binomial and Poisson distributions. *Journal of Statistical Planning and Inference*, 141, 1709–1718.

Krishnamoorthy, K. and Thomson, J. (2004). A more powerful test for comparing two Poisson means. *Journal of Statistical Planning and Inference*, 119, 23–35.

Krishnamoorthy, K. and Tian, L. (2008). Inference on the difference and ratio of the means of two inverse Gaussian distributions. *Journal of Statistical Planning and Inference*, 138, 2082–2089.

Krishnamoorthy, K. and Xia, Y. (2007). Inferences on correlation coefficients: One-sample, independent and correlated cases. *Journal of Statistical Planning and Inferences*, 137, 2362–2379.

Krishnamoorthy, K., Xia, Y., and Xie, F. (2011). A simple approximate procedure for constructing binomial and Poisson tolerance intervals. *Communications in Statistics-Theory and Methods*, 40, 2443–2458.

Krishnamoorthy, K. (2013). Comparison of confidence intervals for correlation coefficients based on incomplete monotone samples and those based on listwise deletion. *Journal of Multivariate Analysis*, 114, 378–388.

Krishnamoorthy, K. and Xia, Y. (2014). Simple closed-form confidence intervals for two-parameter exponential distributions: One- and two-sample cases. Submitted for publication.

Krishnamoorthy, K. and Novelo, L. (2014). Small sample inference for gamma parameters: One-sample and two-sample problems. *Environmetrics*, 25, 107–126.

Krishnamoorthy, K. (2014). Modified normal-based approximation for the percentiles of a linear combination of independent random variables with applications. *Communications in Statistics—Simulation and Computation.* DOI:10.1080/03610918.2014.904342

Kuchenhoff, H. and Thamerus, M. (1996). Extreme value analysis of Munich air pollution data. *Environmental and Ecological Statistics*, 3, 127–141.

Lawless, J. F. (1977). Prediction Intervals for the Two Parameter Exponential Distribution. *Technomterics*, 19, 469–472.

Lawless, J. F. (1982). *Statistical Models and Methods for Lifetime Data.* Hoboken, NJ: John Wiley & Sons.

Lawless, J. F. (2003). *Statistical Models and Methods for Lifetime Data.* Hoboken, NJ: John Wiley & Sons.

Lawson, L. R. and Chen, E. Y. (1999). Fatigue crack coalescence in discontinuously reinforced metal matrix composites: Implications for reliability prediction. *Journal of Composites Technology & Research*, 21, 147–152.

Lenth, R. V. (1989). Cumulative distribution function of the noncentral t distribution. *Applied Statistics*, 38, 185–189.

Li, H.-Q., Tang, M.-L., Poon, W.-Y., and Tang, N.-S. (2011). Confidence intervals for difference between two Poisson rates. *Communications in Statistics—Simulation and Computation*, 40, 1478–1491.

Longin, F. M. (1999). Optimal margin level in futures markets: Extreme price movements. *Journal of Futures Markets*, 19, 127–152.

Looney, S. W., and Gulledge Jr., T. R. (1985). Use of the correlation coefficient with normal probability plots. *American Statistician*, 39, 75–79.

Mantra, A. and Gibbins, C. J. (1999). Modeling of raindrop size distributions from multiwavelength rain attenuation measurements. *Radio Science*, 34, 657–666.

Majumder, K. L. and Bhattacharjee, G. P. (1973a). Algorithm AS 63. The incomplete beta integral. *Applied Statistics*, 22, 409–411.

Majumder, K. L. and Bhattacharjee, G. P. (1973b). Algorithm AS 64. Inverse of the incomplete beta function ratio. *Applied Statistics*, 22, 412–415.

Mathew, T. and Zha, W. (1996). Conservative confidence regions in multivariate calibration. *The Annals of Statistics*, 24, 707–725.

Maxim, L. D., Galvin, J. B., Niebo, R., Segrave, A. M., Kampa, O. A., and Utell, M. J. (2006). Occupational exposure to carbon/coke fibers in plants that produce green or calcined petroleum coke and potential health effects: Fiber concentrations. *Inhalation Toxicology*, 18, 17–32.

Menon, M. V. (1963). Estimation of the shape and scale parameters of the Weibull distribution. *Technometrics*, 5, 175–182.

Min, I. A., Mezic, I., and Leonard, A. (1996). Levy stable distributions for velocity and velocity difference in systems of vortex elements. *Physics of Fluids*, 8, 1169–1180.

Miettinen, O. and Nurminen, M. (1985). Comparative analysis of two rates. *Statistics in Medicine*, 4, 213–226.

Montgomery, D. C. (1996). *Introduction to Statistical Quality Control*, Hoboken, NJ: John Wiley & Sons.

Moser, B. K. and Stevens, G. R. (1992). Homogeneity of variance in the two-sample means test. *The American Statistician*, 46, 19–21.

Muirhead, R. J. (1982). *Aspects of Multivariate Statistical Theory*. Hoboken, NJ: John Wiley & Sons.

Nabe, M., Murata, M., and Miyahara, H. (1998). Analysis and modeling of world wide web traffic for capacity dimensioning of internet access lines. *Performance Evaluation*, 34, 249–271.

Nicas, M. (1994). Modeling respirator penetration values with the beta distribution—an application to occupational tuberculosis transmission. *American Industrial Hygiene Association Journal*, 55, 515–524.

Nieuwenhuijsen, M. J. (1997). Exposure assessment in occupational epidemiology: Measuring present exposures with an example of a study of occupational asthma. *International Archives of Occupational and Environmental Health*, 70, 295–308.

Odeh, R. E., Owen, D. B., Birnbaum, Z. W., and Fisher, L. (1977). *Pocket Book of Statistical Tables*. New York: Marcel Dekker.

Olkin, I. and Pratt, J. (1958). Unbiased estimation of certain correlation coefficients. *The Annals of Mathematical Statistics*, 29, 201–211.

Onoz, B. and Bayazit, M. (1995). Best-fit distributions of largest available flood samples. *Journal of Hydrology*, 167, 195–208.

Oguamanam, D. C. D., Martin, H. R., and Huissoon, J. P. (1995). On the application of the beta distribution to gear damage analysis. *Applied Acoustics*, 45, 247–261.

Owen, D. B. (1964). Control of percentages in both tails of the normal distribution. *Technometrics*, 6, 377–387.

Owen, D. B. (1968). A survey of properties and application of the noncentral t distribution. *Technometrics*, 10, 445–478.

Pareto, V. (1897). Cours dEconomie Politique, .2. F. Rouge, Lausanne, Switzerland.

Parsons, B. L., and Lal, M. (1991). Distribution parameters for flexural strength of ice. *Cold Regions Science and Technology*, 19, 285–293.

Patel J. K., Kapadia, C. H., and Owen, D. B. (1976). *Handbook of Statistical Distributions*. New York: Marcel Dekker.

Patel, J. K. and Read, C. B. (1981). *Handbook of the Normal Distribution*. New York: Marcel Dekker.

Patil, G. P. (1962). Some methods of estimation for the logarithmic series distribution. *Biometrics*, 18, 68–75.

Patil G. P. and Bildikar, S. (1966). On minimum variance unbiased estimation for the logarithmic series distribution. *Sankhya, Ser. A*, 28, 239–250.

Patnaik, P. B. (1949). The noncentral chi-square and F-Distributions and their Applications. *Biometrika*, 36, 202–232.

Peizer, D. B. and Pratt, J. W. (1968). A normal approximation for binomial, F, beta, and other common related tail probabilities. *Journal of the American Statistical Association*, 63, 1416–1483.

Pitman, E. J. G. (1939). A note on normal correlation. *Biometrika*, 39, 9–12.

Press, W. H., Teukolsky, S. A., Vetterling, W. T., and Flannery, B. P. (1997). *Numerical Recipes in C*. Cambridge University Press.

Prince, J. R., Mumma, C. G., and Kouvelis, A. (1988). Applications of the logistic distribution to film sensitometry in nuclear-medicine. *Journal of Nuclear Medicine*, 29, 273–273.

Prochaska, B. J. (1973). A note on the relationship between the geometric and exponential distributions. *The American Statistician*, 27, 27.

Proschan, F. (1963). Theoretical explanation of observed decreasing failure rate. *Technometrics*, 5, 375–383.

Puig, P. and Stephens, M. A. (2000). Tests of fit for the Laplace distribution, with applications. *Technometrics*, 42, 417–424.

Puri, P. S. (1973). On a property of exponential and geometric distributions and its relevance to multivariate failure rate. *Sankhya A*, 35, 61–78.

Rajkai, K., Kabos, S., VanGenuchten, M. T., and Jansson, P. E. (1996). Estimation of water-retention characteristics from the bulk density and particle-size distribution of Swedish soils. *Soil Science*, 161, 832–845.

Rutherford, E. and Geiger, H. (1910). The probability variations in the distribution of α particles. *Philosophical Magazine*, 20, 698–704.

Schmeiser, B. W. and Lal, L. (1980). Squeeze methods for generating gamma variates. *Journal of the American Statistical Association*, 75, 679–682.

Schmeiser, B. W. and Shalaby, M. A. (1980). Acceptance/Rejection methods for beta variate generation. *Journal of the American Statistical Association*, 75, 673–678.

Sahli, A., Trecourt, P., and Robin, S. (1997). One sided acceptance sampling by variables: The case of the Laplace distribution. *Communications in Statistics-Theory and Methods*, 26, 2817–2834.

Saltzman, B. E. (1997). Health risk assessment of fluctuating concentrations using lognormal models. *Journal of the Air & Waste Management Association*, 47, 1152–1160.

Sato, T. (1990). Confidence intervals for effect parameters common in cancer epidemiology. *Environmental Health Perspectives*, 87, 95–101.

Scerri, E. and Farrugia, R. (1996). Wind data evaluation in the Maltese Islands. *Renewable Energy*, 7, 109–114.

Schmee, J., Gladstein, D., and Nelson,W. (1985). Confidence limits of a normal distribution from singly censored samples using maximum likelihood. *Technometrics*, 27, 119–128.

Schmee, J. and Nelson, W.B. (1977). Estimates and approximate confidence limits for (log) normal life distributions from singly censored samples by maximum likelihood. General Electric Co. Corp. Research and Development TIS Report 76CRD250.

Schwarzenberg-Czerny, A. (1997). The correct probability distribution for the phase dispersion minimization periodogram. *Astrophysical Journal*, 489, 941–945.

Schulz, T. W. and Griffin, S. (1999). Estimating risk assessment exposure point concentrations when the data are not normal or lognormal. *Risk Analysis*, 19, 577–584.

Sharma, P., Khare, M., and Chakrabarti, S. P. (1999) Application of extreme value theory for predicting violations of air quality standards for an urban road intersection. *Transportation Research Part D–Transport and Environment*, 4, 201–216.

Shivanagaraju, C., Mahanty, B., Vizayakumar, K., and Mohapatra, P. K. J. (1998). Beta-distributed age in manpower planning models. *Applied Mathematical Modeling*, 22, 23–37.

Shyu, J. C. and Owen, D. B. (1986). One-sided tolerance intervals for the two-parameter double exponential distribution. *Communications in Statistics-Simulation and Computation*, 15, 101–119.

Simpson, J. (1972). Use of the gamma distribution in single-cloud rainfall analysis. *Monthly Weather Review*, 100, 309–312.

Sivapalan, M. and Bloschl, G. (1998). Transformation of point rainfall to areal rainfall: Intensity-duration frequency curves. *Journal of Hydrology*, 204, 150–167.

Smith, W. B. and Hocking, R. R. (1972). Wishart variates generator, Algorithm AS 53. *Applied Statistics*, 21, 341–345.

Stapf, S., Kimmich, R., Seitter, R. O., Maklakov, A. I., and Skid, V. D. (1996). Proton and deuteron field-cycling NMR relaxometry of liquids confined in porous glasses. *Colloids and Surfaces: A Physicochemical and Engineering Aspects*, 115, 107–114.

Stein, C. (1981). Estimation of the mean of a multivariate normal distribution. *The Annals of Statistics*, 9, 1135–1151.

Stevens, W. L. (1950). Fiducial limits of the parameter of a discontinuous distribution. *Biometrika*, 37, 117–129.

Storer, B. E., and Kim, C. (1990). Exact properties of some exact test statistics comparing two binomial proportions. *Journal of the American Statistical Association*, 85, 146—155.

Stephenson, D. B., Kumar, K. R., Doblas-Reyes, F. J., Royer, J. F., Chauvin, E., and Pezzulli, S. (1999). Extreme daily rainfall events and their impact on ensemble forecasts of the Indian monsoon. *Monthly Weather Review*, 127, 1954–1966.

Sulaiman, M. Y., Oo, W. H., Abd Wahab, M., and Zakaria, A. (1999). Application of beta distribution model to Malaysian sunshine data. *Renewable Energy*, 18, 573–579.

Taraldsen, G. and Lindqvist, B. (2005). The multiple roots simulation algorithm, the inverse Gaussian distribution, and the sufficient conditional Monte Carlo method. Preprint Statistics No. 4/2005, Norwegian University of Science and Technology, Trondheim, Norway.

Thatcher, A. R. (1964). Relationships between Bayesian and confidence limits for prediction. *Journal of the Royal Statistical Society*, Ser. B 26, 176–192.

Thomas, D. G. and Gart, J. J. (1977). A table of exact confidence limits for differences and ratios of two proportions and their odds ratios. *Journal of the American Statistical Association*, 72, 73–76.

Thoman, D.R., Bain, L.J., and Antle, C.E. (1969). Inferences on the parameters of the Weibull distribution. *Technometrics*, 11, 445–446.

Thompson, S. K. (1992). *Sampling*. Hoboken, NJ: John Wiley & Sons.

Tuggle, R. M. (1982). Assessment of occupational exposure using one-sided tolerance limits. *American Industrial Hygiene Association Journal*, 43, 338–346.

Vangel, M. G. (1996). Confidence intervals for a normal coefficient of variation. *The American Statistician*, 50, 21–26.

Wald, A. and Wolfowitz, J. (1946). Tolerance limits for a normal distribution. *Annals of the Mathematical Statistics*, 17, 208–215.

Wang, L. J. and Wang, X. G. (1998). Diameter and strength distributions of merino wool in early stage processing. *Textile Research Journal*, 68, 87–93.

Wang, H. and Tsung, F. (2009). Tolerance intervals with improved coverage probabilities for binomial and Poisson variables. *Technometrics*, 51, 25–33.

Wang, H. (2008). Coverage probability of prediction intervals for discrete random variables. *Computational Statistics and Data Analysis*, 53, 17–26.

Wani, J. K. (1975). Clopper-Pearson system of confidence intervals for the logarithmic distributions. *Biometrics*, 31, 771–775.

Weerahandi, S., (1995). *Exact Statistical Methods for Data Analysis*. New York: Springer-Verlag.

Wiens, B. L. (1999). When log-normal and gamma models give different results: A case study. *American Statistician*, 53, 89–93.

Wilson, E. B. and Hilferty, M. M. (1931). The distribution of chi-squares. *Proceedings of the National Academy of Sciences*, 17, 684–688.

Winterton, S. S., Smy, T. J., and Tarr, N. G. (1992). On the source of scatter in contact resistance data. *Journal of Electronic Materials*, 21, 917–921.

Williamson, E. and Bretherton, M. H. (1964). Tables of logarithmic series distribution. *Annals of Mathematical Statistics*, 35, 284–297.

Xu, Y. L. (1995) Model and full-scale comparison of fatigue-related characteristics of wind pressures on the Texas Tech building. *Journal of Wind Engineering and Industrial Aerodynamics*, 58, 147–173.

Yang, Y., Allen, J. C., Knapp, J. L., and Stansly, P. A. (1995). Frequency-distribution of citrus rust mite (acari, eriophyidae) damage on fruit in hamlin orange trees. *Environmental Entomology*, 24, 1018–1023.

Zimmerman, D. W. (2004). Conditional probabilities of rejecting H_0 by pooled and separate-variances t tests given heterogeneity of sample variances. *Communications in Statistics–Simulation and Computation*, 33, 69–81.

Zou, G. Y. and Donner, A. (2008). Construction of confidence limits about effect measures: A general approach. *Statistics in Medicine*, 27, 1693–1702.

Zou, G. Y., Taleban, J., and Huo, C. Y. (2009a). Simple confidence intervals for lognormal means and their differences with environmental applications. *Environmetrics*, 20, 172–180.

Zou, G. Y., Huo, C. Y., and Taleban, J. (2009b). Confidence interval estimation for lognormal data with application to health economics. *Computational Statistics and Data Analysis*, 53, 3755–3764.

Index

Printed in the United States
by Baker & Taylor Publisher Services